Teacher Edition

Eureka Math Grade 8 Modules 1 & 2

Special thanks go to the Gordan A. Cain Center and to the Department of Mathematics at Louisiana State University for their support in the development of *Eureka Math*.

Published by Great Minds

Printed in the U.S.A.
This book may be purchased from the publisher at eureka-math.org
10 9 8 7 6 5 4 3

ISBN 978-1-63255-620-2

Table of Contents[1]

Integer Exponents and Scientific Notation

[1]Each lesson is ONE day, and ONE day is considered a 45-minute period.

Grade 8 • Module 1

Integer Exponents and Scientific Notation

OVERVIEW

In Module 1, students' knowledge of operations on numbers is expanded to include operations on numbers in integer exponents. Module 1 also builds on students' understanding from previous grades with regard to transforming expressions. Students were introduced to exponential notation in Grade 5 as they used whole number exponents to denote powers of ten (**5.NBT.A.2**). In Grade 6, students expanded the use of exponents to include bases other than ten as they wrote and evaluated exponential expressions limited to whole-number exponents (**6.EE.A.1**). Students made use of exponents again in Grade 7 as they learned formulas for the area of a circle (**7.G.B.4**) and volume (**7.G.B.6**).

In this module, students build upon their foundation with exponents as they make conjectures about how zero and negative exponents of a number should be defined and prove the properties of integer exponents (**8.EE.A.1**). These properties are codified into three laws of exponents. They make sense out of very large and very small numbers, using the number line model to guide their understanding of the relationship of those numbers to each other (**8.EE.A.3**).

Having established the properties of integer exponents, students learn to express the magnitude of a positive number through the use of scientific notation and to compare the relative size of two numbers written in scientific notation (**8.EE.A.3**). Students explore the use of scientific notation and choose appropriately sized units as they represent, compare, and make calculations with very large quantities (e.g., the U.S. national debt, the number of stars in the universe, and the mass of planets) and very small quantities, such as the mass of subatomic particles (**8.EE.A.4**).

The Mid-Module Assessment follows Topic A. The End-of-Module Assessment follows Topic B.

Focus Standards

Work with radicals and integer exponents.

8.EE.A.1 Know and apply the properties of integer exponents to generate equivalent numerical expressions. *For example,* $3^2 \times 3^{-5} = 3^{-3} = 1/3^3 = 1/27$.

8.EE.A.3 Use numbers expressed in the form of a single digit times an integer power of 10 to estimate very large or very small quantities, and to express how many times as much one is than the other. *For example, estimate the population of the United States as* 3×10^8 *and the population of the world as* 7×10^9*, and determine that the world population is more than* 20 *times larger.*

EUREKA MATH™

8.EE.A.4 Perform operations with numbers expressed in scientific notation, including problems where both decimal and scientific notation are used. Use scientific notation and choose units of appropriate size for measurements of very large or very small quantities (e.g., use millimeters per year for seafloor spreading). Interpret scientific notation that has been generated by technology.

Foundational Standards

Understand the place value system.

5.NBT.A.2 Explain patterns in the number of zeros of the product when multiplying a number by powers of 10, and explain patterns in the placement of the decimal point when a decimal is multiplied or divided by a power of 10. Use whole-number exponents to denote powers of 10.

Apply and extend previous understandings of arithmetic to algebraic expressions.

6.EE.A.1 Write and evaluate numerical expressions involving whole-number exponents.

Solve real-life and mathematical problems involving angle measure, area, surface area, and volume.

7.G.B.4 Know the formulas for the area and circumference of a circle and use them to solve problems; give an informal derivation of the relationship between the circumference and area of a circle.

7.G.B.6 Solve real-world and mathematical problems involving area, volume and surface area of two- and three-dimensional objects composed of triangles, quadrilaterals, polygons, cubes, and right prisms.

Focus Standards for Mathematical Practice

MP.2 **Reason abstractly and quantitatively.** Students use concrete numbers to explore the properties of numbers in exponential form and then prove that the properties are true for all positive bases and all integer exponents using symbolic representations for bases and exponents. As lessons progress, students use symbols to represent integer exponents and make sense of those quantities in problem situations. Students refer to symbolic notation in order to contextualize the requirements and limitations of given statements (e.g., letting m, n represent positive integers, letting a, b represent all integers, both with respect to the properties of exponents).

MP.3 **Construct viable arguments and critique the reasoning of others.** Students reason through the acceptability of definitions and proofs (e.g., the definitions of x^0 and x^{-b} for all integers b and positive integers x). New definitions, as well as proofs, require students to analyze situations and break them into cases. Further, students examine the implications of these definitions and proofs on existing properties of integer exponents. Students keep the goal of a logical argument in mind while attending to details that develop during the reasoning process.

MP.6 **Attend to precision.** Beginning with the first lesson on exponential notation, students are required to attend to the definitions provided throughout the lessons and the limitations of symbolic statements, making sure to express what they mean clearly. Students are provided a hypothesis, such as $x < y$, for positive integers x, y, and then are asked to evaluate whether a statement, like $-2 < 5$, contradicts this hypothesis.

MP.7 **Look for and make use of structure.** Students understand and make analogies to the distributive law as they develop properties of exponents. Students will know $x^m \cdot x^n = x^{m+n}$ as an analog of $mx + nx = (m + n)x$ and $(x^m)^n = x^{m \cdot n}$ as an analog of $n \cdot (m \cdot x) = (n \cdot m) \cdot x$.

MP.8 **Look for and express regularity in repeated reasoning.** While evaluating the cases developed for the proofs of laws of exponents, students identify when a statement must be proved or if it has already been proven. Students see the use of the laws of exponents in application problems and notice the patterns that are developed in problems.

Terminology

New or Recently Introduced Terms

- **Order of Magnitude** (The *order of magnitude of a finite decimal* is the exponent in the power of 10 when that decimal is expressed in scientific notation.

  For example, the order of magnitude of 192.7 is 2, because when 192.7 is expressed in scientific notation as 1.927×10^2, 2 is the exponent of 10^2.)
- **Scientific Notation** (The *scientific notation for a finite decimal* is the representation of that decimal as the product of a decimal s and a power of 10, where s satisfies the property that its absolute value is at least one but less than ten, or in symbolic notation, $1 \leq |s| < 10$.

  For example, the scientific notation for 192.7 is 1.927×10^2.)

Familiar Terms and Symbols[2]

- Base, Exponent, Power
- Equivalent Fractions
- Expanded Form (of decimal numbers)
- Exponential Notation
- Integer
- Square and Cube (of a number)
- Whole Number

Suggested Tools and Representations

- Scientific Calculator

Rapid White Board Exchanges

Implementing an RWBE requires that each student be provided with a personal white board, a white board marker, and an eraser. An economic choice for these materials is to place two sheets of tag board (recommended) or cardstock, one red and one white, into a sheet protector. The white side is the "paper" side that students write on. The red side is the "signal" side, which can be used for students to indicate they have finished working—"Show red when ready." Sheets of felt cut into small squares can be used as erasers.

An RWBE consists of a sequence of 10 to 20 problems on a specific topic or skill that starts out with a relatively simple problem and progressively gets more difficult. The teacher should prepare the problems in a way that allows the teacher to reveal them to the class one at a time. A flip chart or PowerPoint presentation can be used, or the teacher can write the problems on the board and either cover some with paper or simply write only one problem on the board at a time.

The teacher reveals, and possibly reads aloud, the first problem in the list and announces, "Go." Students work the problem on their personal white boards as quickly as possible. Depending on teacher preference, students can be directed to hold their work up for their teacher to see their answers as soon as they have the answer ready or to turn their white boards face down to show the red side when they have finished. In the latter case, the teacher says, "Hold up your work," once all students have finished. The teacher gives immediate feedback to each student, pointing and/or making eye contact with the student and responding with an affirmation for correct work, such as "Good job!", "Yes!", or "Correct!", or responding with guidance for incorrect work such as "Look again," "Try again," "Check your work," etc. Feedback can also be more specific, such as "Watch your division facts," or "Error in your calculation."

If many students have struggled to get the answer correct, go through the solution of that problem as a class before moving on to the next problem in the sequence. Fluency in the skill has been established when the class is able to go through a sequence of problems leading up to and including the level of the relevant student objective, without pausing to go through the solution of each problem individually.

[2]These are terms and symbols students have seen previously.

Sprints

Sprints are designed to develop fluency. They should be fun, adrenaline-rich activities that intentionally build energy and excitement. A fast pace is essential. During Sprint administration, teachers assume the role of athletic coaches. A rousing routine fuels students' motivation to do their personal best. Student recognition of increasing success is critical, and so every improvement is acknowledged. (See the Sprint Delivery Script for the suggested means of acknowledging and celebrating student success.)

One Sprint has two parts with closely related problems on each. Students complete the two parts of the Sprint in quick succession with the goal of improving on the second part, even if only by one more. The problems on the second Sprint should not be harder, or easier, than the problems on the first Sprint. The problems on a Sprint should progress from easiest to hardest. The first quarter of problems on the Sprint should be simple enough that all students find them accessible (though not all students will finish the first quarter of problems within one minute). The last quarter of problems should be challenging enough that even the strongest students in the class find them challenging.

Sprints scores are not recorded. Thus, there is no need for students to write their names on the Sprints. The low-stakes nature of the exercise means that even students with allowances for extended time can participate. When a particular student finds the experience undesirable, it is reasonable to either give the student a copy of the sprint to practice with the night before, or to allow the student to opt out and take the Sprint home.

With practice, the Sprint routine takes about 8 minutes.

Sprint Delivery Script

Gather the following: stopwatch, a copy of Sprint A for each student, a copy of Sprint B for each student, answers for Sprint A and Sprint B. The following delineates a script for delivery of a pair of Sprints.

This sprint covers: *topic.*

Do not look at the Sprint; keep it turned face down on your desk.

There are xx problems on the Sprint. You will have 60 seconds. Do as many as you can. I do not expect any of you to finish.

On your mark, get set, GO.

60 seconds of silence.

STOP. Circle the last problem you completed.

I will read the answers. You say "YES" if your answer matches. Mark the ones you have wrong by circling the number of the problem. Don't try to correct them.

Energetically, rapid-fire call the answers ONLY.

Stop reading answers after there are no more students answering, "Yes."

Fantastic! Count the number you have correct, and write it on the top of the page. This is your personal goal for Sprint B.

Raise your hand if you have 1 or more correct. 2 or more, 3 or more, ...

Let us all applaud our runner-up, [insert name], with x correct. And let us applaud our winner, [insert name], with x correct.

You have a few minutes to finish up the page and get ready for the next Sprint.

Students are allowed to talk and ask for help; let this part last as long as most are working seriously.

Stop working. I will read the answers again so you can check your work. You say "YES" if your answer matches.

Energetically, rapid-fire call the answers ONLY.

Optionally, ask students to stand, and lead them in an energy-expanding exercise that also keeps the brain going. Examples are jumping jacks or arm circles, etc., while counting by 15's starting at 15, going up to 150 and back down to 0. You can follow this first exercise with a cool down exercise of a similar nature, such as calf raises with counting by one-sixths $\left(\frac{1}{6}, \frac{1}{3}, \frac{1}{2}, \frac{2}{3}, \frac{5}{6}, 1, \ldots\right)$.

Hand out the second Sprint, and continue reading the script.

Keep the Sprint face down on your desk.

There are xx problems on the Sprint. You will have 60 seconds. Do as many as you can. Your goal is to improve your score from the first Sprint.

On your mark, get set, GO.

60 seconds of silence.

STOP. Circle the last problem you completed.

I will read the answers. You say "YES" if your answer matches. Mark the ones you have wrong by circling the number of the problem. Don't try to correct them.

Quickly read the answers ONLY.

Count the number you have correct, and write it on the top of the page. Write the amount by which your score improved at the top of the page and circle it.

Raise your hand if you have 1 or more correct. 2 or more, 3 or more, ...

Let us all applaud our runner-up, [insert name], with x correct. And let us applaud our winner, [insert name], with x correct.

Raise your hand if you improved your score by 1 or more. 2 or more, 3 or more, ...

Let us all applaud our runner-up for most improved, [insert name]. And let us applaud our winner for most improved, [insert name].

You can take the Sprint home and finish it if you want.

Preparing to Teach a Module

Preparation of lessons will be more effective and efficient if there has been an adequate analysis of the module first. Each module in *A Story of Ratios* can be compared to a chapter in a book. How is the module moving the plot, the mathematics, forward? What new learning is taking place? How are the topics and objectives building on one another? The following is a suggested process for preparing to teach a module.

Step 1: Get a preview of the plot.

A: Read the Table of Contents. At a high level, what is the plot of the module? How does the story develop across the topics?

B: Preview the module's Exit Tickets to see the trajectory of the module's mathematics and the nature of the work students are expected to be able to do.

Note: When studying a PDF file, enter "Exit Ticket" into the search feature to navigate from one Exit Ticket to the next.

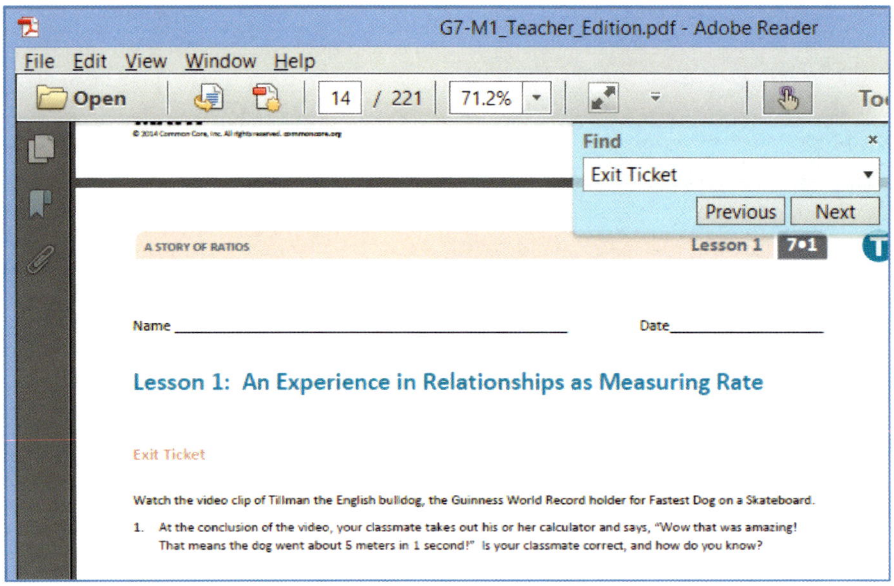

Step 2: Dig into the details.

A: Dig into a careful reading of the Module Overview. While reading the narrative, liberally reference the lessons and Topic Overviews to clarify the meaning of the text—the lessons demonstrate the strategies, show how to use the models, clarify vocabulary, and build understanding of concepts.

B: Having thoroughly investigated the Module Overview, read through the Student Outcomes of each lesson (in order) to further discern the plot of the module. How do the topics flow and tell a coherent story? How do the outcomes move students to new understandings?

Step 3: Summarize the story.

Complete the Mid- and End-of-Module Assessments. Use the strategies and models presented in the module to explain the thinking involved. Again, liberally reference the lessons to anticipate how students who are learning with the curriculum might respond.

Preparing to Teach a Lesson

A three-step process is suggested to prepare a lesson. It is understood that at times teachers may need to make adjustments (customizations) to lessons to fit the time constraints and unique needs of their students. The recommended planning process is outlined below. Note: The ladder of Step 2 is a metaphor for the teaching sequence. The sequence can be seen not only at the macro level in the role that this lesson plays in the overall story, but also at the lesson level, where each rung in the ladder represents the next step in understanding or the next skill needed to reach the objective. To reach the objective, or the top of the ladder, all students must be able to access the first rung and each successive rung.

Step 1: Discern the plot.

A: Briefly review the module's Table of Contents, recalling the overall story of the module and analyzing the role of this lesson in the module.

B: Read the Topic Overview related to the lesson, and then review the Student Outcome(s) and Exit Ticket of each lesson in the topic.

C: Review the assessment following the topic, keeping in mind that assessments can be found midway through the module and at the end of the module.

Step 2: Find the ladder.

A: Work through the lesson, answering and completing each question, example, exercise, and challenge.

B: Analyze and write notes on the new complexities or new concepts introduced with each question or problem posed; these notes on the sequence of new complexities and concepts are the rungs of the ladder.

C: Anticipate where students might struggle, and write a note about the potential cause of the struggle.

D: Answer the Closing questions, always anticipating how students will respond.

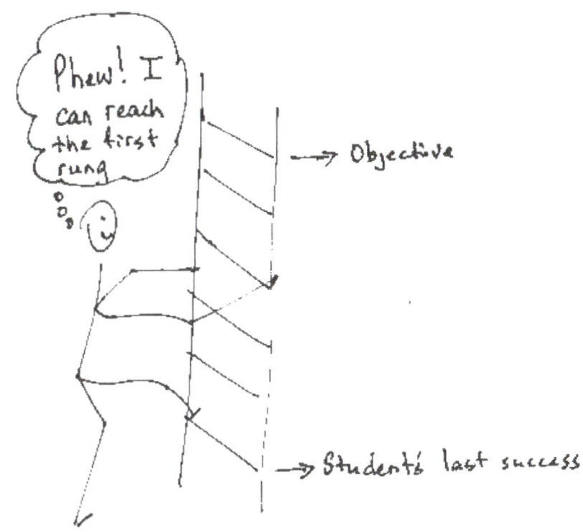

Step 3: Hone the lesson.

Lessons may need to be customized if the class period is not long enough to do all of what is presented and/or if students lack prerequisite skills and understanding to move through the entire lesson in the time allotted. A suggestion for customizing the lesson is to first decide upon and designate each question, example, exercise, or challenge as either "Must Do" or "Could Do."

A: Select "Must Do" dialogue, questions, and problems that meet the Student Outcome(s) while still providing a coherent experience for students; reference the ladder. The expectation should be that the majority of the class will be able to complete the "Must Do" portions of the lesson within the allocated time. While choosing the "Must Do" portions of the lesson, keep in mind the need for a balance of dialogue and conceptual questioning, application problems, and abstract problems, and a balance between students using pictorial/graphical representations and abstract representations. Highlight dialogue to be included in the delivery of instruction so that students have a chance to articulate and consolidate understanding as they move through the lesson.

B: "Must Do" portions might also include remedial work as necessary for the whole class, a small group, or individual students. Depending on the anticipated difficulties, the remedial work might take on different forms as suggested in the chart below.

Anticipated Difficulty	"Must Do" Remedial Problem Suggestion
The first problem of the lesson is too challenging.	Write a short sequence of problems on the board that provides a ladder to Problem 1. Direct students to complete those first problems to empower them to begin the lesson.
There is too big of a jump in complexity between two problems.	Provide a problem or set of problems that bridge student understanding from one problem to the next.
Students lack fluency or foundational skills necessary for the lesson.	Before beginning the lesson, do a quick, engaging fluency exercise, such as a Rapid White Board Exchange or Sprint. Before beginning any fluency activity for the first time, assess that students have conceptual understanding of the problems in the set and that they are poised for success with the easiest problem in the set.
More work is needed at the concrete or pictorial level.	Provide manipulatives or the opportunity to draw solution strategies.
More work is needed at the abstract level.	Add a White Board Exchange of abstract problems to be completed toward the end of the lesson.

C: "Could Do" problems are for students who work with greater fluency and understanding and can, therefore, complete more work within a given time frame.

D: At times, a particularly complex problem might be designated as a "Challenge!" problem to provide to advanced students. Consider creating the opportunity for students to share their "Challenge!" solutions with the class at a weekly session or on video.

E: If the lesson is customized, be sure to carefully select Closing questions that reflect such decisions and adjust the Exit Ticket if necessary.

Assessment Summary

Assessment Type	Administered	Format	Standards Addressed
Mid-Module Assessment Task	After Topic A	Constructed response with rubric	8.EE.A.1
End-of-Module Assessment Task	After Topic B	Constructed response with rubric	8.EE.A.3, 8.EE.A.4

Topic A

Exponential Notation and Properties of Integer Exponents

8.EE.A.1

Focus Standard:	8.EE.A.1	Know and apply the properties of integer exponents to generate equivalent numerical expressions. *For example,* $3^2 \times 3^{-5} = 3^{-3} = 1/3^3 = 1/27$.
Instructional Days:	6	
Lesson 1:	Exponential Notation (S)[1]	
Lesson 2:	Multiplication and Division of Numbers in Exponential Form (S)	
Lesson 3:	Numbers in Exponential Form Raised to a Power (S)	
Lesson 4:	Numbers Raised to the Zeroth Power (E)	
Lesson 5:	Negative Exponents and the Laws of Exponents (S)	
Lesson 6:	Proofs of Laws of Exponents (S)	

In Topic A, students begin by learning the precise definition of exponential notation where the exponent is restricted to being a positive integer. In Lessons 2 and 3, students discern the structure of exponents by relating multiplication and division of expressions with the same base to combining like terms using the distributive property and by relating multiplying three factors using the associative property to raising a power to a power.

Lesson 4 expands the definition of exponential notation to include what it means to raise a nonzero number to a zero power; students verify that the properties of exponents developed in Lessons 2 and 3 remain true. Properties of exponents are extended again in Lesson 5 when a positive integer, raised to a negative exponent, is defined. In Lesson 5, students accept the properties of exponents as true for all integer exponents and are shown the value of learning them; in other words, if the three properties of exponents are known, then facts about dividing numbers in exponential notation with the same base and raising fractions to a power are also known.

[1]Lesson Structure Key: **P**-Problem Set Lesson, **M**-Modeling Cycle Lesson, **E**-Exploration Lesson, **S**-Socratic Lesson

Topic A culminates in Lesson 6 when students work to prove the laws of exponents for all integer exponents. Throughout Topic A, students generate equivalent numerical expressions by applying properties of integer exponents, first with positive integer exponents, then with whole number exponents, and concluding with integer exponents in general.

Lesson 1: Exponential Notation

Student Outcomes

- Students know what it means for a number to be raised to a power and how to represent the repeated multiplication symbolically.
- Students know the reason for some bases requiring parentheses.

Lesson Notes

This lesson is foundational for the topic of properties of integer exponents. For the first time in this lesson, students are seeing the use of exponents with negative valued bases. It is important that students explore and understand the importance of parentheses in such cases, just as with rational base values. It may also be the first time that students are seeing the notation (dots and braces) used in this lesson. If students have already mastered the skills in this lesson, it is optional to move forward and begin with Lesson 2 or provide opportunities for students to explore how to rewrite expressions in a different base, 4^2 as 2^4, for example.

Classwork

Discussion (15 minutes)

> *Scaffolding:*
>
> Remind students of their previous experiences:
>
> - The square of a number (e.g., 3×3 is denoted by 3^2).
> - From the expanded form of a whole number, we also learned that 10^3 stands for $10 \times 10 \times 10$.

MP.2 & MP.7

When we add 5 copies of 3, we devise an abbreviation (i.e., a new notation) for this purpose.

$$3 + 3 + 3 + 3 + 3 = 5 \times 3$$

Now if we multiply 5 factors of 3, how should we abbreviate this?

$$3 \times 3 \times 3 \times 3 \times 3 = ?$$

Allow students to make suggestions (see sidebar for scaffolds).

$$3 \times 3 \times 3 \times 3 \times 3 = 3^5$$

Similarly, we also write $3^3 = 3 \times 3 \times 3$; $3^4 = 3 \times 3 \times 3 \times 3$; etc.

We see that when we add 5 summands of 3, we write 5×3, but when we multiply 5 factors of 3, we write 3^5. Thus, the *multiplication by* 5 in the context of addition corresponds exactly to the superscript 5 in the context of multiplication.

Make students aware of the correspondence between addition and multiplication because what they know about *repeated addition* will help them learn exponents as *repeated multiplication* as we go forward.

5^6 means $5 \times 5 \times 5 \times 5 \times 5 \times 5$, and $\left(\frac{9}{7}\right)^4$ means $\frac{9}{7} \times \frac{9}{7} \times \frac{9}{7} \times \frac{9}{7}$.

You have seen this kind of notation before; it is called exponential notation. In general, for any number x and any positive integer n,

$$x^n = \underbrace{(x \cdot x \cdots x)}_{n \text{ times}}.$$

The number x^n is called x raised to the nth power, where n is the exponent of x in x^n and x is the base of x^n.

Examples 1–5

Work through Examples 1–5 as a group, and supplement with additional examples if needed.

Example 1

$5 \times 5 \times 5 \times 5 \times 5 \times 5 = 5^6$

Example 2

$\frac{9}{7} \times \frac{9}{7} \times \frac{9}{7} \times \frac{9}{7} = \left(\frac{9}{7}\right)^4$

Example 3

$\left(-\frac{4}{11}\right)^3 = \left(-\frac{4}{11}\right) \times \left(-\frac{4}{11}\right) \times \left(-\frac{4}{11}\right)$

Example 4

$(-2)^6 = (-2) \times (-2) \times (-2) \times (-2) \times (-2) \times (-2)$

Example 5

$3.8^4 = 3.8 \times 3.8 \times 3.8 \times 3.8$

- Notice the use of parentheses in Examples 2, 3, and 4. Do you know why we use them?
  - *In cases where the base is either fractional or negative, parentheses tell us what part of the expression is included in the base and, therefore, going to be multiplied repeatedly.*
- Suppose n is a fixed **positive integer**. Then 3^n by definition is $3^n = \underbrace{(3 \times \cdots \times 3)}_{n \text{ times}}$.
- Again, if n is a fixed positive integer, then by definition:

$$7^n = \underbrace{(7 \times \cdots \times 7)}_{n \text{ times}},$$

$$\left(\frac{4}{5}\right)^n = \underbrace{\left(\frac{4}{5} \times \cdots \times \frac{4}{5}\right)}_{n \text{ times}},$$

$$(-2.3)^n = \underbrace{((-2.3) \times \cdots \times (-2.3))}_{n \text{ times}}.$$

If students ask about values of n that are not positive integers, ask them to give an example and to consider what such an exponent would indicate. Let them know that *integer* exponents will be discussed later in this module, so they should continue examining their question as we move forward. Positive and negative fractional exponents are a topic that will be introduced in Algebra II.

MP.6

- In general, for any number x, $x^1 = x$, and for any positive integer $n > 1$, x^n is by definition:

$$x^n = \underbrace{(x \cdot x \cdots x)}_{n \text{ times}}.$$

- The number x^n is called **x raised to the nth power**, where n is the **exponent** of x in x^n, and x is the **base** of x^n.
- x^2 is called the **square** of x, and x^3 is its **cube**.
- You have seen this kind of notation before when you gave the expanded form of a whole number for powers of 10; it is called **exponential notation**.

Students might ask why we use the terms *square* and *cube* to represent exponential expressions with exponents of 2 and 3, respectively. Refer them to earlier grades and finding the area of a square and the volume of a cube. These geometric quantities are obtained by multiplying equal factors. The area of a square with side lengths of 4 units is $4 \text{ units} \times 4 \text{ units} = 4^2 \text{ units}^2$ or 16 units^2. Similarly, the volume of a cube with edge lengths of 4 units is $4 \text{ units} \times 4 \text{ units} \times 4 \text{ units} = 4^3 \text{ units}^3$ or 64 units^3.

Exercises 1–10 (5 minutes)

Have students complete these independently and check their answers before moving on.

Exercise 1

$\underbrace{4 \times \cdots \times 4}_{7 \text{ times}} = 4^7$

Exercise 2

$\underbrace{3.6 \times \cdots \times 3.6}_{\_\_\_\_ \text{ times}} = 3.6^{47}$

47 times

Exercise 3

$\underbrace{(-11.63) \times \cdots \times (-11.63)}_{34 \text{ times}} = (-11.63)^{34}$

Exercise 4

$\underbrace{12 \times \cdots \times 12}_{\_\_\_\_ \text{ times}} = 12^{15}$

15 times

Exercise 5

$\underbrace{(-5) \times \cdots \times (-5)}_{10 \text{ times}} = (-5)^{10}$

Exercise 6

$\underbrace{\frac{7}{2} \times \cdots \times \frac{7}{2}}_{21 \text{ times}} = \left(\frac{7}{2}\right)^{21}$

Exercise 7

$\underbrace{(-13) \times \cdots \times (-13)}_{6 \text{ times}} = (-13)^6$

Exercise 8

$\underbrace{\left(-\frac{1}{14}\right) \times \cdots \times \left(-\frac{1}{14}\right)}_{10 \text{ times}} = \left(-\frac{1}{14}\right)^{10}$

Exercise 9

$\underbrace{x \cdot x \cdots x}_{185 \text{ times}} = x^{185}$

Exercise 10

$\underbrace{x \cdot x \cdots x}_{\_\_\_\_ \text{ times}} = x^n$

n times

Exercises 11–14 (15 minutes)

Allow students to complete Exercises 11–14 individually or in small groups. As an alternative, provide students with several examples of exponential expressions whose bases are negative values, and whose exponents alternate between odd and even whole numbers. Ask students to discern a pattern from their calculations, form a conjecture, and work to justify their conjecture. They should find that a negative value raised to an even exponent results in a positive value since the product of two negative values yields a positive product. They should also find that having an even number of negative factors means each factor pairs with another, resulting in a set of positive products. Likewise, they should conclude that a negative number raised to an odd exponent always results in a negative value. This is because any odd whole number is 1 greater than an even number (or zero). This means that while the even set of negative factors results in a positive value, there will remain one more negative factor to negate the resulting product.

- When a negative number is raised to an odd power, what is the sign of the result?
- When a negative number is raised to an even power, what is the sign of the result?

Point out that when a negative number is raised to an odd power, the sign of the answer is negative. Conversely, if a negative number is raised to an even power, the sign of the answer is positive.

Exercise 11

Will these products be positive or negative? How do you know?

$$\underbrace{(-1) \times (-1) \times \cdots \times (-1)}_{12 \text{ times}} = (-1)^{12}$$

This product will be positive. Students may state that they computed the product and it was positive. If they say that, let them show their work. Students may say that the answer is positive because the exponent is positive; however, this would not be acceptable in view of the next example.

$$\underbrace{(-1) \times (-1) \times \cdots \times (-1)}_{13 \text{ times}} = (-1)^{13}$$

This product will be negative. Students may state that they computed the product and it was negative. If so, ask them to show their work. Based on the discussion of the last problem, you may need to point out that a positive exponent does not always result in a positive product.

The two problems in Exercise 12 force the students to think beyond the computation level. If students struggle, revisit the previous two problems, and have them discuss in small groups what an even number of negative factors yields and what an odd number of negative factors yields.

Exercise 12

Is it necessary to do all of the calculations to determine the sign of the product? Why or why not?

$$\underbrace{(-5) \times (-5) \times \cdots \times (-5)}_{95 \text{ times}} = (-5)^{95}$$

Students should state that an odd number of negative factors yields a negative product.

$$\underbrace{(-1.8) \times (-1.8) \times \cdots \times (-1.8)}_{122 \text{ times}} = (-1.8)^{122}$$

Students should state that an even number of negative factors yields a positive product.

Exercise 13

Fill in the blanks indicating whether the number is positive or negative.

If n is a positive even number, then $(-55)^n$ is *positive.*

If n is a positive odd number, then $(-72.4)^n$ is *negative.*

Exercise 14

Josie says that $\underbrace{(-15) \times \cdots \times (-15)}_{6 \text{ times}} = -15^6$. Is she correct? How do you know?

Students should state that Josie is not correct for the following two reasons: (1) They just stated that an even number of factors yields a positive product, and this conflicts with the answer Josie provided, and (2) the notation is used incorrectly because, as is, the answer is the negative of 15^6, instead of the product of 6 copies of -15. The base is (-15). Recalling the discussion at the beginning of the lesson, when the base is negative it should be written clearly by using parentheses. Have students write the answer correctly.

Closing (5 minutes)

- Why should we bother with exponential notation? Why not just write out the multiplication?

Engage the class in discussion, but make sure to address at least the following two reasons:

1. Like all good notation, exponential notation saves writing.
2. Exponential notation is used for recording scientific measurements of very large and very small quantities. It is indispensable for the clear indication of the magnitude of a number (see Lessons 10–13).
   - Here is an example of the labor-saving aspect of the exponential notation: Suppose a colony of bacteria doubles in size every 8 hours for a few days under tight laboratory conditions. If the initial size is B, what is the size of the colony after 2 days?
     - *In 2 days, there are six 8-hour periods; therefore, the size will be 2^6B.*

If time allows, give more examples as a lead in to Lesson 2. Example situations: (1) exponential decay with respect to heat transfer, vibrations, ripples in a pond, or (2) exponential growth with respect to interest on a bank deposit after some years have passed.

Exit Ticket (5 minutes)

Name ______________________________ Date ______________

Lesson 1: Exponential Notation

Exit Ticket

1.

a. Express the following in exponential notation:

$$\underbrace{(-13) \times \cdots \times (-13)}_{35 \text{ times}}.$$

b. Will the product be positive or negative? Explain.

2. Fill in the blank:

$$\underbrace{\frac{2}{3} \times \cdots \times \frac{2}{3}}_{\_\_\_\_\text{times}} = \left(\frac{2}{3}\right)^4$$

3. Arnie wrote:

$$\underbrace{(-3.1) \times \cdots \times (-3.1)}_{4 \text{ times}} = -3.1^4$$

Is Arnie correct in his notation? Why or why not?

Exit Ticket Sample Solutions

1.

a. Express the following in exponential notation:

$$\underbrace{(-13) \times \cdots \times (-13)}_{35 \text{ times}}$$

$(-13)^{35}$

b. Will the product be positive or negative? Explain.

The product will be negative. The expanded form shows 34 negative factors plus one more negative factor. Any even number of negative factors yields a positive product. The remaining 35th negative factor negates the resulting product.

2. Fill in the blank:

$$\underbrace{\frac{2}{3} \times \cdots \times \frac{2}{3}}_{\_\_\_\text{ times}} = \left(\frac{2}{3}\right)^4$$

4 times

3. Arnie wrote:

$$\underbrace{(-3.1) \times \cdots \times (-3.1)}_{4 \text{ times}} = -3.1^4$$

Is Arnie correct in his notation? Why or why not?

Arnie is not correct. The base, -3.1, should be in parentheses to prevent ambiguity. At present the notation is not correct.

Problem Set Sample Solutions

1. Use what you know about exponential notation to complete the expressions below.

$$\underbrace{(-5) \times \cdots \times (-5)}_{17 \text{ times}} = (-5)^{17}$$

$$\underbrace{3.7 \times \cdots \times 3.7}_{\_\_\_\text{ times}} = 3.7^{19}$$

19 times

$$\underbrace{7 \times \cdots \times 7}_{\_\_\_\text{ times}} = 7^{45}$$

45 times

$$\underbrace{6 \times \cdots \times 6}_{4 \text{ times}} = 6^4$$

$$\underbrace{4.3 \times \cdots \times 4.3}_{13 \text{ times}} = 4.3^{13}$$

$$\underbrace{(-1.1) \times \cdots \times (-1.1)}_{9 \text{ times}} = (-1.1)^9$$

$$\underbrace{\left(\frac{2}{3}\right) \times \cdots \times \left(\frac{2}{3}\right)}_{19 \text{ times}} = \left(\frac{2}{3}\right)^{19}$$

$$\underbrace{\left(-\frac{11}{5}\right) \times \cdots \times \left(-\frac{11}{5}\right)}_{\_\_\_\text{ times}} = \left(-\frac{11}{5}\right)^x$$

x times

$\underbrace{(-12) \times \cdots \times (-12)}_{\_\_\_\text{ times}} = (-12)^{15}$ $\underbrace{a \times \cdots \times a}_{m\text{ times}} = a^m$

15 times

2. Write an expression with (-1) as its base that will produce a positive product, and explain why your answer is valid.

   Accept any answer with (-1) to an exponent that is even.

3. Write an expression with (-1) as its base that will produce a negative product, and explain why your answer is valid.

   Accept any answer with (-1) to an exponent that is odd.

4. Rewrite each number in exponential notation using 2 as the base.

   $8 = 2^3$ $16 = 2^4$ $32 = 2^5$

   $64 = 2^6$ $128 = 2^7$ $256 = 2^8$

5. Tim wrote 16 as $(-2)^4$. Is he correct? Explain.

   Tim is correct that $16 = (-2)^4$. $(-2)(-2)(-2)(-2) = (4)(4) = 16$.

6. Could -2 be used as a base to rewrite 32? 64? Why or why not?

   A base of -2 cannot be used to rewrite 32 because $(-2)^5 = -32$. A base of -2 can be used to rewrite 64 because $(-2)^6 = 64$. If the exponent, n, is even, $(-2)^n$ will be positive. If the exponent, n, is odd, $(-2)^n$ cannot be a positive number.

Lesson 2: Multiplication of Numbers in Exponential Form

Student Outcomes

- Students use the definition of exponential notation to make sense of the first law of exponents.
- Students see a rule for simplifying exponential expressions involving division as a consequence of the first law of exponents.
- Students write equivalent numerical and symbolic expressions using the first law of exponents.

Lesson Notes

In this lesson, students learn their first rule for exponents and apply it to problems that contain only positive integer exponents. The laws of exponents are presented in a slow, methodical way. Specifically, students first learn how to multiply and divide expressions with positive integer exponents. Next, they extend their understanding of the laws to whole numbers (Lesson 4) and then to all integers (Lesson 5). For this reason, for positive integers m and n, we apply the restriction that $m > n$ for expressions of the form $\frac{x^m}{x^n}$. This is a temporary restriction that eliminates the possibility of students arriving at an answer with a negative exponent, something they have yet to learn.

Ultimately, the goal of the work in this lesson is to develop students' fluency generating equivalent expressions; however, it is unlikely this can be achieved in one period. It is perfectly acceptable for students to use their knowledge of exponential notation to generate those equivalent expressions until they build intuition of the behavior of exponents and are ready to use the laws fluently and accurately. This is the reason that answers are in the form of a sum or difference of two integers. The instructional value of answers left in this form far outweighs the instructional value of answers that have been added or subtracted. When it is appropriate, transition students into the normal form of the answer.

For some classes, it may be necessary to split this lesson over two periods. Consider delivering instruction through Exercise 20 on day one and beginning with the discussion that follows Exercise 20 on day two. Another possible customization of the lesson may include providing opportunities for students to discover the properties of exponents prior to giving the mathematical rationale as to why they are true. For example, present students with the problems in Example 1 and allow them to share their thinking about what the answer should be, and then provide the mathematical reasoning behind their correct solutions. Finally, the exercises in this lesson go from simple to complex. Every student should be able to complete the simple exercises, and many students will be challenged by the complex problems. It is not necessary that all students achieve mastery over the complex problems, but they should master those directly related to the standard (e.g., Exercises 1–13 in the first part of the lesson).

Knowing and applying the properties of integer exponents to generate equivalent expressions is the primary goal of this lesson. Students should be exposed to general arguments as to why the properties are true and be able to explain them on their own with concrete numbers; however, the relationship between the laws of exponents and repeated addition, a concept that is introduced in Grade 6 Module 4, is not as important and could be omitted if time is an issue.

Classwork

Discussion (8 minutes)

We have to find out the basic properties of this new concept of raising a number to a power. There are three simple ones, and we will discuss them in this and the next lesson.

- (1) How to multiply different powers of the same number x: If m, n are positive integers, what is $x^m \cdot x^n$?

Let students explore on their own and then in groups: $3^5 \times 3^7$.

- *Answer:* $3^5 \times 3^7 = \underbrace{(3 \times \cdots \times 3)}_{5 \text{ times}} \times \underbrace{(3 \times \cdots \times 3)}_{7 \text{ times}} = \underbrace{(3 \times \cdots \times 3)}_{5+7 \text{ times}} = 3^{5+7}$

In general, if x is any number and m, n are positive integers, then

$$x^m \cdot x^n = x^{m+n}$$

because

$$x^m \times x^n = \underbrace{(x \cdots x)}_{m \text{ times}} \times \underbrace{(x \cdots x)}_{n \text{ times}} = \underbrace{(x \cdots x)}_{m+n \text{ times}} = x^{m+n}.$$

> *Scaffolding:*
> - Use concrete numbers for x, m, and n.
> - This property was seen informally in Grade 7 through the use of units of measure for distance, area, and volume. For instance, in calculating the volume of a right prism, students multiplied the area of the prism's base in units^2 times the height of the prism in units to get its volume in units^3 (i.e., x units$^2 \times$ y units $=$ xy units^{2+1} $=$ xy units^3).

In general, if x is any number and m, n are positive integers, then

$$x^m \cdot x^n = x^{m+n}$$

because

$$x^m \times x^n = \underbrace{(x \cdots x)}_{m \text{ times}} \times \underbrace{(x \cdots x)}_{n \text{ times}} = \underbrace{(x \cdots x)}_{m+n \text{ times}} = x^{m+n}.$$

> *Scaffolding:*
> Advanced students may ask why m and n are restricted to positive integers. If so, ask them to consider some examples and what those examples might mean. For instance, if $m = 2$ and $n = 0$, then
> $$\begin{aligned} 3^m \cdot 3^n &= 3^2 \cdot 3^0 \\ &= 3^{2+0} \\ &= 3^2. \end{aligned}$$
> Interestingly, this means that 3^0 acts like the multiplicative identity 1. This idea is explored in Lesson 4.

Examples 1–2

Work through Examples 1 and 2 in the manner just shown. (Supplement with additional examples if needed.)

It is preferable to write the answers as an addition of exponents to emphasize the use of the identity. That step should not be left out. That is, $5^2 \times 5^4 = 5^6$ does not have the same instructional value as $5^2 \times 5^4 = 5^{2+4}$.

> *Scaffolding:*
> Remind students that to remove ambiguity, bases that contain fractions or negative numbers require parentheses.

Example 1

$5^2 \times 5^4 = 5^{2+4}$

Example 2

$$\left(-\frac{2}{3}\right)^4 \times \left(-\frac{2}{3}\right)^5 = \left(-\frac{2}{3}\right)^{4+5}$$

MP.2 & MP.7

- What is the analog of $x^m \cdot x^n = x^{m+n}$ in the context of repeated addition of a number x?

Allow time for a brief discussion.

- *If we add m copies of x and then add to it another n copies of x, we end up adding $m+n$ copies of x. By the distributive law:*

$$mx + nx = (m+n)x .$$

This is further confirmation of what we observed at the beginning of Lesson 1: The exponent $m+n$ in x^{m+n} in the context of repeated multiplication corresponds exactly to the $m+n$ in $(m+n)x$ in the context of repeated addition.

Exercises 1–20 (11 minutes)

Have students complete Exercises 1–8 independently. Check their answers, and then have them complete Exercises 9–20.

Exercise 1

$14^{23} \times 14^{8} = 14^{23+8}$

Exercise 2

$(-72)^{10} \times (-72)^{13} = (-72)^{10+13}$

Exercise 3

$5^{94} \times 5^{78} = 5^{94+78}$

Exercise 4

$(-3)^{9} \times (-3)^{5} = (-3)^{9+5}$

Exercise 5

Let a be a number.

$a^{23} \cdot a^{8} = a^{23+8}$

Exercise 6

Let f be a number.

$f^{10} \cdot f^{13} = f^{10+13}$

Exercise 7

Let b be a number.

$b^{94} \cdot b^{78} = b^{94+78}$

Exercise 8

Let x be a positive integer. If $(-3)^9 \times (-3)^x = (-3)^{14}$, what is x?

$x = 5$

In Exercises 9–16, students need to think about how to rewrite some factors so the bases are the same. Specifically, $2^4 \times 8^2 = 2^4 \times 2^6 = 2^{4+6}$ and $3^7 \times 9 = 3^7 \times 3^2 = 3^{7+2}$. Make clear that these expressions can only be combined into a single base because the bases are the same. Also included is a non-example, $5^4 \times 2^{11}$, that cannot be combined into a single base using this identity. Exercises 17–20 offer further applications of the identity.

What would happen if there were more terms with the same base? Write an equivalent expression for each problem.

do as example

Exercise 9

$9^4 \times 9^6 \times 9^{13} = 9^{4+6+13}$

Exercise 10

$2^3 \times 2^5 \times 2^7 \times 2^9 = 2^{3+5+7+9}$

Can the following expressions be written in simpler form? If so, write an equivalent expression. If not, explain why not.

Exercise 11

$6^5 \times 4^9 \times 4^3 \times 6^{14} = 4^{9+3} \times 6^{5+14}$

Exercise 12

$(-4)^2 \cdot 17^5 \cdot (-4)^3 \cdot 17^7 = (-4)^{2+3} \cdot 17^{5+7}$

Exercise 13

$15^2 \cdot 7^2 \cdot 15 \cdot 7^4 = 15^{2+1} \cdot 7^{2+4}$

Exercise 14

$2^4 \times 8^2 = 2^4 \times 2^6 = 2^{4+6}$

Exercise 15

$3^7 \times 9 = 3^7 \times 3^2 = 3^{7+2}$

Exercise 16

$5^4 \times 2^{11} =$

Cannot be simplified. Bases are different and cannot be rewritten in the same base.

> *Scaffolding:*
>
> Some of these exercises can be written in several forms. The sample responses only provide one such form.

Exercise 17

Let x be a number. Rewrite the expression in a simpler form.

$(2x^3)(17x^7) = 34x^{10}$

Exercise 18

Let a and b be numbers. Use the distributive law to rewrite the expression in a simpler form.

$a(a+b) = a^2 + ab$

Exercise 19

Let a and b be numbers. Use the distributive law to rewrite the expression in a simpler form.

$b(a+b) = ab + b^2$

Exercise 20

Let a and b be numbers. Use the distributive law to rewrite the expression in a simpler form.

$(a+b)(a+b) = a^2 + ab + ba + b^2 = a^2 + 2ab + b^2$

> *Scaffolding:*
>
> Remind students of the rectangular array used in Grade 7 Module 6 to multiply expressions of this form:
>
> 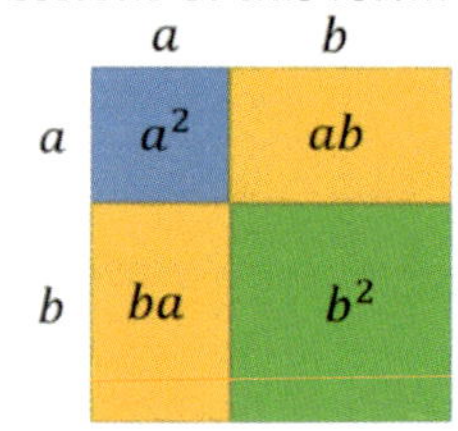
>
>
> $(a+b)(a+b)$
> $= a^2 + ab + ba + b^2$
> $= a^2 + 2ab + b^2$

Discussion (9 minutes)

Now that we know something about multiplication, we actually know a little about how to divide numbers in exponential notation too. This is not a new law of exponents but a (good) consequence of knowing the first law of exponents. Make this clear to students.

- (2) We have just learned how to multiply two different positive integer powers of the same number x. It is time to ask how to *divide* different powers of a number x. If m and n are positive integers, what is $\frac{x^m}{x^n}$?

> *Scaffolding:*
>
> Use concrete numbers for x, m, and n.

Allow time for a brief discussion.

- What is $\frac{3^7}{3^5}$? (Observe: The power of 7 in the numerator is bigger than the power of 5 in the denominator. The general case of arbitrary positive integer exponents will be addressed in Lesson 5, so all problems in this lesson will have greater exponents in the numerator than in the denominator.)
  - *Expect students to write* $\frac{3^7}{3^5} = \frac{3\cdot3\cdot3\cdot3\cdot3\cdot3\cdot3}{3\cdot3\cdot3\cdot3\cdot3}$. *However, we should nudge them to see how the formula* $x^m \cdot x^n = x^{m+n}$ *comes into play.*
  - *Answer:*

$$\frac{3^7}{3^5} = \frac{3^5 \cdot 3^2}{3^5} \qquad \textit{By } x^m x^n = x^{m+n}$$

$$= 3^2 \qquad \textit{By equivalent fractions}$$

$$= 3^{7-5}$$

> *Scaffolding:*
>
> Advanced students may ask about a case of the numerator and denominator having the same base, but the exponent of the denominator is greater than the exponent of the numerator. Write the numerator and denominator in expanded form, and then divide out their common factors. Students see that the remaining factors are in the denominator. This serves as an excellent opportunity to develop intuition with regard to the meaning of negative integer exponents.

- Observe that the exponent 2 in 3^2 *is* the difference of 7 and 5 (see the numerator $3^5 \cdot 3^2$ on the first line).
- In general, if x is nonzero and m, n are positive integers, then:

$$\frac{x^m}{x^n} = x^{m-n}$$

The restriction on m and n given below is to prevent negative exponents from coming up in problems before students learn about them. If advanced students want to consider the remaining cases, $m = n$ and $m < n$, they can gain some insight to the meaning of the zeroth power and negative integer exponents. In general instruction however, these cases are reserved for Lessons 4 and 5.

- Let's restrict (for now) $m > n$. Then there is a positive integer l, so that $m = n + l$. Then, we can rewrite the identity as follows:

$$\frac{x^m}{x^n} = \frac{x^{n+l}}{x^n}$$

$$= \frac{x^n \cdot x^l}{x^n} \qquad \text{By } x^m x^n = x^{m+n}$$

$$= x^l \qquad \text{By equivalent fractions}$$

$$= x^{m-n} \qquad \text{Because } m = n + l \text{ implies } l = m - n$$

Therefore, $\frac{x^m}{x^n} = x^{m-n}$, if $m > n$.

> **In general, if x is nonzero and m, n are positive integers, then**
>
> $$\frac{x^m}{x^n} = x^{m-n}.$$

This formula is as far as we can go for now. The reason is that $\frac{3^5}{3^7}$ in terms of exponents is $3^{5-7} = 3^{-2}$, and that answer makes no sense at the moment since we have no meaning for a negative exponent. This motivates our search for a definition of negative exponent, as we shall do in Lesson 5.

MP.7

- What is the analog of $\frac{x^m}{x^n} = x^{m-n}$, if $m > n$ in the context of repeated addition of a number x?
  - *Division is to multiplication as subtraction is to addition, so if n copies of a number x is subtracted from m copies of x, and $m > n$, then $(mx) - (nx) = (m - n)x$ by the distributive law. (Incidentally, observe once more how the exponent $m - n$ in x^{m-n}, in the context of repeated multiplication, corresponds exactly to the $m - n$ in $(m - n)x$ in the context of repeated addition.)*

Examples 3–4

Work through Examples 3 and 4 in the manner shown. (Supplement with additional examples if needed.)

It is preferable to write the answers as a subtraction of exponents to emphasize the use of the identity.

Example 3

$$\frac{\left(\frac{3}{5}\right)^8}{\left(\frac{3}{5}\right)^6} = \left(\frac{3}{5}\right)^{8-6}$$

Example 4

$$\frac{4^5}{4^2} = 4^{5-2}$$

Scaffolding:

In Grade 3, students began recognizing division problems as *missing factor* problems. Students can relate that work to this work. For example: $\frac{4^5}{4^2} = 4^?$ is equivalent to missing factor problem $4^2 \cdot 4^? = 4^5$. Using the first law of exponents, this means $4^{2+?} = 4^5$, and $? = 3$.

Exercises 21–32 (11 minutes)

Students complete Exercises 21–24 independently. Check their answers, and then have them complete Exercises 25–32 in pairs or small groups.

Exercise 21

$$\frac{7^9}{7^6} = \mathbf{7^{9-6}}$$

Exercise 22

$$\frac{(-5)^{16}}{(-5)^7} = \mathbf{(-5)^{16-7}}$$

Exercise 23

$$\frac{\left(\frac{8}{5}\right)^9}{\left(\frac{8}{5}\right)^2} = \mathbf{\left(\frac{8}{5}\right)^{9-2}}$$

Exercise 24

$$\frac{13^5}{13^4} = \mathbf{13^{5-4}}$$

Exercise 25

Let a, b be nonzero numbers. What is the following number?

$$\frac{\left(\frac{a}{b}\right)^9}{\left(\frac{a}{b}\right)^2} = \left(\frac{a}{b}\right)^{9-2}$$

> *Scaffolding:*
>
> Try Exercise 25 as a missing factor problem (see scaffold box above) using knowledge of dividing fractions.

Exercise 26

Let x be a nonzero number. What is the following number?

$$\frac{x^5}{x^4} = x^{5-4}$$

Can the following expressions be written in simpler forms? If yes, write an equivalent expression for each problem. If not, explain why not.

Exercise 27

$$\frac{2^7}{4^2} = \frac{2^7}{2^4} = 2^{7-4}$$

Exercise 28

$$\frac{3^{23}}{27} = \frac{3^{23}}{3^3} = 3^{23-3}$$

Exercise 29

$$\frac{3^5 \cdot 2^8}{3^2 \cdot 2^3} = 3^{5-2} \cdot 2^{8-3}$$

Exercise 30

$$\frac{(-2)^7 \cdot 95^5}{(-2)^5 \cdot 95^4} = (-2)^{7-5} \cdot 95^{5-4}$$

Exercise 31

Let x be a number. Write each expression in a simpler form.

a. $\frac{5}{x^3}(3x^8) = 15x^5$

b. $\frac{5}{x^3}(-4x^6) = -20x^3$

c. $\frac{5}{x^3}(11x^4) = 55x$

Exercise 32

Anne used an online calculator to multiply $2\ 000\ 000\ 000 \times 2\ 000\ 000\ 000\ 000$. The answer showed up on the calculator as $4\text{e} + 21$, as shown below. Is the answer on the calculator correct? How do you know?

$2\ 000\ 000\ 000 \times 2\ 000\ 000\ 000\ 000 = 4\ 000\ 000\ 000\ 000\ 000\ 000\ 000$.

The answer must mean 4 followed by 21 zeros. That means that the answer on the calculator is correct.

This problem is hinting at scientific notation (i.e., $(2 \times 10^9)(2 \times 10^{12}) = 4 \times 10^{9+12}$*). Accept any reasonable explanation of the answer.*

Closing (3 minutes)

Summarize, or have students summarize, the lesson.

- State the two identities and how to write equivalent expressions for each.

Optional Fluency Exercise (2 minutes)

This exercise is not an expectation of the standard, but it may prepare students for work with squared numbers in Module 2 with respect to the Pythagorean theorem. Therefore, this is an optional fluency exercise.

Have students chorally respond to numbers squared and cubed that you provide. For example, you say "1 squared," and students respond, "1." Next, you say, "2 squared," and students respond "4." Have students respond to all squares, in order, up to 15. When squares are finished, start with "1 cubed," and students respond "1." Next, say "2 cubed," and students respond "8." Have students respond to all cubes, in order, up to 10. If time allows, have students respond to random squares and cubes.

Exit Ticket (3 minutes)

Name ______________________________ Date ______________

Lesson 2: Multiplication of Numbers in Exponential Form

Exit Ticket

Write each expression using the fewest number of bases possible.

1. Let a and b be positive integers. $23^a \times 23^b =$

2. $5^3 \times 25 =$

3. Let x and y be positive integers and $x > y$. $\frac{11^x}{11^y} =$

4. $\frac{2^{13}}{2^3} =$

Exit Ticket Sample Solutions

Note to Teacher: Accept both forms of the answer; in other words, accept an answer that shows the exponents as a sum or difference as well as an answer where the numbers are actually added or subtracted.

Write each expression using the fewest number of bases possible.

1. **Let a and b be positive integers.** $23^a \times 23^b =$

   $23^a \times 23^b = 23^{a+b}$

2. $5^3 \times 25 =$

   $$5^3 \times 25 = 5^3 \times 5^2$$
   $$= 5^{3+2}$$
   $$= 5^5$$

3. **Let x and y be positive integers and $x > y$.** $\frac{11^x}{11^y} =$

   $\frac{11^x}{11^y} = 11^{x-y}$

4. $\frac{2^{13}}{8} =$

   $\frac{2^{13}}{2^3} = 2^{13-3} = 2^{10}$

EUREKA MATH™

Problem Set Sample Solutions

To ensure success with Problems 1 and 2, students should complete at least bounces 1–4 with support in class. Consider working on Problem 1 as a class activity and assigning Problem 2 for homework.

Students may benefit from a simple drawing of the scenario. It will help them see why the factor of 2 is necessary when calculating the distance traveled for each bounce. Make sure to leave the total distance traveled in the format shown so that students can see the pattern that is developing. Simplifying at any step will make it difficult to write the general statement for n number of bounces.

1. **A certain ball is dropped from a height of x feet. It always bounces up to $\frac{2}{3}x$ feet. Suppose the ball is dropped from 10 feet and is stopped exactly when it touches the ground after the 30th bounce. What is the total distance traveled by the ball? Express your answer in exponential notation.**

Bounce	Computation of Distance Traveled in Previous Bounce	Total Distance Traveled (in feet)
1	$2\left(\frac{2}{3}\right)10$	$10+2\left(\frac{2}{3}\right)10$
2	$2\left[\frac{2}{3}\left(\frac{2}{3}\right)10\right]$ $=2\left(\frac{2}{3}\right)^2 10$	$10+2\left(\frac{2}{3}\right)10+2\left(\frac{2}{3}\right)^2 10$
3	$2\left[\frac{2}{3}\left(\frac{2}{3}\right)^2 10\right]$ $=2\left(\frac{2}{3}\right)^3 10$	$10+2\left(\frac{2}{3}\right)10+2\left(\frac{2}{3}\right)^2 10+2\left(\frac{2}{3}\right)^3 10$
4	$2\left[\frac{2}{3}\left(\frac{2}{3}\right)^3 10\right]$ $=2\left(\frac{2}{3}\right)^4 10$	$10+2\left(\frac{2}{3}\right)10+2\left(\frac{2}{3}\right)^2 10+2\left(\frac{2}{3}\right)^3 10+2\left(\frac{2}{3}\right)^4 10$
30	$2\left(\frac{2}{3}\right)^{30} 10$	$10+2\left(\frac{2}{3}\right)10+2\left(\frac{2}{3}\right)^2 10+2\left(\frac{2}{3}\right)^3 10+2\left(\frac{2}{3}\right)^4 10+\cdots+2\left(\frac{2}{3}\right)^{30} 10$
n	$2\left(\frac{2}{3}\right)^n 10$	$10+20\left(\frac{2}{3}\right)\left(1+\left(\frac{2}{3}\right)+\left(\frac{2}{3}\right)^2+\cdots+\left(\frac{2}{3}\right)^n\right)$

2. **If the same ball is dropped from 10 feet and is stopped exactly at the highest point after the 25th bounce, what is the total distance traveled by the ball? Use what you learned from the last problem.**

Based on the last problem, we know that each bounce causes the ball to travel $2\left(\frac{2}{3}\right)^n 10$ feet. If the ball is stopped at the highest point of the 25th bounce, then the distance traveled on that last bounce is just $\left(\frac{2}{3}\right)^{25} 10$ feet because it does not make the return trip to the ground. Therefore, the total distance traveled by the ball in feet in this situation is

$$10+2\left(\frac{2}{3}\right)10+2\left(\frac{2}{3}\right)^2 10+2\left(\frac{2}{3}\right)^3 10+2\left(\frac{2}{3}\right)^4 10+\cdots+2\left(\frac{2}{3}\right)^{23} 10+2\left(\frac{2}{3}\right)^{24} 10.$$

3. Let a and b be numbers and $b \neq 0$, and let m and n be positive integers. Write each expression using the fewest number of bases possible.

$(-19)^5 \cdot (-19)^{11} = (-19)^{5+11}$	$2.7^5 \times 2.7^3 = 2.7^{5+3}$
$\frac{7^{10}}{7^3} = 7^{10-3}$	$\left(\frac{1}{5}\right)^2 \cdot \left(\frac{1}{5}\right)^{15} = \left(\frac{1}{5}\right)^{2+15}$
$\left(-\frac{9}{7}\right)^m \cdot \left(-\frac{9}{7}\right)^n = \left(-\frac{9}{7}\right)^{m+n}$	$\frac{ab^3}{b^2} = ab^{3-2}$

4. Let the dimensions of a rectangle be $(4 \times (871\,209)^5 + 3 \times 49\,762\,105)$ ft. by $(7 \times (871\,209)^3 - (49\,762\,105)^4)$ ft. Determine the area of the rectangle. (Hint: You do not need to expand all the powers.)

$$\text{Area} = (4 \times (871\,209)^5 + 3 \times 49\,762\,105)\text{ ft.}(7 \times (871\,209)^3 - (49\,762\,105)^4)\text{ ft.}$$

$$= (28 \times (871\,209)^8 - 4 \times (871\,209)^5(49\,762\,105)^4 + 21 \times (871\,209)^3(49\,762\,105) - 3 \times (49\,762\,105)^5)\text{ sq. ft.}$$

5. A rectangular area of land is being sold off in smaller pieces. The total area of the land is 2^{15} square miles. The pieces being sold are 8^3 square miles in size. How many smaller pieces of land can be sold at the stated size? Compute the actual number of pieces.

$8^3 = 2^9$ $\qquad \frac{2^{15}}{2^9} = 2^{15-9} = 2^6 = 64$ $\qquad$ *64 pieces of land can be sold.*

EUREKA MATH

Lesson 3: Numbers in Exponential Form Raised to a Power

Student Outcomes

- Students know how to take powers of powers. Students know that when a product is raised to a power, each factor of the product is raised to that power.
- Students write simplified, equivalent numeric, and symbolic expressions using this new knowledge of powers.

Lesson Notes

As with Lesson 2, consider providing opportunities for students to discover the property of exponents introduced in this lesson prior to giving the mathematical rationale as to why it is true. For example, you may present students with the problems in Examples 1 and 2 and allow them to share their thinking about what the answer should be and then provide the mathematical reasoning behind their correct solutions.

We continue the work of knowing and applying the properties of integer exponents to generate equivalent expressions in this lesson. As with Lesson 2, students should be exposed to general arguments as to why the properties are true and be able to explain them on their own with concrete numbers. However, the relationship between the laws of exponents and repeated addition is not as important and could be omitted if time is an issue. The discussion that relates taking a power to a power and the four arithmetic operations may also be omitted, but do allow time for students to consider the relationship demonstrated in the concrete problems $(5 \times 8)^{17}$ and $5^{17} \times 8^{17}$.

Classwork

Discussion (10 minutes)

MP.2 & MP.7

Suppose we add 4 copies of 3, thereby getting $(3 + 3 + 3 + 3)$ and then add 5 copies of the sum. We get

$$(3+3+3+3)+(3+3+3+3)+(3+3+3+3)+(3+3+3+3)+(3+3+3+3).$$

Now, by the definition of multiplication, adding 4 copies of 3 is denoted by (4×3),

$$(4\times 3)+(4\times 3)+(4\times 3)+(4\times 3)+(4\times 3),$$

and also by definition of multiplication, adding 5 copies of this product is then denoted by $5 \times (4 \times 3)$. So,

$$5\times(4\times 3)=(3+3+3+3)+(3+3+3+3)+(3+3+3+3)+(3+3+3+3)+(3+3+3+3).$$

A closer examination of the right side of the above equation reveals that we are adding 3 to itself 20 times (i.e., adding 3 to itself (5×4) times). Therefore,

$$5\times(4\times 3)=(5\times 4)\times 3.$$

So ultimately, because multiplying can be considered as repeated addition, multiplying three numbers is really repeated addition of a value represented by repeated addition.

MP.2 & MP.7

Now, let us consider repeated multiplication.

$$\text{(For example, } \underbrace{(3\times3\times3\times3)\times(3\times3\times3\times3)\cdots\times(3\times3\times3\times3)}_{5\text{ times}} = \underbrace{3^4\times3^4\cdots\times3^4}_{5\text{ times}}.)$$

- What is multiplying 4 copies of 3 and then multiplying 5 copies of the product?
  - *Multiplying* 4 *copies of* 3 *is* 3^4*, and multiplying* 5 *copies of the product is* $(3^4)^5$*. We wish to say this is equal to* 3^x *for some positive integer* x*. By the analogy initiated in Lesson 1, the* 5×4 *in* $(5\times4)\times3$ *should correspond to the exponent* x *in* 3^x*; therefore, the answer should be*

$$(3^4)^5 = 3^{5\times4}.$$

This is correct because

$$\begin{aligned}(3^4)^5 &= (3\times3\times3\times3)^5\\ &= \underbrace{(3\times3\times3\times3)\times\cdots\times(3\times3\times3\times3)}_{5\text{ times}}\\ &= \underbrace{3\times3\times\cdots\times3}_{5\times4\text{ times}}\\ &= 3^{5\times4}.\end{aligned}$$

Examples 1–2

Work through Examples 1 and 2 in the same manner. (Supplement with additional examples if needed.) Have students calculate the resulting exponent; however, emphasis should be placed on the step leading to the resulting exponent, which is the product of the exponents.

Example 1

$$\begin{aligned}(7^2)^6 &= (7\times7)^6\\ &= \underbrace{(7\times7)\times\cdots\times(7\times7)}_{6\text{ times}}\\ &= \underbrace{7\times\cdots\times7}_{6\times2\text{ times}}\\ &= 7^{6\times2}\end{aligned}$$

Example 2

$$\begin{aligned}(1.3^3)^{10} &= (1.3\times1.3\times1.3)^{10}\\ &= \underbrace{(1.3\times1.3\times1.3)\times\cdots\times(1.3\times1.3\times1.3)}_{10\text{ times}}\\ &= \underbrace{1.3\times\cdots\times1.3}_{10\times3\text{ times}}\\ &= 1.3^{10\times3}\end{aligned}$$

In the same way, we have

For any number x and any positive integers m and n,

$$(x^m)^n = x^{nm}$$

because

$$\begin{aligned}(x^m)^n &= (\underbrace{x\cdot x\cdots x}_{m\text{ times}})^n\\ &= \underbrace{\underbrace{(x\cdot x\cdots x)}_{m\text{ times}}\times\cdots\times\underbrace{(x\cdot x\cdots x)}_{m\text{ times}}}_{n\text{ times}}\\ &= x^{nm}.\end{aligned}$$

Exercises 1–6 (10 minutes)

Have students complete Exercises 1–4 independently. Check their answers, and then have students complete Exercises 5–6.

Exercise 1

$(15^3)^9 = 15^{9\times 3}$

Exercise 2

$((-2)^5)^8 = (-2)^{8\times 5}$

Exercise 3

$(3.4^{17})^4 = 3.4^{4\times 17}$

Exercise 4

Let s be a number.

$(s^{17})^4 = s^{4\times 17}$

Exercise 5

Sarah wrote $(3^5)^7 = 3^{12}$. Correct her mistake. Write an exponential expression using a base of 3 and exponents of 5, 7, and 12 that would make her answer correct.

Correct way: $(3^5)^7 = 3^{35}$; *Rewritten Problem:* $3^5 \times 3^7 = 3^{5+7} = 3^{12}$.

Exercise 6

A number y satisfies $y^{24} - 256 = 0$. What equation does the number $x = y^4$ satisfy?

Since $x = y^4$, then $(x)^6 = (y^4)^6$. Therefore, $x = y^4$ would satisfy the equation $x^6 - 256 = 0$.

Discussion (10 minutes)

MP.7

From the point of view of algebra and arithmetic, the most basic question about raising a number to a power has to be the following: How is this operation related to the four arithmetic operations? In other words, for two numbers x, y and a positive integer n,

1. How is $(xy)^n$ related to x^n and y^n?
2. How is $\left(\frac{x}{y}\right)^n$ related to x^n and y^n, $y \neq 0$?
3. How is $(x + y)^n$ related to x^n and y^n?
4. How is $(x - y)^n$ related to x^n and y^n?

Scaffolding:

As an alternative to the discussion, provide students with the four questions shown and encourage them to work with partners or in small groups to find a relationship in each case *if one exists*. You might consider assigning one case to each group and present their findings to the class. Students will find the latter two cases to be much more complicated.

The answers to the last two questions turn out to be complicated; students learn about this in high school under the heading of the **binomial theorem**. However, they should at least be aware that, in general,

$(x + y)^n \neq x^n + y^n$, *unless* $n = 1$. *For example,* $(2 + 3)^2 \neq 2^2 + 3^2$.

Allow time for discussion of Problem 1. Students can begin by talking in partners or small groups and then share with the class.

Scaffolding:

Provide a numeric example for students to work on

$(5 \times 8)^{17} = 5^{17} \times 8^{17}$.

Some students may want to simply multiply 5×8, but remind them to focus on the above-stated goal, which is to relate $(5 \times 8)^{17}$ to 5^{17} and 8^{17}. Therefore, we want to see 17 copies of 5 and 17 copies of 8 on the right side. Multiplying 5×8 would take us in a different direction.

$$\begin{aligned}(5 \times 8)^{17} &= \underbrace{(5 \times 8) \times \cdots \times (5 \times 8)}_{17 \text{ times}} \\ &= \underbrace{(5 \times \cdots \times 5)}_{17 \text{ times}} \times \underbrace{(8 \times \cdots \times 8)}_{17 \text{ times}} \\ &= 5^{17} \times 8^{17}\end{aligned}$$

The following computation is a different way of proving the equality.

$$\begin{aligned}5^{17} \times 8^{17} &= \underbrace{(5 \times \cdots \times 5)}_{17 \text{ times}} \times \underbrace{(8 \times \cdots \times 8)}_{17 \text{ times}} \\ &= \underbrace{(5 \times 8) \times \cdots \times (5 \times 8)}_{17 \text{ times}} \\ &= (5 \times 8)^{17}\end{aligned}$$

Answer to Problem 1:

Because in $(xy)^n$, the factors x and y are repeatedly multiplied n times, resulting in factors of x^n and y^n:

$$(xy)^n = x^n y^n$$

because

$(xy)^n = \underbrace{(xy)\cdots(xy)}_{n \text{ times}}$	By definition of raising a number to the nth power
$= \underbrace{(x \cdot x \cdots x)}_{n \text{ times}} \cdot \underbrace{(y \cdot y \cdots y)}_{n \text{ times}}$	By commutative and associative properties
$= x^n y^n$	By definition of x^n

Scaffolding:

Advanced learners may ask about cases in which n is not a positive integer. At this point in the module, some students may have begun to develop an intuition about what other integer exponents mean. Encourage them to continue thinking as we begin examining zero exponents in Lesson 4 and negative integer exponents in Lesson 5.

For any numbers x and y, and positive integer n,

$$(xy)^n = x^n y^n$$

because

$$\begin{aligned}(xy)^n &= \underbrace{(xy)\cdots(xy)}_{n \text{ times}} \\ &= \underbrace{(x \cdot x \cdots x)}_{n \text{ times}} \cdot \underbrace{(y \cdot y \cdots y)}_{n \text{ times}} \\ &= x^n y^n.\end{aligned}$$

Exercises 7–13 (10 minutes)

Have students complete Exercises 17–12 independently and then check their answers.

Exercise 7

$(11 \times 4)^9 = 11^{9\times1} \times 4^{9\times1}$

Exercise 8

$(3^2 \times 7^4)^5 = 3^{5\times2} \times 7^{5\times4}$

Exercise 9

Let a, b, and c be numbers.

$(3^2a^4)^5 = 3^{5\times2}a^{5\times4}$

Exercise 10

Let x be a number.

$(5x)^7 = 5^{7\times1} \cdot x^{7\times1}$

Exercise 11

Let x and y be numbers.

$(5xy^2)^7 = 5^{7\times1} \cdot x^{7\times1} \cdot y^{7\times2}$

Exercise 12

Let a, b, and c be numbers.

$(a^2bc^3)^4 = a^{4\times2} \cdot b^{4\times1} \cdot c^{4\times3}$

Have students work in pairs or small groups on Exercise 13 after you present the problem.

First ask students to explain why we must assume $y \neq 0$. They should say that if the denominator were zero then the value of the fraction would be undefined.

- The answer to the fourth question is similar to the third: If x, y are any two numbers, such that $y \neq 0$ and n is a positive integer, then

$$\left(\frac{x}{y}\right)^n = \frac{x^n}{y^n}.$$

Scaffolding:

- Have students review problems just completed.
- Remind students to begin with the definition of a number raised to a power.

Exercise 13

Let x and y be numbers, $y \neq 0$, and let n be a positive integer. How is $\left(\frac{x}{y}\right)^n$ related to x^n and y^n?

$$\left(\frac{x}{y}\right)^n = \frac{x^n}{y^n}$$

Because

$$\left(\frac{x}{y}\right)^n = \underbrace{\frac{x}{y} \times \cdots \times \frac{x}{y}}_{n \text{ times}} \quad \textit{By definition}$$

$$= \frac{\overbrace{x \cdot x \cdots x}^{n \text{ times}}}{\underbrace{y \cdot y \cdots y}_{n \text{ times}}} \quad \textit{By the product formula}$$

$$= \frac{x^n}{y^n} \quad \textit{By definition}$$

Let students know that this type of reasoning is required to prove facts in mathematics. They should always supply a reason for each step or at least know the reason the facts are connected. Further, it is important to keep in mind what we already know in order to figure out what we do not know. Students are required to write two proofs for the Problem Set that are extensions of the proofs they have done in class.

Closing (2 minutes)

Summarize, or have students summarize, the lesson. Students should state that they now know how to take powers of powers.

Exit Ticket (3 minutes)

Name ______________________________ Date______________

Lesson 3: Numbers in Exponential Form Raised to a Power

Exit Ticket

Write each expression as a base raised to a power or as the product of bases raised to powers that is equivalent to the given expression.

1. $(9^3)^6 =$

2. $(113^2 \times 37 \times 51^4)^3 =$

3. Let x, y, z be numbers. $(x^2yz^4)^3 =$

4. Let x, y, z be numbers and let m, n, p, q be positive integers. $(x^m y^n z^p)^q =$

5. $\frac{4^8}{5^8} =$

Exit Ticket Sample Solutions

Write each expression as a base raised to a power or as the product of bases raised to powers that is equivalent to the given expression.

1. $(9^3)^6 =$

   $(9^3)^6 = 9^{6\times 3} = 9^{18}$

2. $(113^2 \times 37 \times 51^4)^3 =$

   $(113^2 \times 37 \times 51^4)^3 = \left((113^2 \times 37) \times 51^4\right)^3$ *By associative law*

   $= (113^2 \times 37)^3 \times (51^4)^3$ *Because $(xy)^n = x^n y^n$ for all numbers x, y*

   $= (113^2)^3 \times 37^3 \times (51^4)^3$ *Because $(xy)^n = x^n y^n$ for all numbers x, y*

   $= 113^6 \times 37^3 \times 51^{12}$ *Because $(x^m)^n = x^{mn}$ for all numbers x*

3. Let x, y, z be numbers. $(x^2yz^4)^3 =$

   $(x^2yz^4)^3 = \left((x^2 \times y) \times z^4\right)^3$ *By associative law*

   $= (x^2 \times y)^3 \times (z^4)^3$ *Because $(xy)^n = x^n y^n$ for all numbers x, y*

   $= (x^2)^3 \times y^3 \times (z^4)^3$ *Because $(xy)^n = x^n y^n$ for all numbers x, y*

   $= x^6 \times y^3 \times z^{12}$ *Because $(x^m)^n = x^{mn}$ for all numbers x*

   $= x^6y^3z^{12}$

4. Let x, y, z be numbers and let m, n, p, q be positive integers. $(x^m y^n z^p)^q =$

   $(x^m y^n z^p)^q = \left((x^m \times y^n) \times z^p\right)^q$ *By associative law*

   $= (x^m \times y^n)^q \times (z^p)^q$ *Because $(xy)^n = x^n y^n$ for all numbers x, y*

   $= (x^m)^q \times (y^n)^q \times (z^p)^q$ *Because $(xy)^n = x^n y^n$ for all numbers x, y*

   $= x^{mp} \times y^{nq} \times z^{pq}$ *Because $(x^m)^n = x^{mn}$ for all numbers x*

   $= x^{mq}y^{nq}z^{pq}$

5. $\frac{4^8}{5^8} =$

   $\frac{4^8}{5^8} = \left(\frac{4}{5}\right)^8$

Problem Set Sample Solutions

1. Show (prove) in detail why $(2 \cdot 3 \cdot 7)^4 = 2^4 3^4 7^4$.

$(2 \cdot 3 \cdot 7)^4 = (2 \cdot 3 \cdot 7)(2 \cdot 3 \cdot 7)(2 \cdot 3 \cdot 7)(2 \cdot 3 \cdot 7)$ — *By definition*

$= (2 \cdot 2 \cdot 2 \cdot 2)(3 \cdot 3 \cdot 3 \cdot 3)(7 \cdot 7 \cdot 7 \cdot 7)$ — *By repeated use of the commutative and associative properties*

$= 2^4 3^4 7^4$ — *By definition*

2. Show (prove) in detail why $(xyz)^4 = x^4 y^4 z^4$ for any numbers x, y, z.

The left side of the equation $(xyz)^4$ means $(xyz)\ (xyz)\ (xyz)\ (xyz)$. Using the commutative and associative properties of multiplication, we can write $(xyz)\ (xyz)\ (xyz)\ (xyz)$ as $(xxxx)(yyyy)\ (zzzz)$, which in turn can be written as $x^4 y^4 z^4$, which is what the right side of the equation states.

3. Show (prove) in detail why $(xyz)^n = x^n y^n z^n$ for any numbers x, y, and z and for any positive integer n.

Beginning with the left side of the equation, $(xyz)^n$ means $\underbrace{(xyz) \cdot (xyz) \cdots (xyz)}_{n \text{ times}}$. Using the commutative and associative properties of multiplication, $\underbrace{(xyz) \cdot (xyz) \cdots (xyz)}_{n \text{ times}}$ can be rewritten as $\underbrace{(x \cdot x \cdots x)}_{n \text{ times}} \underbrace{(y \cdot y \cdots y)}_{n \text{ times}} \underbrace{(z \cdot z \cdots z)}_{n \text{ times}}$ and, finally, $x^n y^n z^n$, which is what the right side of the equation states. We can also prove this equality by a different method, as follows. Beginning with the right side $x^n y^n z^n$ means $\underbrace{(x \cdot x \cdots x)}_{n \text{ times}} \underbrace{(y \cdot y \cdots y)}_{n \text{ times}} \underbrace{(z \cdot z \cdots z)}_{n \text{ times}}$, which by the commutative property of multiplication can be rewritten as $\underbrace{(xyz) \cdot (xyz) \cdots (xyz)}_{n \text{ times}}$. Using exponential notation, $\underbrace{(xyz) \cdot (xyz) \cdots (xyz)}_{n \text{ times}}$ can be rewritten as $(xyz)^n$, which is what the left side of the equation states.

Lesson 4: Numbers Raised to the Zeroth Power

Student Outcomes

- Students know that a number raised to the zeroth power is equal to one.
- Students recognize the need for the definition to preserve the properties of exponents.

Lesson Notes

In this lesson we introduce the zeroth power and its definition. Most of the time in this lesson should be spent having students work through possible meanings of numbers raised to the zeroth power and then checking the validity of their claims. For that reason, focus should be placed on the Exploratory Challenge in this lesson. Encourage students to share their thinking about what a number raised to the zeroth power could mean. It may be necessary to guide students through the work of developing cases to check the definition $x^0 = 1$ in Exercise 1 and the check of the first law of exponents in Exercise 2. Exercises 3 and 4 may be omitted if time is an issue.

Classwork

Concept Development (5 minutes): Let us summarize our main conclusions about exponents. For any numbers x, y and any positive integers m, n, the following holds

$$x^m \cdot x^n = x^{m+n} \quad (1)$$

$$(x^m)^n = x^{mn} \quad (2)$$

$$(xy)^n = x^n y^n. \quad (3)$$

We have shown that for any numbers x, y, and any positive integers m, n, the following holds

$$x^m \cdot x^n = x^{m+n} \quad \textbf{(1)}$$

$$(x^m)^n = x^{mn} \quad \textbf{(2)}$$

$$(xy)^n = x^n y^n. \quad \textbf{(3)}$$

Definition: ____

If we assume $x \neq 0$ in equation (4) and $y \neq 0$ in equation (5) below, then we also have

$$\frac{x^m}{x^n} = x^{m-n} \quad (4)$$

$$\left(\frac{x}{y}\right)^n = \frac{x^n}{y^n}. \quad (5)$$

G8-M1-B1-v1.3.0_05/04/15

There is an obvious reason why the x in (4) and the y in (5) must be nonzero: We cannot divide by 0. Please note that in high school it is further necessary to restrict the values of x and y to nonnegative numbers when defining rational number exponents. Our examination of exponents in Grade 8 is limited to the integers, however, so restricting the base values to nonnegative numbers should not be a concern for students at this time.

We group equations (1)–(3) together because they are the foundation on which all the results about exponents rest. When they are suitably generalized, as they are above, they imply (4) and (5). Therefore, we concentrate on (1)–(3).

The most important feature of (1)–(3) is that they are *simple* and *formally (symbolically) natural*. Mathematicians want these three identities to continue to hold for *all* exponents m and n, without the restriction that m and n be positive integers because of these two desirable qualities. We should do it one step at a time. Our goal in this grade is to extend the validity of (1)–(3) to all *integers* m and n.

Exploratory Challenge (20 minutes)

The first step in this direction is to introduce the definition of the 0th exponent of a number and to then use it to prove that (1)–(3) remain valid when m and n are not just positive integers but all whole numbers (including 0). Since our goal is to make sure (1)–(3) remain valid even when m and n may be 0, the very definition of the 0th exponent of a number must pose no *obvious* contradiction to (1)–(3). With this in mind, let us consider what it means to raise a number x to the zeroth power. For example, what should 3^0 mean?

- *Students will likely respond that* 3^0 *should equal* 0. When they do, demonstrate why that would contradict our existing understanding of properties of exponents using (1). Specifically, if m is a positive integer and we let $3^0 = 0$, then

$$3^m \cdot 3^0 = 3^{m+0},$$

but since we let $3^0 = 0$, it means that the left side of the equation would equal zero. That creates a contradiction because

$$0 \neq 3^{m+0}.$$

Therefore, letting $3^0 = 0$ will not help us to extend (1)–(3) to all whole numbers m and n.

- *Next, students may say that we should let* $3^0 = 3$. Show the two problematic issues this would create. First, in Lesson 1, we learned that, by definition, $x^1 = x$, and we do not want to have two powers that yield the same result. Second, it would violate the existing rules we have developed: Looking specifically at (1) again, if we let $3^0 = 3$, then

$$3^m \cdot 3^0 = 3^{m+0},$$

but

$$3^m \cdot 3^0 = \underbrace{3 \times \cdots \times 3}_{m \text{ times}} \cdot 3$$
$$= 3^{m+1},$$

which again is a contradiction.

> *Scaffolding:*
>
> Ask struggling students to use what they know about the laws of exponents to rewrite $3^m \cdot 3^0$. Their resulting expression should be 3^m. Then ask: If a number, 3, is multiplied by another number, and the product is 3, what does that mean about the other number? It means the other number must be 1. Then have students apply this thinking to the equation $3^m \cdot 3^0 = 3^m$ to determine the value of 3^0.

If we believe that equation (1) should hold even when $n = 0$, then, for example, $3^{2+0} = 3^2 \times 3^0$, which is the same as $3^2 = 3^2 \times 3^0$; therefore, after multiplying both sides by the number $\frac{1}{3^2}$, we get $1 = 3^0$. In the same way, our belief that (1) should hold when either m or n is 0, would lead us to conclude that we should define $x^0 = 1$ for any nonzero x. Therefore, we give the following definition:

MP.6 **Definition:** For any positive number x, we define $x^0 = 1$.

Students should write this definition of x^0 in the lesson summary box on their classwork paper.

Now that x^n is defined for all whole numbers n, check carefully that (1)–(3) remain valid for all whole numbers m and n.

MP.3 Have students independently complete Exercise 1; provide correct values for m and n before proceeding to the development of cases (A)–(C).

Exercise 1

List all possible cases of whole numbers m and n for identity (1). More precisely, when $m > 0$ and $n > 0$, we already know that (1) is correct. What are the other possible cases of m and n for which (1) is yet to be verified?

Case (A): $m > 0$ and $n = 0$

Case (B): $m = 0$ and $n > 0$

Case (C): $m = n = 0$

Model how to check the validity of a statement using Case (A) with equation (1) as part of Exercise 2. Have students work independently or in pairs to check the validity of (1) in Case (B) and Case (C) to complete Exercise 2. Next, have students check the validity of equations (2) and (3) using Cases (A)–(C) for Exercises 3 and 4.

Exercise 2

Check that equation (1) is correct for each of the cases listed in Exercise 1.

Case (A): $x^m \cdot x^0 = x^m$? *Yes, because $x^m \cdot x^0 = x^m \cdot 1 = x^m$.*

Case (B): $x^0 \cdot x^n = x^n$? *Yes, because $x^0 \cdot x^n = 1 \cdot x^n = x^n$.*

Case (C): $x^0 \cdot x^0 = x^0$? *Yes, because $x^0 \cdot x^0 = 1 \cdot 1 = x^0$.*

Exercise 3

Do the same with equation (2) by checking it case-by-case.

Case (A): $(x^m)^0 = x^{0 \times m}$? Yes, because x^m is a number, and a number raised to a zero power is 1. $1 = x^0 = x^{0 \times m}$. So, the left side is 1. The right side is also 1 because $x^{0 \times m} = x^0 = 1$.

Case (B): $(x^0)^n = x^{n \times 0}$? Yes, because, by definition $x^0 = 1$ and $1^n = 1$, the left side is equal to 1. The right side is equal to $x^0 = 1$, so both sides are equal.

Case (C): $(x^0)^0 = x^{0 \times 0}$? Yes, because, by definition of the zeroth power of x, both sides are equal to 1.

Exercise 4

Do the same with equation (3) by checking it case-by-case.

Case (A): $(xy)^0 = x^0y^0$? *Yes, because the left side is* 1 *by the definition of the zeroth power, while the right side is* $1 \times 1 = 1$.

Case (B): Since $n > 0$, *we already know that (3) is valid.*

Case (C): This is the same as Case (A), which we have already shown to be valid.

Exploratory Challenge 2 (5 minutes)

Students practice writing numbers in expanded form in Exercises 5 and 6. Students use the definition of x^0, for any number x, learned in this lesson.

Clearly state that you want to see the ones digit multiplied by 10^0. That is the important part of the expanded notation because it leads to the use of negative powers of 10 for decimals in Lesson 5.

> *Scaffolding:*
>
> You may need to remind students how to write numbers in expanded form with Exercise 5.

Exercise 5

Write the expanded form of $8{,}374$ using exponential notation.

$8374 = (8 \times 10^3) + (3 \times 10^2) + (7 \times 10^1) + (4 \times 10^0)$

Exercise 6

Write the expanded form of $6{,}985{,}062$ using exponential notation.

$6\,985\,062 = (6 \times 10^6) + (9 \times 10^5) + (8 \times 10^4) + (5 \times 10^3) + (0 \times 10^2) + (6 \times 10^1) + (2 \times 10^0)$

> *Scaffolding:*
>
> Ask advanced learners to consider the expanded form of a number such as 945.78. Ask them if it is possible to write this number in expanded form using exponential notation. If so, what exponents would represent the decimal place values? This question serves as a sneak peek into the meaning of negative integer exponents.

Closing (3 minutes)

Summarize, or have students summarize, the lesson.

- The rules of exponents that we have worked on prior to today only work for positive integer exponents; now those same exponent rules have been extended to all whole numbers.
- The next logical step is to attempt to extend these rules to all integer exponents.

Exit Ticket (2 minutes)

Fluency Exercise (10 minutes)

Sprint: Rewrite expressions with the same base for positive exponents only. Make sure to tell the students that all letters within the problems of the Sprint are meant to denote numbers. This exercise can be administered at any point during the lesson. Refer to the Sprints and Sprint Delivery Script sections in the Module Overview for directions to administer a Sprint.

Name ______________________________ Date ______________

Lesson 4: Numbers Raised to the Zeroth Power

Exit Ticket

1. Simplify the following expression as much as possible.

$$\frac{4^{10}}{4^{10}} \cdot 7^0 =$$

2. Let a and b be two numbers. Use the distributive law and then the definition of zeroth power to show that the numbers $(a^0 + b^0)a^0$ and $(a^0 + b^0)b^0$ are equal.

Exit Ticket Sample Solutions

1. Simplify the following expression as much as possible.

$$\frac{4^{10}}{4^{10}} \cdot 7^0 = 4^{10-10} \cdot 1 = 4^0 \cdot 1 = 1 \cdot 1 = 1$$

2. Let a and b be two numbers. Use the distributive law and then the definition of zeroth power to show that the numbers $(a^0 + b^0)a^0$ and $(a^0 + b^0)b^0$ are equal.

$$\begin{aligned}(a^0 + b^0)a^0 &= a^0 \cdot a^0 + b^0 \cdot a^0 \\ &= a^{0+0} + a^0b^0 \\ &= a^0 + a^0b^0 \\ &= 1 + 1 \cdot 1 \\ &= 1 + 1 \\ &= 2\end{aligned}$$

$$\begin{aligned}(a^0 + b^0)b^0 &= a^0 \cdot b^0 + b^0 \cdot b^0 \\ &= a^0b^0 + b^{0+0} \\ &= a^0b^0 + b^0 \\ &= 1 \cdot 1 + 1 \\ &= 1 + 1 \\ &= 2\end{aligned}$$

Since both numbers are equal to 2, they are equal.

Problem Set Sample Solutions

Let x, y be numbers $(x, y \neq 0)$. Simplify each of the following expressions.

1.
$$\begin{aligned}\frac{y^{12}}{y^{12}} &= y^{12-12} \\ &= y^0 \\ &= 1\end{aligned}$$

2.
$$\begin{aligned}9^{15} \cdot \frac{1}{9^{15}} &= \frac{9^{15}}{9^{15}} \\ &= 9^{15-15} \\ &= 9^0 \\ &= 1\end{aligned}$$

3.
$$\begin{aligned}(7(123456.789)^4)^0 &= \\ &= 7^0(123456.789)^{4\times 0} \\ &= 7^0(123456.789)^0 \\ &= 1\end{aligned}$$

4.
$$\begin{aligned}2^2 \cdot \frac{1}{2^5} \cdot 2^5 \cdot \frac{1}{2^2} &= \frac{2^2}{2^2} \cdot \frac{2^5}{2^5} \\ &= 2^{2-2} \cdot 2^{5-5} \\ &= 2^0 \cdot 2^0 \\ &= 1\end{aligned}$$

5.
$$\begin{aligned}\frac{x^{41}}{y^{15}} \cdot \frac{y^{15}}{x^{41}} &= \frac{x^{41} \cdot y^{15}}{y^{15} \cdot x^{41}} \\ &= \frac{x^{41}}{x^{41}} \cdot \frac{y^{15}}{y^{15}} \\ &= x^{41-41} \cdot y^{15-15} \\ &= x^0 \cdot y^0 \\ &= 1\end{aligned}$$

Number Correct: ______

Applying Properties of Exponents to Generate Equivalent Expressions—Round 1

Directions: Simplify each expression using the laws of exponents. Use the least number of bases possible and only positive exponents. All letters denote numbers.

1.	$2^2 \cdot 2^3 =$		23.	$6^3 \cdot 6^2 =$	
2.	$2^2 \cdot 2^4 =$		24.	$6^2 \cdot 6^3 =$	
3.	$2^2 \cdot 2^5 =$		25.	$(-8)^3 \cdot (-8)^7 =$	
4.	$3^7 \cdot 3^1 =$		26.	$(-8)^7 \cdot (-8)^3 =$	
5.	$3^8 \cdot 3^1 =$		27.	$(0.2)^3 \cdot (0.2)^7 =$	
6.	$3^9 \cdot 3^1 =$		28.	$(0.2)^7 \cdot (0.2)^3 =$	
7.	$7^6 \cdot 7^2 =$		29.	$(-2)^{12} \cdot (-2)^1 =$	
8.	$7^6 \cdot 7^3 =$		30.	$(-2.7)^{12} \cdot (-2.7)^1 =$	
9.	$7^6 \cdot 7^4 =$		31.	$1.1^6 \cdot 1.1^9 =$	
10.	$11^{15} \cdot 11 =$		32.	$57^6 \cdot 57^9 =$	
11.	$11^{16} \cdot 11 =$		33.	$x^6 \cdot x^9 =$	
12.	$2^{12} \cdot 2^2 =$		34.	$2^7 \cdot 4 =$	
13.	$2^{12} \cdot 2^4 =$		35.	$2^7 \cdot 4^2 =$	
14.	$2^{12} \cdot 2^6 =$		36.	$2^7 \cdot 16 =$	
15.	$99^5 \cdot 99^2 =$		37.	$16 \cdot 4^3 =$	
16.	$99^6 \cdot 99^3 =$		38.	$3^2 \cdot 9 =$	
17.	$99^7 \cdot 99^4 =$		39.	$3^2 \cdot 27 =$	
18.	$5^8 \cdot 5^2 =$		40.	$3^2 \cdot 81 =$	
19.	$6^8 \cdot 6^2 =$		41.	$5^4 \cdot 25 =$	
20.	$7^8 \cdot 7^2 =$		42.	$5^4 \cdot 125 =$	
21.	$r^8 \cdot r^2 =$		43.	$8 \cdot 2^9 =$	
22.	$s^8 \cdot s^2 =$		44.	$16 \cdot 2^9 =$	

Applying Properties of Exponents to Generate Equivalent Expressions—Round 1 [KEY]

Directions: Simplify each expression using the laws of exponents. Use the least number of bases possible and only positive exponents. All letters denote numbers.

1.	$2^2 \cdot 2^3 =$	2^5
2.	$2^2 \cdot 2^4 =$	2^6
3.	$2^2 \cdot 2^5 =$	2^7
4.	$3^7 \cdot 3^1 =$	3^8
5.	$3^8 \cdot 3^1 =$	3^9
6.	$3^9 \cdot 3^1 =$	3^{10}
7.	$7^6 \cdot 7^2 =$	7^8
8.	$7^6 \cdot 7^3 =$	7^9
9.	$7^6 \cdot 7^4 =$	7^{10}
10.	$11^{15} \cdot 11 =$	11^{16}
11.	$11^{16} \cdot 11 =$	11^{17}
12.	$2^{12} \cdot 2^2 =$	2^{14}
13.	$2^{12} \cdot 2^4 =$	2^{16}
14.	$2^{12} \cdot 2^6 =$	2^{18}
15.	$99^5 \cdot 99^2 =$	99^7
16.	$99^6 \cdot 99^3 =$	99^9
17.	$99^7 \cdot 99^4 =$	99^{11}
18.	$5^8 \cdot 5^2 =$	5^{10}
19.	$6^8 \cdot 6^2 =$	6^{10}
20.	$7^8 \cdot 7^2 =$	7^{10}
21.	$r^8 \cdot r^2 =$	r^{10}
22.	$s^8 \cdot s^2 =$	s^{10}

23.	$6^3 \cdot 6^2 =$	6^5
24.	$6^2 \cdot 6^3 =$	6^5
25.	$(-8)^3 \cdot (-8)^7 =$	$(-8)^{10}$
26.	$(-8)^7 \cdot (-8)^3 =$	$(-8)^{10}$
27.	$(0.2)^3 \cdot (0.2)^7 =$	$(0.2)^{10}$
28.	$(0.2)^7 \cdot (0.2)^3 =$	$(0.2)^{10}$
29.	$(-2)^{12} \cdot (-2)^1 =$	$(-2)^{13}$
30.	$(-2.7)^{12} \cdot (-2.7)^1 =$	$(-2.7)^{13}$
31.	$1.1^6 \cdot 1.1^9 =$	1.1^{15}
32.	$57^6 \cdot 57^9 =$	57^{15}
33.	$x^6 \cdot x^9 =$	x^{15}
34.	$2^7 \cdot 4 =$	2^9
35.	$2^7 \cdot 4^2 =$	2^{11}
36.	$2^7 \cdot 16 =$	2^{11}
37.	$16 \cdot 4^3 =$	4^5
38.	$3^2 \cdot 9 =$	3^4
39.	$3^2 \cdot 27 =$	3^5
40.	$3^2 \cdot 81 =$	3^6
41.	$5^4 \cdot 25 =$	5^6
42.	$5^4 \cdot 125 =$	5^7
43.	$8 \cdot 2^9 =$	2^{12}
44.	$16 \cdot 2^9 =$	2^{13}

Number Correct: ______

Improvement: ______

Applying Properties of Exponents to Generate Equivalent Expressions—Round 2

Directions: Simplify each expression using the laws of exponents. Use the least number of bases possible and only positive exponents. All letters denote numbers.

1.	$5^2 \cdot 5^3 =$	
2.	$5^2 \cdot 5^4 =$	
3.	$5^2 \cdot 5^5 =$	
4.	$2^7 \cdot 2^1 =$	
5.	$2^8 \cdot 2^1 =$	
6.	$2^9 \cdot 2^1 =$	
7.	$3^6 \cdot 3^2 =$	
8.	$3^6 \cdot 3^3 =$	
9.	$3^6 \cdot 3^4 =$	
10.	$7^{15} \cdot 7 =$	
11.	$7^{16} \cdot 7 =$	
12.	$11^{12} \cdot 11^2 =$	
13.	$11^{12} \cdot 11^4 =$	
14.	$11^{12} \cdot 11^6 =$	
15.	$23^5 \cdot 23^2 =$	
16.	$23^6 \cdot 23^3 =$	
17.	$23^7 \cdot 23^4 =$	
18.	$13^7 \cdot 13^3 =$	
19.	$15^7 \cdot 15^3 =$	
20.	$17^7 \cdot 17^3 =$	
21.	$x^7 \cdot x^3 =$	
22.	$y^7 \cdot y^3 =$	

23.	$7^3 \cdot 7^2 =$	
24.	$7^2 \cdot 7^3 =$	
25.	$(-4)^3 \cdot (-4)^{11} =$	
26.	$(-4)^{11} \cdot (-4)^3 =$	
27.	$(0.2)^3 \cdot (0.2)^{11} =$	
28.	$(0.2)^{11} \cdot (0.2)^3 =$	
29.	$(-2)^9 \cdot (-2)^5 =$	
30.	$(-2.7)^5 \cdot (-2.7)^9 =$	
31.	$3.1^6 \cdot 3.1^6 =$	
32.	$57^6 \cdot 57^6 =$	
33.	$z^6 \cdot z^6 =$	
34.	$4 \cdot 2^9 =$	
35.	$4^2 \cdot 2^9 =$	
36.	$16 \cdot 2^9 =$	
37.	$16 \cdot 4^3 =$	
38.	$9 \cdot 3^5 =$	
39.	$3^5 \cdot 9 =$	
40.	$3^5 \cdot 27 =$	
41.	$5^7 \cdot 25 =$	
42.	$5^7 \cdot 125 =$	
43.	$2^{11} \cdot 4 =$	
44.	$2^{11} \cdot 16 =$	

EUREKA MATH

Applying Properties of Exponents to Generate Equivalent Expressions—Round 2 [KEY]

Directions: Simplify each expression using the laws of exponents. Use the least number of bases possible and only positive exponents. All letters denote numbers.

1.	$5^2 \cdot 5^3 =$	5^5	23.	$7^3 \cdot 7^2 =$	7^5
2.	$5^2 \cdot 5^4 =$	5^6	24.	$7^2 \cdot 7^3 =$	7^5
3.	$5^2 \cdot 5^5 =$	5^7	25.	$(-4)^3 \cdot (-4)^{11} =$	$(-4)^{14}$
4.	$2^7 \cdot 2^1 =$	2^8	26.	$(-4)^{11} \cdot (-4)^3 =$	$(-4)^{14}$
5.	$2^8 \cdot 2^1 =$	2^9	27.	$(0.2)^3 \cdot (0.2)^{11} =$	$(0.2)^{14}$
6.	$2^9 \cdot 2^1 =$	2^{10}	28.	$(0.2)^{11} \cdot (0.2)^3 =$	$(0.2)^{14}$
7.	$3^6 \cdot 3^2 =$	3^8	29.	$(-2)^9 \cdot (-2)^5 =$	$(-2)^{14}$
8.	$3^6 \cdot 3^3 =$	3^9	30.	$(-2.7)^5 \cdot (-2.7)^9 =$	$(-2.7)^{14}$
9.	$3^6 \cdot 3 =$	3^{10}	31.	$3.1^6 \cdot 3.1^6 =$	3.1^{12}
10.	$7^{15} \cdot 7 =$	7^{16}	32.	$57^6 \cdot 57^6 =$	57^{12}
11.	$7^{16} \cdot 7 =$	7^{17}	33.	$z^6 \cdot z^6 =$	z^{12}
12.	$11^{12} \cdot 11^2 =$	11^{14}	34.	$4 \cdot 2^9 =$	2^{11}
13.	$11^{12} \cdot 11^4 =$	11^{16}	35.	$4^2 \cdot 2^9 =$	2^{13}
14.	$11^{12} \cdot 11^6 =$	11^{18}	36.	$16 \cdot 2^9 =$	2^{13}
15.	$23^5 \cdot 23^2 =$	23^7	37.	$16 \cdot 4^3 =$	4^5
16.	$23^6 \cdot 23^3 =$	23^9	38.	$9 \cdot 3^5 =$	3^7
17.	$23^7 \cdot 23^4 =$	23^{11}	39.	$3^5 \cdot 9 =$	3^7
18.	$13^7 \cdot 13^3 =$	13^{10}	40.	$3^5 \cdot 27 =$	3^8
19.	$15^7 \cdot 15^3 =$	15^{10}	41.	$5^7 \cdot 25 =$	5^9
20.	$17^7 \cdot 17^3 =$	17^{10}	42.	$5^7 \cdot 125 =$	5^{10}
21.	$x^7 \cdot x^3 =$	x^{10}	43.	$2^{11} \cdot 4 =$	2^{13}
22.	$y^7 \cdot y^3 =$	y^{10}	44.	$2^{11} \cdot 16 =$	2^{15}

Lesson 5: Negative Exponents and the Laws of Exponents

Student Outcomes

- Students know the definition of a number raised to a negative exponent.
- Students simplify and write equivalent expressions that contain negative exponents.

Lesson Notes

We are now ready to extend the existing laws of exponents to include all integers. As with previous lessons, have students work through a concrete example, such as those in the first Discussion, before giving the mathematical rationale. Note that in this lesson the symbols used to represent the exponents change from m and n to a and b. This change is made to clearly highlight that we are now working with all integer exponents, not just positive integers or whole numbers as in the previous lessons.

In line with previous implementation suggestions, it is important that students are shown the symbolic arguments in this lesson, but less important for students to reproduce them on their own. Students should learn to fluently and accurately apply the laws of exponents as in Exercises 5–10, 11, and 12. Use discretion to omit other exercises.

Classwork

Discussion (10 minutes)

This lesson, and the next, refers to several of the equations used in the previous lessons. It may be helpful if students have some way of referencing these equations quickly (e.g., a poster in the classroom or handout). For convenience, an equation reference sheet has been provided on page 61.

Let x and y be positive numbers throughout this lesson. Recall that we have the following three identities (6)–(8).

For all whole numbers m and n:

$$x^m \cdot x^n = x^{m+n} \tag{6}$$

$$(x^m)^n = x^{mn} \tag{7}$$

$$(xy)^n = x^n y^n \tag{8}$$

Make clear that we want (6)–(8) to remain true even when m and n are *integers*. Before we can say that, we have to first decide what something like 3^{-5} should mean.

Allow time for the class to discuss the question, "What should 3^{-5} mean?" As in Lesson 4, where we introduced the concept of the zeroth power of a number, the overriding idea here is that the negative power of a number should be defined in a way to ensure that (6)–(8) continue to hold when m and n are integers and not just whole numbers. Students will likely say that it should mean -3^5. Tell students that if that is what it meant, that is what we would write.

When they get stuck, ask students this question, "Using equation (6), what *should* $3^5 \cdot 3^{-5}$ equal?" Students should respond that they want to believe that equation (6) is still correct even when m and n are integers, and therefore, they *should* have $3^5 \cdot 3^{-5} = 3^{5+(-5)} = 3^0 = 1$.

> *Scaffolding:*
>
> Ask students, "If x is a number, then what value of x would make the following true: $3^5 x = 1$?"

- What does this say about the value 3^{-5}?
  - *The value 3^{-5} must be a fraction because $3^5 \cdot 3^{-5} = 1$, specifically the reciprocal of 3^5.*
- Then, would it not be reasonable to define 3^{-n}, in general, as $\frac{1}{3^n}$?

MP.6 **Definition:** For any nonzero number x and for any positive integer n, we define x^{-n} as $\frac{1}{x^n}$.

> *Scaffolding:*
>
> As an alternative to providing the consequence of the definition, ask advanced learners to consider what would happen if we removed the restriction that n is a positive integer. Allow them time to reach the conclusion shown in equation (9).

Note that this definition of negative exponents says x^{-1} is just the reciprocal, $\frac{1}{x}$, of x. In particular, x^{-1} would make no sense if $x = 0$. This explains why we must restrict x to being nonzero at this juncture.

The definition has the following consequence:

For a nonzero x, $x^{-b} = \frac{1}{x^b}$ for all *integers* b. (9)

Note that (9) contains more information than the definition of negative exponent. For example, it implies that, with $b = -3$ in (9), $5^3 = \frac{1}{5^{-3}}$.

Proof of (9): There are three possibilities for b: $b > 0$, $b = 0$, and $b < 0$. If the b in (9) is positive, then (9) is just the definition of x^{-b}, and there is nothing to prove. If $b = 0$, then both sides of (9) are seen to be equal to 1 and are, therefore, equal to each other. Again, (9) is correct. Finally, in general, let b be negative. Then $b = -n$ for some *positive* integer n. The left side of (9) is $x^{-b} = x^{-(-n)}$. The right side of (9) is equal to

$$\frac{1}{x^{-n}} = \frac{1}{\frac{1}{x^n}} = 1 \times \frac{x^n}{1} = x^n$$

where we have made use of invert and multiply to simplify the complex fraction. Hence, the left side of (9) is again equal to the right side. The proof of (9) is complete.

Definition: For any nonzero number x, and for any positive integer n, we define x^{-n} as $\frac{1}{x^n}$.

Note that this definition of negative exponents says x^{-1} is just the reciprocal, $\frac{1}{x}$, of x.

As a consequence of the definition, for a nonnegative x and all *integers* b, we get

$$x^{-b} = \frac{1}{x^b}.$$

Allow time to discuss why we need to understand negative exponents.

- Answer: As we have indicated in Lesson 4, the basic impetus for the consideration of negative (and, in fact, arbitrary) exponents is the fascination with identities (1)–(3) (Lesson 4), which are valid only for positive integer exponents. Such nice looking identities *should be* valid for all exponents. These identities are the starting point for the consideration of all other exponents beyond the positive integers. Even without knowing this aspect of identities (1)–(3), one can see the benefit of having negative exponents by looking at the **complete expanded form of a decimal**. For example, the complete expanded form of 328.5403 is

$$(3 \times 10^2) + (2 \times 10^1) + (8 \times 10^0) + (5 \times 10^{-1}) + (4 \times 10^{-2}) + (0 \times 10^{-3}) + (3 \times 10^{-4}).$$

By writing the place value of the decimal digits in negative powers of 10, one gets a sense of the *naturalness* of the complete expanded form as the sum of whole number multiples of *descending* powers of 10.

Exercises 1–10 (10 minutes)

Students complete Exercise 1 independently or in pairs. Provide the correct solution. Then have students complete Exercises 2–10 independently.

Exercise 1

Verify the general statement $x^{-b} = \frac{1}{x^b}$ for $x = 3$ and $b = -5$.

If b were a positive integer, then we have what the definition states. However, b is a negative integer, specifically $b = -5$, so the general statement in this case reads

$$3^{-(-5)} = \frac{1}{3^{-5}}.$$

The right side of this equation is

$$\frac{1}{3^{-5}} = \frac{1}{\frac{1}{3^5}} = 1 \times \frac{3^5}{1} = 3^5.$$

Since the left side is also 3^5, both sides are equal.

$$3^{-(-5)} = \frac{1}{3^{-5}} = 3^5$$

Exercise 2

What is the value of (3×10^{-2})?

$$(3 \times 10^{-2}) = 3 \times \frac{1}{10^2} = \frac{3}{10^2} = 0.03$$

Exercise 3

What is the value of (3×10^{-5})?

$$(3 \times 10^{-5}) = 3 \times \frac{1}{10^5} = \frac{3}{10^5} = 0.00003$$

Exercise 4

Write the complete expanded form of the decimal 4.728 in exponential notation.

$$4.728 = (4 \times 10^0) + (7 \times 10^{-1}) + (2 \times 10^{-2}) + (8 \times 10^{-3})$$

For Exercises 5–10, write an equivalent expression, in exponential notation, to the one given, and simplify as much as possible.

Exercise 5

$5^{-3} = \frac{1}{5^3}$

Exercise 6

$\frac{1}{8^9} = 8^{-9}$

Exercise 7

$3 \cdot 2^{-4} = 3 \cdot \frac{1}{2^4} = \frac{3}{2^4}$

Exercise 8

Let x be a nonzero number.

$x^{-3} = \frac{1}{x^3}$

Exercise 9

Let x be a nonzero number.

$\frac{1}{x^9} = x^{-9}$

Exercise 10

Let x, y be two nonzero numbers.

$xy^{-4} = x \cdot \frac{1}{y^4} = \frac{x}{y^4}$

Discussion (5 minutes)

We now state our main objective: *For any nonzero numbers* x and y *and for all* integers a and b,

$$x^a \cdot x^b = x^{a+b} \tag{10}$$

$$(x^b)^a = x^{ab} \tag{11}$$

$$(xy)^a = x^a y^a \tag{12}$$

We accept that for nonzero numbers x and y and all integers a and b,

$$x^a \cdot x^b = x^{a+b}$$

$$(x^b)^a = x^{ab}$$

$$(xy)^a = x^a y^a.$$

We claim

$$\frac{x^a}{x^b} = x^{a-b} \qquad \text{for all integers } a, b.$$

$$\left(\frac{x}{y}\right)^a = \frac{x^a}{y^a} \qquad \text{for any integer } a.$$

Identities (10)–(12) are called the **laws of exponents** for integer exponents. They clearly generalize (6)–(8).

Consider mentioning that (10)–(12) are valid even when a and b are rational numbers. (Make sure they know rational numbers refer to positive and negative fractions.) The fact that they are true also for all real numbers can only be proved in college.

The laws of exponents will be proved in the next lesson. For now, we want to use them effectively.

MP.2

In the process, we will get a glimpse of why they are worth learning. We will show that knowing (10)–(12) means also knowing (4) and (5) automatically. Thus, it is enough to know only three facts, (10)–(12), rather than *five* facts, (10)–(12) and (4) and (5). Incidentally, the preceding sentence demonstrates why it is essential to learn how to use symbols because if (10)–(12) were stated in terms of explicit numbers, the preceding sentence would not even make sense.

We reiterate the following: The discussion below assumes the validity of (10)–(12) for the time being. We claim

$$\frac{x^a}{x^b} = x^{a-b} \quad \text{for all integers } a, b. \qquad (13)$$

$$\left(\frac{x}{y}\right)^a = \frac{x^a}{y^a} \quad \text{for any integer } a. \qquad (14)$$

Note that identity (13) says much more than (4): Here, a and b can be integers, rather than positive integers and, moreover, there is no requirement that $a > b$. Similarly, unlike (5), the a in (14) is an integer rather than just a positive integer.

Tell students that the need for formulas about complex fractions will be obvious in subsequent lessons and will not be consistently pointed out. Ask students to explain why these must be considered complex fractions.

Exercises 11 and 12 (4 minutes)

Students complete Exercises 11 and 12 independently or in pairs in preparation of the proof of (13) in general.

Exercise 11

$$\frac{19^2}{19^5} = 19^{2-5}$$

Exercise 12

$$\frac{17^{16}}{17^{-3}} = 17^{16} \times \frac{1}{17^{-3}} = 17^{16} \times 17^3 = 17^{16+3}$$

Proof of (13):

$$\frac{x^a}{x^b} = x^a \cdot \frac{1}{x^b} \quad \text{By the product formula for complex fractions}$$

$$= x^a \cdot x^{-b} \quad \text{By } x^{-b} = \frac{1}{x^b} \text{ (9)}$$

$$= x^{a+(-b)} \quad \text{By } x^a \cdot x^b = x^{a+b} \text{ (10)}$$

$$= x^{a-b}$$

Exercises 13 and 14 (8 minutes)

Students complete Exercise 13 in preparation for the proof of (14). Check before continuing to the general proof of (14).

Exercise 13

If we let $b = -1$ in (11), a be any integer, and y be any nonzero number, what do we get?

$$(y^{-1})^a = y^{-a}$$

Exercise 14

Show directly that $\left(\frac{7}{5}\right)^{-4} = \frac{7^{-4}}{5^{-4}}$.

$$\left(\frac{7}{5}\right)^{-4} = \left(7 \cdot \frac{1}{5}\right)^{-4} \quad \textit{By the product formula}$$

$$= (7 \cdot 5^{-1})^{-4} \quad \textit{By definition}$$

$$= 7^{-4} \cdot (5^{-1})^{-4} \quad \textit{By } (xy)^a = x^a y^a \textit{ (12)}$$

$$= 7^{-4} \cdot 5^4 \quad \textit{By } (x^b)^a = x^{ab} \textit{ (11)}$$

$$= 7^{-4} \cdot \frac{1}{5^{-4}} \quad \textit{By } x^{-b} = \frac{1}{x^b} \textit{ (9)}$$

$$= \frac{7^{-4}}{5^{-4}} \quad \textit{By product formula}$$

Proof of (14):

$$\left(\frac{x}{y}\right)^a = \left(x \cdot \frac{1}{y}\right)^a \quad \text{By the product formula for complex fractions}$$

$$= (xy^{-1})^a \quad \text{By definition}$$

$$= x^a (y^{-1})^a \quad \text{By } (xy)^a = x^a y^a \text{ (12)}$$

$$= x^a y^{-a} \quad \text{By } (x^b)^a = x^{ab} \text{ (11), also see Exercise 13}$$

$$= x^a \cdot \frac{1}{y^a} \quad \text{By } x^{-b} = \frac{1}{x^b} \text{ (9)}$$

$$= \frac{x^a}{y^a}$$

Students complete Exercise 14 independently. Provide the solution when they are finished.

Closing (3 minutes)

Summarize, or have students summarize, the lesson.

- By assuming (10)–(12) were true for integer exponents, we see that (4) and (5) would also be true.
- (10)–(12) are worth remembering because they are so useful and allow us to limit what we need to memorize.

Exit Ticket (5 minutes)

Name ______________________________ Date______________

Lesson 5: Negative Exponents and the Laws of Exponents

Exit Ticket

Write each expression in a simpler form that is equivalent to the given expression.

1. $76543^{-4} =$

2. Let f be a nonzero number. $f^{-4} =$

3. $671 \times 28796^{-1} =$

4. Let a, b be numbers $(b \neq 0)$. $ab^{-1} =$

5. Let g be a nonzero number. $\frac{1}{g^{-1}} =$

EUREKA MATH™

Exit Ticket Sample Solutions

Write each expression in a simpler form that is equivalent to the given expression.

1. $76543^{-4} = \frac{1}{76543^4}$

2. Let f be a nonzero number. $f^{-4} = \frac{1}{f^4}$

3. $671 \times 28796^{-1} = 671 \times \frac{1}{28796} = \frac{671}{28796}$

4. Let a, b be numbers $(b \neq 0)$. $ab^{-1} = a \cdot \frac{1}{b} = \frac{a}{b}$

5. Let g be a nonzero number. $\frac{1}{g^{-1}} = g$

Problem Set Sample Solutions

1. Compute: $3^3 \times 3^2 \times 3^1 \times 3^0 \times 3^{-1} \times 3^{-2} = 3^3 = 27$

   Compute: $5^2 \times 5^{10} \times 5^8 \times 5^0 \times 5^{-10} \times 5^{-8} = 5^2 = 25$

   Compute for a nonzero number, a: $a^m \times a^n \times a^l \times a^{-n} \times a^{-m} \times a^{-l} \times a^0 = a^0 = 1$

2. Without using (10), show directly that $(17.6^{-1})^8 = 17.6^{-8}$.

$$(17.6^{-1})^8 = \left(\frac{1}{17.6}\right)^8 \quad \textit{By definition}$$

$$= \frac{1^8}{17.6^8} \quad \textit{By } \left(\frac{x}{y}\right)^n = \frac{x^n}{y^n} \ (5)$$

$$= \frac{1}{17.6^8}$$

$$= 17.6^{-8} \quad \textit{By definition}$$

3. Without using (10), show (prove) that for any whole number n and any nonzero number y, $(y^{-1})^n = y^{-n}$.

$$(y^{-1})^n = \left(\frac{1}{y}\right)^n \quad \textit{By definition}$$

$$= \frac{1^n}{y^n} \quad \textit{By } \left(\frac{x}{y}\right)^n = \frac{x^n}{y^n} \ (5)$$

$$= \frac{1}{y^n}$$

$$= y^{-n} \quad \textit{By definition}$$

4. Without using (13), show directly that $\frac{2.8^{-5}}{2.8^{7}} = 2.8^{-12}$.

$$\frac{2.8^{-5}}{2.8^{7}} = 2.8^{-5} \times \frac{1}{2.8^{7}}$$ *By the product formula for complex fractions*

$$= \frac{1}{2.8^{5}} \times \frac{1}{2.8^{7}}$$ *By definition*

$$= \frac{1}{2.8^{5} \times 2.8^{7}}$$ *By the product formula for complex fractions*

$$= \frac{1}{2.8^{5+7}}$$ *By $x^{a} \cdot x^{b} = x^{a+b}$ (10)*

$$= \frac{1}{2.8^{12}}$$

$$= 2.8^{-12}$$ *By definition*

Equation Reference Sheet

For any numbers x, y [$x \neq 0$ in (4) and $y \neq 0$ in (5)] and any positive integers m, n, the following holds:	
$x^m \cdot x^n = x^{m+n}$	(1)
$(x^m)^n = x^{mn}$	(2)
$(xy)^n = x^n y^n$	(3)
$\frac{x^m}{x^n} = x^{m-n}$	(4)
$\left(\frac{x}{y}\right)^n = \frac{x^n}{y^n}$	(5)
For any numbers x, y and for all whole numbers m, n, the following holds:	
$x^m \cdot x^n = x^{m+n}$	(6)
$(x^m)^n = x^{mn}$	(7)
$(xy)^n = x^n y^n$	(8)
For any nonzero number x and all integers b, the following holds:	
$x^{-b} = \frac{1}{x^b}$	(9)
For any numbers x, y and all integers a, b, the following holds:	
$x^a \cdot x^b = x^{a+b}$	(10)
$(x^b)^a = x^{ab}$	(11)
$(xy)^a = x^a y^a$	(12)
$\frac{x^a}{x^b} = x^{a-b} \qquad x \neq 0$	(13)
$\left(\frac{x}{y}\right)^a = \frac{x^a}{y^a} \qquad x, y \neq 0$	(14)

Lesson 6: Proofs of Laws of Exponents

Student Outcomes

- Students extend the previous laws of exponents to include all integer exponents.
- Students base symbolic proofs on concrete examples to show that $(x^b)^a = x^{ab}$ is valid for all integer exponents.

Lesson Notes

This lesson is not designed for all students, but rather for those who would benefit from a lesson that enriches their existing understanding of the laws of exponents. For that reason, this is an optional lesson that can be used with students who have demonstrated mastery over concepts in Topic A.

Classwork

Discussion (8 minutes)

The goal of this lesson is to show why the laws of exponents, (10)–(12), are correct for all integers a and b and for all $x, y \neq 0$. We recall (10)–(12):

For all $x, y \neq 0$ and for all integers a and b, we have

$$x^a \cdot x^b = x^{a+b} \qquad (10)$$

$$\left(x^b\right)^a = x^{ab} \qquad (11)$$

$$(xy)^a = x^a y^a. \qquad (12)$$

Scaffolding:

Ask advanced learners to consider why it is necessary to restrict the values of x and y to nonzero numbers. They should be able to respond that if a or b is a negative integer, the value of the expression could depend on division by zero, which is undefined.

MP.7 & MP.8 This is a tedious process as the proofs for all three are somewhat similar. The proof of (10) is the most complicated of the three, but students who understand the proof of the easier identity (11) should get a good idea of how all three proofs go. Therefore, we will only prove (11) completely.

We have to first decide on a strategy to prove (11). Ask students what we already know about (11).

MP.1 Elicit the following from students

- Equation (7) of Lesson 5 says for any nonzero x, $(x^m)^n = x^{mn}$ for all whole numbers m and n.

  How does this help us? It tells us that:

  (A) (11) is already known to be true when the integers a and b, in addition, satisfy $a \geq 0$, $b \geq 0$.

Scaffolding:

Keep statements (A), (B), and (C) visible throughout the lesson for reference purposes.

EUREKA MATH™

MP.1

- Equation (9) of Lesson 5 says that the following holds:

  (B) $x^{-m} = \frac{1}{x^m}$ for any whole number m.

  How does this help us? As we shall see from an exercise below, (B) is the statement that another special case of (11) is known.

- We also know that if x is nonzero, then

  (C) $\left(\frac{1}{x}\right)^m = \frac{1}{x^m}$ for any whole number m.

  This is because if m is a positive integer, (C) is implied by equation (5) of Lesson 4, and if $m = 0$, then both sides of (C) are equal to 1.

  How does this help us? We will see from another exercise below that (C) is in fact another special case of (11), which is already known to be true.

The Laws of Exponents

For $x, y \neq 0$, and all integers a, b, the following holds:

$$x^a \cdot x^b = x^{a+b}$$
$$(x^b)^a = x^{ab}$$
$$(xy)^a = x^a y^a.$$

Facts we will use to prove (11):

(A) **(11) is already known to be true when the integers a and b satisfy $a \geq 0, b \geq 0$.**

(B) $x^{-m} = \frac{1}{x^m}$ **for any whole number m.**

(C) $\left(\frac{1}{x}\right)^m = \frac{1}{x^m}$ **for any whole number m.**

Exercises 1–3 (6 minutes)

Students complete Exercises 1–3 in small groups.

Exercise 1

Show that (C) is implied by equation (5) of Lesson 4 when $m > 0$, and explain why (C) continues to hold even when $m = 0$.

Equation (5) says for any numbers x, y, $(y \neq 0)$ and any positive integer n, the following holds: $\left(\frac{x}{y}\right)^n = \frac{x^n}{y^n}$. So,

$$\left(\frac{1}{x}\right)^m = \frac{1^m}{x^m} \quad \text{By } \left(\frac{x}{y}\right)^n = \frac{x^n}{y^n} \text{ for positive integer } n \text{ and nonzero } y \text{ (5)}$$

$$= \frac{1}{x^m} \quad \text{Because } 1^m = 1$$

If $m = 0$, then the left side is

$$\left(\frac{1}{x}\right)^m = \left(\frac{1}{x}\right)^0$$

$$= 1 \qquad \textit{By definition of } x^0,$$

and the right side is

$$\frac{1}{x^m} = \frac{1}{x^0}$$

$$= \frac{1}{1} \qquad \textit{By definition of } x^0$$

$$= 1.$$

Exercise 2

Show that (B) is in fact a special case of (11) by rewriting it as $(x^m)^{-1} = x^{(-1)m}$ for any whole number m, so that if $b = m$ (where m is a whole number) and $a = -1$, (11) becomes (B).

(B) says $x^{-m} = \frac{1}{x^m}$.

The left side of (B), x^{-m} is equal to $x^{(-1)m}$.

The right side of (B), $\frac{1}{x^m}$, is equal to $(x^m)^{-1}$ by the definition of $(x^m)^{-1}$ in Lesson 5.

Therefore, (B) says exactly that $(x^m)^{-1} = x^{(-1)m}$.

Exercise 3

Show that (C) is a special case of (11) by rewriting (C) as $(x^{-1})^m = x^{m(-1)}$ for any whole number m. Thus, (C) is the special case of (11) when $b = -1$ and $a = m$, where m is a whole number.

(C) says $\left(\frac{1}{x}\right)^m = \frac{1}{x^m}$ for any whole number m.

The left side of (C) is equal to

$$\left(\frac{1}{x}\right)^m = (x^{-1})^m \qquad \textit{By definition of } x^{-1},$$

and the right side of (C) is equal to

$$\frac{1}{x^m} = x^{-m} \qquad \textit{By definition of } x^{-m},$$

and the latter is equal to $x^{m(-1)}$. Therefore, (C) says $(x^{-1})^m = x^{m(-1)}$ for any whole number m.

EUREKA MATH™

Discussion (4 minutes)

MP.3 In view of the fact that the reasoning behind the proof of **(A)** (Lesson 4) clearly cannot be extended to a case in which a and/or b is negative, it may be time to consider proving (11) in several separate cases so that, at the end, these cases *together* cover all possibilities. **(A)** suggests that we consider the following four separate cases of identity (11):

(i) $a, b \geq 0$

(ii) $a \geq 0, b < 0$

(iii) $a < 0,\ b \geq 0$

(iv) $a, b < 0$.

- Why are there are no other possibilities?
- Do we need to prove case (*i*)?
  - *No, because **(A)** corresponds to case (i) of (11).*

We will prove the three remaining cases in succession.

> *Scaffolding:*
> - Have students think about the four quadrants of the plane.
> - Read the meaning of the four cases aloud as you write them symbolically.

Discussion (10 minutes)

Case (ii): We have to prove that for any nonzero x, $(x^b)^a = x^{ab}$, when the integers a and b satisfy $a \geq 0, b < 0$. For example, we have to show that $(5^{-3})^4 = 5^{(-3)4}$, or $(5^{-3})^4 = 5^{-12}$. The following is the proof:

$(5^{-3})^4 = \left(\frac{1}{5^3}\right)^4$ By definition

$= \frac{1}{\left(5^3\right)^4}$ By $\left(\frac{1}{x}\right)^m = \frac{1}{x^m}$ for any whole number m (C)

$= \frac{1}{5^{12}}$ By $(x^m)^n = x^{mn}$ for all whole numbers m and n (A)

$= 5^{-12}$ By definition

In general, we just imitate this argument. Let $b = -c$, where c is a positive integer. We now show that the left side and the right side of $(x^b)^a = x^{ab}$ are equal. The left side is

$(x^b)^a = (x^{-c})^a$

$= \left(\frac{1}{x^c}\right)^a$ By $x^{-m} = \frac{1}{x^m}$ for any whole number m (B)

$= \frac{1}{(x^c)^a}$ By $\left(\frac{1}{x}\right)^m = \frac{1}{x^m}$ for any whole number m (C)

$= \frac{1}{x^{ac}}$ By $(x^m)^n = x^{mn}$ for all whole numbers m and n (A)

> *Scaffolding:*
> - Keep the example, done previously with concrete numbers, visible so students can relate the symbolic argument to the work just completed.
> - Remind students that when using concrete numbers, we can push through computations and show that the left side and right side are the same. For symbolic arguments, we must look at each side separately and show that the two sides are equal.

The right side is

$$x^{ab} = x^{a(-c)}$$
$$= x^{-(ac)}$$
$$= \frac{1}{x^{ac}}. \qquad \text{By } x^{-m} = \frac{1}{x^m} \text{ for any whole number } m \text{ (B)}$$

The left and right sides are equal; thus, *case (ii)* is done.

Case (iii): We have to prove that for any nonzero x, $(x^b)^a = x^{ab}$, when the integers a and b satisfy $a < 0$ and $b \geq 0$. This is very similar to case *(ii)*, so it will be left as an exercise.

Exercise 4 (4 minutes)

Students complete Exercise 4 independently or in pairs.

Exercise 4

Proof of Case (iii): **Show that when $a < 0$ and $b \geq 0$, $(x^b)^a = x^{ab}$ is still valid. Let $a = -c$ for some positive integer c. Show that the left and right sides of $(x^b)^a = x^{ab}$ are equal.**

The left side is

$$(x^b)^a = (x^b)^{-c}$$
$$= \frac{1}{(x^b)^c} \qquad \text{By } x^{-m} = \frac{1}{x^m} \text{ for any whole number } m \text{ (B)}$$
$$= \frac{1}{x^{cb}}. \qquad \text{By } (x^m)^n = x^{mn} \text{ for all whole numbers } m \text{ and } n \text{ (A)}$$

The right side is

$$x^{ab} = x^{(-c)b}$$
$$= x^{-(cb)}$$
$$= \frac{1}{x^{cb}}. \qquad \text{By } x^{-m} = \frac{1}{x^m} \text{ for any whole number } m \text{ (B)}$$

So, the two sides are equal.

Discussion (8 minutes)

The only case remaining in the proof of (11) is *case (iv)*. Thus, we have to prove that for any nonzero x, $(x^b)^a = x^{ab}$ when the integers a and b satisfy $a < 0$ and $b < 0$. For example, $(7^{-5})^{-8} = 7^{5\cdot 8}$ because

$$(7^{-5})^{-8} = \frac{1}{(7^{-5})^8} \qquad \text{By } x^{-m} = \frac{1}{x^m} \text{ for any whole number } m \text{ (B)}$$
$$= \frac{1}{7^{-(5\cdot 8)}} \qquad \text{By case (ii)}$$
$$= 7^{5\cdot 8}. \qquad \text{By } x^{-m} = \frac{1}{x^m} \text{ for any whole number } m \text{ (B)}$$

In general, we can imitate this explicit argument with numbers as we did in case *(ii)*. Let $a = -c$ and $b = -d$, where c and d are positive integers. Then, the left side is

$$\begin{aligned}(x^b)^a &= (x^{-c})^{-d} \\ &= \frac{1}{(x^{-c})^d} && \text{By } x^{-m} = \frac{1}{x^m} \text{ for any whole number } m \text{ (B)} \\ &= \frac{1}{x^{-cd}} && \text{By case } (ii) \\ &= \frac{1}{\frac{1}{x^{cd}}} && \text{By } x^{-m} = \frac{1}{x^m} \text{ for any whole number } m \text{ (B)} \\ &= x^{cd}. && \text{By invert-and-multiply for division of complex fractions}\end{aligned}$$

> *Scaffolding:*
>
> Students may ask why the exponent c remained negative in the second line while d became positive. Reconcile this through the use of a concrete example or by pointing to the previous problem $(7^{-5})^{-8} = \frac{1}{(7^{-5})^8}$ first line.

The right side is

$$\begin{aligned}x^{ab} &= x^{(-c)(-d)} \\ &= x^{cd}.\end{aligned}$$

The left side is equal to the right side; thus, *case (iv)* is finished. Putting all of the cases together, the proof of (11) is complete. We now know that (11) is true for any nonzero integer x and any integers a, b.

Closing (2 minutes)

Summarize, or have students summarize, the lesson.

- We have proven the laws of exponents are valid for any integer exponent.

Exit Ticket (3 minutes)

Name ______________________________ Date____________

Lesson 6: Proofs of Laws of Exponents

Exit Ticket

1. Show directly that for any nonzero integer x, $x^{-5} \cdot x^{-7} = x^{-12}$.

2. Show directly that for any nonzero integer x, $(x^{-2})^{-3} = x^{6}$.

Exit Ticket Sample Solutions

1. Show directly that for any nonzero integer x, $x^{-5} \cdot x^{-7} = x^{-12}$.

$x^{-5} \cdot x^{-7} = \frac{1}{x^5} \cdot \frac{1}{x^7}$ — *By $x^{-m} = \frac{1}{x^m}$ for any whole number m (B)*

$= \frac{1}{x^5 \cdot x^7}$ — *By the product formula for complex fractions*

$= \frac{1}{x^{5+7}}$ — *By $x^m \cdot x^n = x^{m+n}$ for whole numbers m and n (6)*

$= \frac{1}{x^{12}}$

$= x^{-12}$ — *By $x^{-m} = \frac{1}{x^m}$ for any whole number m (B)*

2. Show directly that for any nonzero integer x, $(x^{-2})^{-3} = x^6$.

$(x^{-2})^{-3} = \frac{1}{(x^{-2})^3}$ — *By $x^{-m} = \frac{1}{x^m}$ for any whole number m (B)*

$= \frac{1}{x^{-(2\cdot 3)}}$ — *By case (ii) of (11)*

$= \frac{1}{x^{-6}}$

$= x^6$ — *By $x^{-m} = \frac{1}{x^m}$ for any whole number m (B)*

Problem Set Sample Solutions

1. You sent a photo of you and your family on vacation to seven Facebook friends. If each of them sends it to five of their friends, and each of those friends sends it to five of their friends, and those friends send it to five more, how many people (not counting yourself) will see your photo? No friend received the photo twice. Express your answer in exponential notation.

# of New People to View Your Photo	*Total # of People to View Your Photo*
7	7
5×7	$7 + (5 \times 7)$
$5 \times 5 \times 7$	$7 + (5 \times 7) + (5^2 \times 7)$
$5 \times 5 \times 5 \times 7$	$7 + (5 \times 7) + (5^2 \times 7) + (5^3 \times 7)$

The total number of people who viewed the photo is $(5^0 + 5^1 + 5^2 + 5^3) \times 7$.

2. Show directly, without using (11), that $(1.27^{-36})^{85} = 1.27^{-36\cdot 85}$.

$(1.27^{-36})^{85} = \left(\frac{1}{1.27^{36}}\right)^{85}$ *By definition*

$= \frac{1}{(1.27^{36})^{85}}$ *By $\left(\frac{1}{x}\right)^m = \frac{1}{x^m}$ for any whole number m (C)*

$= \frac{1}{1.27^{36\cdot 85}}$ *By $(x^m)^n = x^{mn}$ for whole numbers m and n (7)*

$= 1.27^{-36\cdot 85}$ *By $x^{-m} = \frac{1}{x^m}$ for any whole number m (B)*

3. Show directly that $\left(\frac{2}{13}\right)^{-127} \cdot \left(\frac{2}{13}\right)^{-56} = \left(\frac{2}{13}\right)^{-183}$.

$\left(\frac{2}{13}\right)^{-127} \cdot \left(\frac{2}{13}\right)^{-56} = \frac{1}{\left(\frac{2}{13}\right)^{127}} \cdot \frac{1}{\left(\frac{2}{13}\right)^{56}}$ *By definition*

$= \frac{1}{\left(\frac{2}{13}\right)^{127} \cdot \left(\frac{2}{13}\right)^{56}}$ *By the product formula for complex fractions*

$= \frac{1}{\left(\frac{2}{13}\right)^{127+56}}$ *By $x^m \cdot x^n = x^{m+n}$ for whole numbers m and n (6)*

$= \frac{1}{\left(\frac{2}{13}\right)^{183}}$

$= \left(\frac{2}{13}\right)^{-183}$ *By $x^{-m} = \frac{1}{x^m}$ for any whole number m (B)*

4. Prove for any nonzero number x, $x^{-127} \cdot x^{-56} = x^{-183}$.

$x^{-127} \cdot x^{-56} = \frac{1}{x^{127}} \cdot \frac{1}{x^{56}}$ *By definition*

$= \frac{1}{x^{127} \cdot x^{56}}$ *By the product formula for complex fractions*

$= \frac{1}{x^{127+56}}$ *By $x^m \cdot x^n = x^{m+n}$ for whole numbers m and n (6)*

$= \frac{1}{x^{183}}$

$= x^{-183}$ *By $x^{-m} = \frac{1}{x^m}$ for any whole number m (B)*

5. Prove for any nonzero number x, $x^{-m} \cdot x^{-n} = x^{-m-n}$ for positive integers m and n.

$x^{-m} \cdot x^{-n} = \frac{1}{x^m} \cdot \frac{1}{x^n}$ *By definition*

$= \frac{1}{x^m \cdot x^n}$ *By the product formula for complex fractions*

$= \frac{1}{x^{m+n}}$ *By $x^m \cdot x^n = x^{m+n}$ for whole numbers m and n (6)*

$= x^{-(m+n)}$ *By $x^{-m} = \frac{1}{x^m}$ for any whole number m (B)*

$= x^{-m-n}$

6. **Which of the preceding four problems did you find easiest to do? Explain.**

   Students will likely say that $x^{-m} \cdot x^{-n} = x^{-m-n}$ (Problem 5) was the easiest problem to do. It requires the least amount of writing because the symbols are easier to write than decimal or fraction numbers.

7. **Use the properties of exponents to write an equivalent expression that is a product of distinct primes, each raised to an integer power.**

$$\frac{10^5 \cdot 9^2}{6^4} = \frac{(2 \cdot 5)^5 \cdot (3 \cdot 3)^2}{(2 \cdot 3)^4} = \frac{2^5 \cdot 5^5 \cdot 3^2 \cdot 3^2}{2^4 \cdot 3^4} = 2^{5-4} \cdot 3^{4-4} \cdot 5^5 = 2^1 \cdot 3^0 \cdot 5^5 = 2^1 \cdot 1 \cdot 5^5 = 2 \cdot 5^5$$

G8-M1-B1-v1.3.0_05/04/15

Name ______________________________ Date ______________

1. The number of users of social media has increased significantly since the year 2001. In fact, the approximate number of users has tripled each year. It was reported that in 2005 there were 3 million users of social media.

   a. Assuming that the number of users continues to triple each year, for the next three years, determine the number of users in 2006, 2007, and 2008.

   b. Assume the trend in the numbers of users tripling each year was true for all years from 2001 to 2009. Complete the table below using 2005 as year 1 with 3 million as the number of users that year.

Year	-3	-2	-1	0	1	2	3	4	5
# of users in millions					3				

   c. Given only the number of users in 2005 and the assumption that the number of users triples each year, how did you determine the number of users for years 2, 3, 4, and 5?

   d. Given only the number of users in 2005 and the assumption that the number of users triples each year, how did you determine the number of users for years 0, -1, -2, and -3?

e. Write an equation to represent the number of users in millions, N, for year t, $t \geq -3$.

f. Using the context of the problem, explain whether or not the formula $N = 3^t$ would work for finding the number of users in millions in year t, for all $t \leq 0$.

g. Assume the total number of users continues to triple each year after 2009. Determine the number of users in 2012. Given that the world population at the end of 2011 was approximately 7 billion, is this assumption reasonable? Explain your reasoning.

2. Let m be a whole number.

   a. Use the properties of exponents to write an equivalent expression that is a product of unique primes, each raised to an integer power.

$$\frac{6^{21} \cdot 10^7}{30^7}$$

   b. Use the properties of exponents to prove the following identity:

$$\frac{6^{3m} \cdot 10^m}{30^m} = 2^{3m} \cdot 3^{2m}$$

   c. What value of m could be substituted into the identity in part (b) to find the answer to part (a)?

3.

a. Jill writes $2^3 \cdot 4^3 = 8^6$ and the teacher marked it wrong. Explain Jill's error.

b. Find n so that the number sentence below is true:

$$2^3 \cdot 4^3 = 2^3 \cdot 2^n = 2^9$$

c. Use the definition of exponential notation to demonstrate why $2^3 \cdot 4^3 = 2^9$ is true.

d. You write $7^5 \cdot 7^{-9} = 7^{-4}$. Keisha challenges you, "Prove it!" Show directly why your answer is correct without referencing the laws of exponents for integers; in other words, $x^a \cdot x^b = x^{a+b}$ for positive numbers x and integers a and b.

A Progression Toward Mastery					
Assessment Task Item		**STEP 1** Missing or incorrect answer and little evidence of reasoning or application of mathematics to solve the problem.	**STEP 2** Missing or incorrect answer but evidence of some reasoning or application of mathematics to solve the problem.	**STEP 3** A correct answer with some evidence of reasoning or application of mathematics to solve the problem, OR an incorrect answer with substantial evidence of solid reasoning or application of mathematics to solve the problem.	**STEP 4** A correct answer supported by substantial evidence of solid reasoning or application of mathematics to solve the problem.
1	**a–d** 8.EE.A.1	Student answers 0–1 parts of (a)–(d) correctly. Student is able to complete the table for at least values of 0–5 for part (b). Student is unable to respond to questions or left items blank.	Student answers 2–3 parts of (a)–(d) correctly. Student is able to complete the table in part (b) correctly for 5 or more entries, including at least one value on each side of the value given for year 1. Student provides a limited expression of reasoning in parts (c) and (d).	Student answers 3–4 parts of (a)–(d) correctly. Student provides correct answers with some reasoning for making calculations. OR Student has a few miscalculations but provides substantial reasoning with proper use of grade-level vocabulary.	Student answers all parts of (a)–(d) correctly. Student provides solid reasoning for making calculations with proper use of grade-level vocabulary.
	e–g 8.EE.A.1	Student answers 0–1 parts of (e)–(g) correctly. Student is unable to relate the pattern in the problem to exponential growth.	Student answers 1–2 parts of (e)–(g) correctly. Student is able to relate the pattern in the problem to exponential growth by writing an equation. Student justifications are incomplete.	Student answers 2–3 parts of (e)–(g) correctly. Equation given is correct, and student is able to answer questions, but justifications are incomplete. OR The equation given relates the pattern to exponential growth but is incomplete or contains a minor error, and student is able to answer questions using solid reasoning based on the information provided.	Student answers all parts of (e)–(g) correctly. Student justifies answers and makes accurate conclusions based on the information provided in the problem. Student is able to explain limitations of the equation when looking ahead in time and back in time.

2	a 8.EE.A.1	Student answers incorrectly. No evidence of use of properties of exponents.	Student answers incorrectly. Properties of exponents are used incorrectly.	Student answers correctly. Some evidence of use of properties of exponents is shown in calculations.	Student answers correctly. Student provides substantial evidence of the use of properties of exponents to simplify the expression to distinct primes.
	b–c 8.EE.A.1	Student answers parts (b)–(c) incorrectly. No evidence of use of properties of exponents.	Student answers parts (b)–(c) incorrectly. Properties of exponents are used incorrectly.	Student answers part (b) and/or part (c) correctly. Some evidence of use of properties of exponents is shown in calculations.	Student answers both parts (b) and (c) correctly. Student provides substantial evidence of the use of properties of exponents to prove the identity.
3	a 8.EE.A.1	Student states that Jill's response is correct. OR Student is unable to identify the mistake and provides no additional information.	Student states that Jill's answer is incorrect. Student is unable to identify the mistake of multiplying unlike bases. Student may have used what he or she knows about exponential notation to multiply numbers to show the answer was incorrect.	Student identifies Jill's error as "multiplied unlike bases."	Student identifies Jill's error as "multiplied unlike bases." Student provides a thorough explanation as to how unlike bases can be rewritten so that properties of exponents can be used properly.
	b 8.EE.A.1	Student is unable to identify the correct value for n.	Student correctly answers $n = 6$. No explanation is provided as to why the answer is correct.	Student correctly answers $n = 6$. Student states that $4^3 = 2^6$ with little or no explanation or work shown.	Student correctly answers $n = 6$. Student *clearly shows* that 4^3 is equivalent to 2^6.
	c 8.EE.A.1	Student uses the definition of exponential notation to rewrite 4^3 as $4 \times 4 \times 4$. Student is unable to complete the problem.	Student multiplies 4^3 to get 64 and is able to rewrite it as a number with a base of 2 but has the wrong exponent.	Student correctly rewrites 4^3 as 2^6 and then uses the first property of exponents to show that the answer is correct.	Student correctly rewrites 4^3 as 2^6. Student uses definition of exponential notation to rewrite each number as repeated multiplication. Student clearly shows how/why the exponents are added to simplify such expressions.

	d 8.EE.A.1	Student may be able to rewrite 7^{-9} as a fraction but is unable to operate with fractions.	Student is unable to show why part (d) is correct but uses a property of exponents to state that the given answer is correct.	Student answers part (d) but misuses or leaves out definitions in explanations and proofs.	Student answers part (d) correctly and uses definitions and properties to thoroughly explain and prove the answer. Answer shows strong evidence that student understands exponential notation and can use the properties of exponents proficiently.

Name ______________________________ Date ______________

1. The number of users of social media has increased significantly since the year 2001. In fact, the approximate number of users has tripled each year. It was reported that in 2005 there were 3 million users of social media.

   a. Assuming that the number of users continues to triple each year, for the next three years, determine the number of users in 2006, 2007, and 2008.

   2006 – 9 MILLION
   2007 – 27 MILLION
   2008 - 81 MILLION

   b. Assume the trend in the numbers of users tripling each year was true for all years from 2001 to 2009. Complete the table below using 2005 as year 1 with 3 million as the number of users that year.

Year	-3	-2	-1	0	1	2	3	4	5
# of users in millions	$\frac{1}{27}$	$\frac{1}{9}$	$\frac{1}{3}$	1	3	9	27	81	243

   c. Given only the number of users in 2005 and the assumption that the number of users triples each year, how did you determine the number of users for years 2, 3, 4, and 5?

   I MULTIPLIED THE PRECEDING YEAR'S NUMBER OF USERS BY 3.

   d. Given only the number of users in 2005 and the assumption that the number of users triples each year, how did you determine the number of users for years 0, −1, −2, and −3?

   I DIVIDED THE NEXT YEAR'S NUMBER OF USERS BY 3.

e. Write an equation to represent the number of users in millions, N, for year t, $t \geq -3$.

$N = 3^t$

f. Using the context of the problem, explain whether or not the formula $N = 3^t$ would work for finding the number of users in millions in year t, for all $t \leq 0$.

WE ONLY KNOW THAT THE NUMBER OF USERS HAS TRIPLED EACH YEAR IN THE TIME FRAME OF 2001 TO 2009. FOR THAT REASON, WE CANNOT RELY ON THE FORMULA, $N=3^t$, TO WORK FOR ALL $t \leq 0$, JUST TO $t=-3$, WHICH IS THE YEAR 2001.

g. Assume the total number of users continues to triple each year after 2009. Determine the number of users in 2012. Given that the world population at the end of 2011 was approximately 7 billion, is this assumption reasonable? Explain your reasoning.

2012 IS $t=8$, SO WHEN $t=8$ IN $N=3^t$, $N=6{,}561{,}000{,}000$. THE NUMBER OF USERS IN 2012, 6,561,000,000 DOES NOT EXCEED THE WORLD POPULATION OF 7 BILLION, THEREFORE IT IS POSSIBLE TO HAVE THAT NUMBER OF USERS. BUT 6,561,000,000 IS APPROXIMATELY 94% OF THE WORLD'S POPULATION. THE NUMBER OF USERS IS LIKELY LESS THAN THAT DUE TO POVERTY, ILLNESS, INFANCY, ETC. THE ASSUMPTION IS POSSIBLE, BUT NOT REASONABLE.

2. Let m be a whole number.

a. Use the properties of exponents to write an equivalent expression that is a product of unique primes, each raised to an integer power.

$$\frac{6^{21} \cdot 10^7}{30^7}$$

$$= \frac{(3\cdot 2)^{21}\cdot 10^7}{(3\cdot 10)^7} = \frac{3^{21}\cdot 2^{21}\cdot 10^7}{3^7\cdot 10^7}$$

$$= 3^{21-7}\cdot 2^{21}\cdot 10^{7-7}$$

$$= 3^{14}\cdot 2^{21}\cdot 10^0$$

$$= 3^{14}\cdot 2^{21}$$

b. Use the properties of exponents to prove the following identity:

$$\frac{6^{3m} \cdot 10^m}{30^m} = 2^{3m} \cdot 3^{2m}$$

$$\frac{6^{3m}\cdot 10^m}{30^m} = \frac{(3\cdot 2)^{3m}\cdot 10^m}{(3\cdot 10)^m}$$

$$= \frac{3^{3m}\cdot 2^{3m}\cdot 10^m}{3^m\cdot 10^m}$$

$$= 3^{3m-m}\cdot 2^{3m}\cdot 10^{m-m} = 3^{2m}\cdot 2^{3m} = 2^{3m}\cdot 3^{2m}$$

c. What value of m could be substituted into the identity in part (b) to find the answer to part (a)?

$$2^{3m}\cdot 3^{2m} = 2^{21}\cdot 3^{14}$$

$3m = 21$, $m = 7$ $\quad$ $2m = 14$, $m = 7$

THEREFORE, $m = 7$.

3.

a. Jill writes $2^3 \cdot 4^3 = 8^6$ and the teacher marked it wrong. Explain Jill's error.

JILL MULTIPLIED THE BASES, 2 AND 4, AND ADDED THE EXPONENTS. YOU CAN ONLY ADD THE EXPONENTS WHEN THE BASES BEING MULTIPLIED ARE THE SAME.

b. Find n so that the number sentence below is true:

$$2^3 \cdot 4^3 = 2^3 \cdot 2^n = 2^9$$

$4^3 = 4 \cdot 4 \cdot 4$

$= (2 \cdot 2)(2 \cdot 2)(2 \cdot 2)$

$= 2^6$

THEREFORE:

$2^3 \cdot 4^3 = 2^3 \cdot 2^6 = 2^9$

SO $n = 6$

c. Use the definition of exponential notation to demonstrate why $2^3 \cdot 4^3 = 2^9$ is true.

$4^3 = 2^6$, SO $2^3 \cdot 4^3 = 2^9$ IS EQUIVALENT TO $2^3 \cdot 2^6 = 2^9$.

BY DEFINITION OF EXPONENTIAL NOTATION:

$$2^3 \cdot 2^6 = \underbrace{(2 \times \cdots \times 2)}_{3 \text{ times}} \times \underbrace{(2 \times \cdots \times 2)}_{6 \text{ times}} = \underbrace{(2 \times \cdots \times 2)}_{3+6 \text{ times}} = 2^{3+6} = 2^9$$

d. You write $7^5 \cdot 7^{-9} = 7^{-4}$. Keisha challenges you, "Prove it!" Show directly why your answer is correct without referencing the laws of exponents for integers; in other words, $x^a \cdot x^b = x^{a+b}$ for positive numbers x and integers a and b.

$7^5 \cdot 7^{-9} = 7^5 \cdot \frac{1}{7^9}$ BY DEFINITION

$= \frac{7^5}{7^9}$ BY PRODUCT FORMULA

$= \frac{7^5}{7^5 \cdot 7^4}$ BY $x^m \cdot x^n = x^{m+n}$ for $x > 0$, $m, n \geq 0$

$= \frac{1}{7^4}$ BY EQUIVALENT FRACTIONS

$= 7^{-4}$ BY DEFINITION.

Topic B

Magnitude and Scientific Notation

8.EE.A.3, 8.EE.A.4

Focus Standards:	8.EE.A.3	Use numbers expressed in the form of a single digit times an integer power of 10 to estimate very large or very small quantities, and to express how many times as much one is than the other. *For example, estimate the population of the United States as* 3 *times* 10^8 *and the population of the world as* 7 *times* 10^9*, and determine that the world population is more than* 20 *times larger.*
	8.EE.A.4	Perform operations with numbers expressed in scientific notation, including problems where both decimal and scientific notation are used. Use scientific notation and choose units of appropriate size for measurements of very large or very small quantities (e.g., use millimeters per year for seafloor spreading). Interpret scientific notation that has been generated by technology.
Instructional Days:	7	
Lesson 7:	Magnitude (P)[1]	
Lesson 8:	Estimating Quantities (P)	
Lesson 9:	Scientific Notation (P)	
Lesson 10:	Operations with Numbers in Scientific Notation (P)	
Lesson 11:	Efficacy of Scientific Notation (S)	
Lesson 12:	Choice of Unit (E)	
Lesson 13:	Comparison of Numbers Written in Scientific Notation and Interpreting Scientific Notation Using Technology (E)	

[1]Lesson Structure Key: **P**-Problem Set Lesson, **M**-Modeling Cycle Lesson, **E**-Exploration Lesson, **S**-Socratic Lesson.

In Topic B, students' understanding of integer exponents is expanded to include the concept of magnitude as a measurement. Students learn to estimate how big or how small a number is using magnitude. In Lesson 7, students learn that positive powers of 10 are large numbers and negative powers of 10 are very small numbers. In Lesson 8, students express large numbers in the form of a single digit times a positive power of 10 and express how many times as much one of these numbers is compared to another. Students estimate and compare national to household debt and use estimates of the number of stars in the universe to compare with the number of stars an average human can see.

Lessons 9–13 immerse students in scientific notation. Each lesson demonstrates the need for such a notation and then how to compare and compute with numbers in scientific notation. In Lesson 9, students learn how to write numbers in scientific notation and the importance of the exponent with respect to magnitude. The number line is used to illustrate different magnitudes of 10, and students estimate where a particular number, written in scientific notation, belongs on the number line. Also, in this set of lessons, students use what they know about exponential notation, properties of exponents, and scientific notation to interpret results that have been generated by technology.

Continuing with magnitude, Lesson 10 shows students how to operate with numbers in scientific notation by making numbers have the same magnitude. In Lessons 11–13, students reason quantitatively with scientific notation to understand several instances of how the notation is used in science. For example, students compare masses of protons and electrons written in scientific notation and then compute how many times heavier one is than the other by using their knowledge of ratio and properties of exponents. Students use the population of California and their knowledge of proportions to estimate the population of the U.S. assuming population density is the same. Students calculate the average lifetime of subatomic particles and rewrite very small quantities (e.g., 1.6×10^{-27} kg) in a power-of-ten unit of kilograms that supports easier comparisons of the mass.

It is the direct relationship to science in Lesson 12 that provides an opportunity for students to understand why certain units were developed, like the gigaelectronvolt. Given a list of very large numbers, students choose a unit of appropriate size and then rewrite numbers in the new unit to make comparisons easier. In Lesson 13, students combine all the skills of Module 1 as they compare numbers written in scientific notation by rewriting the given numbers as numbers with the same magnitude, using the properties of exponents. By the end of this topic, students are able to compare and perform operations on numbers given in both decimal and scientific notation.

Lesson 7: Magnitude

Student Outcomes

- Students know that positive powers of 10 are very large numbers, and negative powers of 10 are very small numbers.
- Students know that the exponent of an expression provides information about the magnitude of a number.

Classwork

Discussion (5 minutes)

In addition to applications within mathematics, exponential notation is indispensable in science. It is used to clearly display the **magnitude** of a measurement (e.g., *How big? How small?*). We will explore this aspect of exponential notation in the next seven lessons.

Understanding magnitude demands an understanding of the *integer* powers of 10. Therefore, we begin with two fundamental facts about the integer powers of 10. What does it mean to say that 10^n for large positive integers n are *big* numbers? What does it mean to say that 10^{-n} for large positive integers n are *small* numbers? The examples and exercises in this lesson are intended to highlight exactly those facts. It is not the intent of the examples and exercises to demonstrate how we think about magnitude, rather to provide experience with incredibly large and incredibly small numbers in context.

Scaffolding:

Remind students that special cases are cases when concrete numbers are used. They provide a way to become familiar with the mathematics before moving to the general case.

MP.6 *Fact 1:* The numbers 10^n for arbitrarily large positive integers n are big numbers; given a number M (no matter how big it is), there is a power of 10 that exceeds M.

MP.8 *Fact 2:* The numbers 10^{-n} for arbitrarily large positive integers n are small numbers; given a positive number S (no matter how small it is), there is a (negative) power of 10 that is smaller than S.

Fact 2 is a consequence of *Fact 1*. We address *Fact 1* first. The following two special cases illustrate why this is true.

> Fact 1: *The number 10^n, for arbitrarily large positive integers n, is a* big number in the sense that given a number M (no matter how big it is) there is a power of 10 that exceeds M.
>
> Fact 2: *The number 10^{-n}, for arbitrarily large positive integers n, is a* small number in the sense that given a positive number S (no matter how small it is), there is a (negative) power of 10 that is smaller than S.

Examples 1–2 (5 minutes)

Example 1: Let M be the world population as of March 23, 2013. Approximately, $M = 7{,}073{,}981{,}143$. It has 10 digits and is, therefore, smaller than any whole number with 11 digits, such as $10{,}000{,}000{,}000$. But $10\,000\,000\,000 = 10^{10}$, so $M < 10^{10}$ (i.e., the 10th power of 10 exceeds this M).

Example 2: Let M be the U.S. national debt on March 23, 2013. $M = 16\,755\,133\,009\,522$ to the nearest dollar. It has 14 digits. The largest 14-digit number is $99{,}999{,}999{,}999{,}999$. Therefore,

$$M < 99\,999\,999\,999\,999 < 100\,000\,000\,000\,000 = 10^{14}.$$

That is, the 14th power of 10 exceeds M.

Exercises 1–2 (5 minutes)

Students complete Exercises 1 and 2 independently.

Exercise 1

Let $M = 993{,}456{,}789{,}098{,}765$. Find the smallest power of 10 that will exceed M.

$M = 993\,456\,789\,098\,765 < 999\,999\,999\,999\,999 < 1\,000\,000\,000\,000\,000 = 10^{15}$. Because M has 15 digits, 10^{15} will exceed it.

Exercise 2

Let $M = 78{,}491\frac{899}{987}$. Find the smallest power of 10 that will exceed M.

$M = 78491\frac{899}{987} < 78492 < 99999 < 100\,000 = 10^5$.

Therefore, 10^5 will exceed M.

Example 3 (8 minutes)

This example set is for the general case of *Fact 1*.

<u>Case 1</u>: Let M be a positive integer with n digits.

As we know, the integer $99 \cdots 99$ (with n 9s) is $\geq M$.

Therefore, $100 \cdots 00$ (with n 0s) exceeds M.

Since $10^n = 100 \cdots 00$ (with n 0s), we have $10^n > M$.

Symbolically,

$$M < \underbrace{99 \cdots 9}_{n} < 1\underbrace{00 \cdots 0}_{n} = 10^n.$$

Therefore, for an n-digit positive integer M, the nth power of 10 (i.e., 10^n) always exceeds M.

Case 2: In Case 1, M was a positive integer. For this case, let M be a non-integer. We know M must lie between two consecutive points on a number line (i.e., there is some *integer* N so that M lies between N and $N+1$). In other words, $N < M < N+1$. For example, the number $45{,}572.384$ is between $45{,}572$ and $45{,}473$.

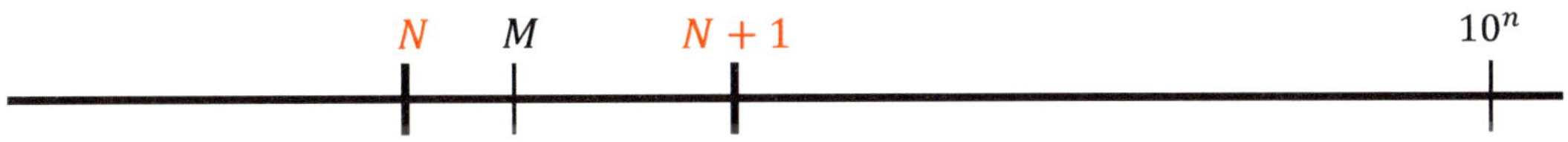

Consider the positive integer $N+1$: According to the reasoning above, there is a positive integer n so that $10^n > N+1$. Since $N+1 > M$, we have $10^n > M$ again. Consequently, for this number M, 10^n exceeds it.

We have now shown why *Fact 1* is correct.

Exercise 3 (2 minutes)

Students discuss Exercise 3 and record their explanations with a partner.

> **Exercise 3**
>
> **Let M be a positive integer. Explain how to find the smallest power of 10 that exceeds it.**
>
> ***If M is a positive integer, then the power of 10 that exceeds it will be equal to the number of digits in M. For example, if M were a 10-digit number, then 10^{10} would exceed M. If M is a positive number, but not an integer, then the power of 10 that would exceed it would be the same power of 10 that would exceed the integer to the right of M on a number line. For example, if $M = 5678.9$, the integer to the right of M is $5{,}679$. Then based on the first explanation, 10^4 exceeds both this integer and M; this is because $M = 5678.9 < 5679 < 10\,000 = 10^4$.***

Example 4 (5 minutes)

- The average ant weighs about 0.0003 grams.

Observe that this number is less than 1 and is very small. We want to express this number as a power of 10. We know that $10^0 = 1$, so we must need a power of 10 that is less than zero. Based on our knowledge of decimals and fractions, we can rewrite the weight of the ant as $\frac{3}{10{,}000}$, which is the same as $\frac{3}{10^4}$. Our work with the laws of exponents taught us that $\frac{3}{10^4} = 3 \times \frac{1}{10^4} = 3 \times 10^{-4}$. Therefore, we can express the weight of an average ant as 3×10^{-4} grams.

- The mass of a neutron is $0.000\,000\,000\,000\,000\,000\,000\,000\,001\,674\,9$ kilograms.

Let's look at an approximated version of the mass of a neutron, $0.000\,000\,000\,000\,000\,000\,000\,000\,001$ kilograms. We already know that 10^{-4} takes us to the place value four digits to the right of the decimal point (ten-thousandths). We need a power of 10 that would take us 27 places to the right of the decimal, which means that we can represent the simplified mass of a neutron as 10^{-27}.

In general, numbers with a value less than 1 but greater than 0 can be expressed using a negative power of 10. The closer a number is to zero, the smaller the power of 10 that will be needed to express it.

Exercises 4–6 (8 minutes)

Students complete Exercises 4–6 independently.

Exercise 4

The chance of you having the same DNA as another person (other than an identical twin) is approximately 1 in 10 trillion (one trillion is a 1 followed by 12 zeros). Given the fraction, express this very small number using a negative power of 10.

$$\frac{1}{10\,000\,000\,000\,000} = \frac{1}{10^{13}} = 10^{-13}$$

Exercise 5

The chance of winning a big lottery prize is about 10^{-8}, and the chance of being struck by lightning in the U.S. in any given year is about $0.000\,001$. Which do you have a greater chance of experiencing? Explain.

$$0.000\,001 = 10^{-6}$$

There is a greater chance of experiencing a lightning strike. On a number line, 10^{-8} is to the left of 10^{-6}. Both numbers are less than one (one signifies 100% probability of occurring). Therefore, the probability of the event that is greater is 10^{-6}—that is, getting struck by lightning.

Exercise 6

There are about 100 million smartphones in the U.S. Your teacher has one smartphone. What share of U.S. smartphones does your teacher have? Express your answer using a negative power of 10.

$$\frac{1}{100\,000\,000} = \frac{1}{10^{8}} = 10^{-8}$$

Closing (2 minutes)

Summarize the lesson for students.

- No matter what number is given, we can find the smallest power of 10 that exceeds that number.
- Very large numbers have a positive power of 10.
- We can use negative powers of 10 to represent very small numbers that are less than one but greater than zero.

Exit Ticket (5 minutes)

Name ______________________________ Date ______________

Lesson 7: Magnitude

Exit Ticket

1. Let $M = 118{,}526.65902$. Find the smallest power of 10 that will exceed M.

2. Scott said that 0.09 was a bigger number than 0.1. Use powers of 10 to show that he is wrong.

Exit Ticket Sample Solutions

1. Let $M = 118,526.65902$. Find the smallest power of 10 that will exceed M.

   Since $M = 118,526.65902 < 118,527 < 1,000,000 < 10^6$, then 10^6 will exceed M.

2. Scott said that 0.09 was a bigger number than 0.1. Use powers of 10 to show that he is wrong.

   We can rewrite 0.09 as $\frac{9}{10^2} = 9 \times 10^{-2}$ and rewrite 0.1 as $\frac{1}{10^1} = 1 \times 10^{-1}$. Because 0.09 has a smaller power of 10, 0.09 is closer to zero and is smaller than 0.1.

Problem Set Sample Solutions

1. What is the smallest power of 10 that would exceed $987,654,321,098,765,432$?

   $987\,654\,321\,098\,765\,432 < 999\,999\,999\,999\,999\,999 < 1\,000\,000\,000\,000\,000\,000 = 10^{18}$

2. What is the smallest power of 10 that would exceed $999,999,999,991$?

   $999\,999\,999\,991 < 999\,999\,999\,999 < 1\,000\,000\,000\,000 = 10^{12}$

3. Which number is equivalent to $0.000\,000\,1$: 10^7 or 10^{-7}? How do you know?

   $0.000\,000\,1 = 10^{-7}$. Negative powers of 10 denote numbers greater than zero but less than 1. Also, the decimal $0.000\,000\,1$ is equal to the fraction $\frac{1}{10^7}$, which is equivalent to 10^{-7}.

4. Sarah said that $0.000\,01$ is bigger than 0.001 because the first number has more digits to the right of the decimal point. Is Sarah correct? Explain your thinking using negative powers of 10 and the number line.

   $0.000\,01 = \frac{1}{100\,000} = 10^{-5}$ and $0.001 = \frac{1}{1\,000} = 10^{-3}$. On a number line, 10^{-5} is closer to zero than 10^{-3}; therefore, 10^{-5} is the smaller number, and Sarah is incorrect.

5. Order the following numbers from least to greatest:

   $10^5 \quad 10^{-99} \quad 10^{-17} \quad 10^{14} \quad 10^{-5} \quad 10^{30}$

   $10^{-99} < 10^{-17} < 10^{-5} < 10^5 < 10^{14} < 10^{30}$

Lesson 8: Estimating Quantities

Student Outcomes

- Students compare and estimate quantities in the form of a single digit times a power of 10.
- Students use their knowledge of ratios, fractions, and laws of exponents to simplify expressions.

Classwork

Discussion (1 minute)

Now that we know about positive and negative powers of 10, we can compare numbers and estimate how many times greater one quantity is compared to another. Note that in the first and subsequent examples when we compare two values a and b, we immediately write the value of the ratio $\frac{a}{b}$ as opposed to writing the ratio $a:b$ first.

With our knowledge of the laws of integer exponents, we can also do other computations to estimate quantities.

Example 1 (4 minutes)

In 1723, the population of New York City was approximately $7{,}248$. By 1870, almost 150 years later, the population had grown to $942{,}292$. We want to determine approximately how many times greater the population was in 1870 compared to 1723.

The word *approximately* in the question lets us know that we do not need to find a precise answer, so we approximate both populations as powers of 10.

- Population in 1723: $7248 < 9999 < 10000 = 10^4$
- Population in 1870: $942\,292 < 999\,999 < 1\,000\,000 = 10^6$

We want to compare the population in 1870 to the population in 1723:

$$\frac{10^6}{10^4}$$

Now we can use what we know about the laws of exponents to simplify the expression and answer the question:

$$\frac{10^6}{10^4} = 10^2.$$

Therefore, there were approximately 100 *times more people in New York City in* 1870 *compared to* 1723.

Exercise 1 (3 minutes)

Have students complete Exercise 1 independently.

> **Exercise 1**
>
> **The Federal Reserve states that the average household in January of 2013 had $\$7,122$ in credit card debt. About how many times greater is the U.S. national debt, which is $\$16,755,133,009,522$? Rewrite each number to the nearest power of 10 that exceeds it, and then compare.**
>
> *Household debt* $= 7122 < 9999 < 10000 = 10^4$.
>
> *U.S. debt* $= 16\,755\,133\,009\,522 < 99\,999\,999\,999\,999 < 100\,000\,000\,000\,000 = 10^{14}$.
>
> $\frac{10^{14}}{10^4} = 10^{14-4} = 10^{10}$. *The U.S. national debt is* 10^{10} *times greater than the average household's credit card debt.*

Discussion (3 minutes)

If our calculations were more precise in the last example, we would have seen that by 1870 the population of New York City actually increased by about 130 times from what it was in 1723.

In order to be more precise, we need to use estimations of our original numbers that are more precise than just powers of 10.

For example, instead of estimating the population of New York City in 1723 (7,248 people) to be 10^4, we can use a more precise estimation: 7×10^3. Using a single-digit integer times a power of ten is more precise because we are rounding the population to the nearest thousand. Conversely, using only a power of ten, we are rounding the population to the nearest ten thousand.

> *Scaffolding:*
> - Make it clear to students that more precise estimations lead to more precise answers.
> - In a later lesson, we learn how to be even more precise than in Example 2.

Consider that the actual population is 7,248.

- $10^4 = 10000$
- $7 \times 10^3 = 7 \times 1000 = 7000$

Which of these two estimations is closer to the actual population?

Clearly, 7×10^3 is a more precise estimation.

Example 2 (4 minutes)

Let's compare the population of New York City to the population of New York State. Specifically, let's find out how many times greater the population of New York State is compared to that of New York City.

The population of New York City is 8,336,697. Let's round this number to the nearest million; this gives us 8,000,000. Written as single-digit integer times a power of 10:

$$8\,000\,000 = 8 \times 10^6.$$

The population of New York State is 19,570,261. Rounding to the nearest million gives us 20,000,000. Written as a single-digit integer times a power of 10:

$$20\,000\,000 = 2 \times 10^7.$$

To estimate the difference in size we compare state population to city population:

$$\frac{2 \times 10^7}{8 \times 10^6}.$$

Now we simplify the expression to find the answer:

$$\frac{2\times10^7}{8\times10^6} = \frac{2}{8} \times \frac{10^7}{10^6} \quad \text{By the product formula}$$

$$= \frac{1}{4} \times 10 \quad \text{By equivalent fractions and the first law of exponents}$$

$$= 0.25 \times 10$$

$$= 2.5$$

Therefore, the population of the state is 2.5 *times that of the city.*

Example 3 (4 minutes)

There are about 9 billion devices connected to the Internet. If a wireless router can support 300 devices, about how many wireless routers are necessary to connect all 9 billion devices wirelessly?

Because 9 billion is a very large number, we should express it as a single-digit integer times a power of 10.

$$9\,000\,000\,000 = 9 \times 10^9$$

The laws of exponents tells us that our calculations will be easier if we also express 300 as a single-digit integer times a power of 10, even though 300 is much smaller.

$$300 = 3 \times 10^2$$

We want to know how many wireless routers are necessary to support 9 billion devices, so we must divide

$$\frac{9 \times 10^9}{3 \times 10^2}.$$

Now, we can simplify the expression to find the answer:

$$\frac{9\times10^9}{3\times10^2} = \frac{9}{3} \times \frac{10^9}{10^2} \quad \text{By the product formula}$$

$$= 3 \times 10^7 \quad \text{By equivalent fractions and the first law of exponents}$$

$$= 30\,000\,000$$

About 30 *million routers are necessary to connect all devices wirelessly.*

Exercises 2–4 (5 minutes)

Have students complete Exercises 2–4 independently or in pairs.

Exercise 2

There are about $3,000,000$ students attending school, kindergarten through Grade 12, in New York. Express the number of students as a single-digit integer times a power of 10.

$$3\,000\,000 = 3 \times 10^6$$

The average number of students attending a middle school in New York is 8×10^2. How many times greater is the overall number of K–12 students compared to the average number of middle school students?

$$\begin{aligned}\frac{3 \times 10^6}{8 \times 10^2} &= \frac{3}{8} \times \frac{10^6}{10^2}\\ &= \frac{3}{8} \times 10^4\\ &= 0.375 \times 10^4\\ &= 3750\end{aligned}$$

There are about $3,750$ times more students in K–12 compared to the number of students in middle school.

Exercise 3

A conservative estimate of the number of stars in the universe is 6×10^{22}. The average human can see about $3,000$ stars at night with his naked eye. About how many times more stars are there in the universe compared to the stars a human can actually see?

$$\frac{6 \times 10^{22}}{3 \times 10^3} = \frac{6}{3} \times \frac{10^{22}}{10^3} = 2 \times 10^{22-3} = 2 \times 10^{19}$$

There are about 2×10^{19} times more stars in the universe compared to the number we can actually see.

Exercise 4

The estimated world population in 2011 was 7×10^9. Of the total population, 682 million of those people were left-handed. Approximately what percentage of the world population is left-handed according to the 2011 estimation?

$$682\,000\,000 \approx 700\,000\,000 = 7 \times 10^8$$

$$\begin{aligned}\frac{7 \times 10^8}{7 \times 10^9} &= \frac{7}{7} \times \frac{10^8}{10^9}\\ &= 1 \times \frac{1}{10}\\ &= \frac{1}{10}\end{aligned}$$

About one-tenth of the population is left-handed, which is equal to 10%.

Example 4 (3 minutes)

The average American household spends about $\$40{,}000$ each year. If there are about 1×10^8 households, what is the total amount of money spent by American households in one year?

Let's express $\$40{,}000$ as a single-digit integer times a power of 10.

$$40000 = 4 \times 10^4$$

The question asks us how much money *all* American households spend in one year, which means that we need to multiply the amount spent by one household by the total number of households:

$(4 \times 10^4)(1 \times 10^8) = (4 \times 1)(10^4 \times 10^8)$	By repeated use of associative and commutative properties
$= 4 \times 10^{12}$	By the first law of exponents

Therefore, American households spend about $\$4{,}000{,}000{,}000{,}000$ *each year altogether!*

Exercise 5 (2 minutes)

Have students complete Exercise 5 independently.

Exercise 5

The average person takes about $30{,}000$ breaths per day. Express this number as a single-digit integer times a power of 10.

$$30000 = 3 \times 10^4$$

If the average American lives about 80 years (or about $30{,}000$ days), how many total breaths will a person take in her lifetime?

$$(3 \times 10^4) \times (3 \times 10^4) = 9 \times 10^8$$

The average American takes about $900{,}000{,}000$ breaths in a lifetime.

Closing (2 minutes)

Summarize the lesson for the students.

- In general, close approximation of quantities will lead to more precise answers.
- We can multiply and divide numbers that are written in the form of a single-digit integer times a power of 10.

Exit Ticket (4 minutes)

Fluency Exercise (10 minutes)

Sprint: Practice the laws of exponents. Instruct students to write answers using positive exponents only. This exercise can be administered at any point during the lesson. Refer to the Sprints and Sprint Delivery Script sections in the Module Overview for directions to administer a Sprint.

Name ______________________________ Date__________________

Lesson 8: Estimating Quantities

Exit Ticket

Most English-speaking countries use the short-scale naming system, in which a trillion is expressed as $1{,}000{,}000{,}000{,}000$. Some other countries use the long-scale naming system, in which a trillion is expressed as $1{,}000{,}000{,}000{,}000{,}000{,}000$. Express each number as a single-digit integer times a power of ten. How many times greater is the long-scale naming system than the short-scale?

Exit Ticket Sample Solution

Most English-speaking countries use the short-scale naming system, in which a trillion is expressed as $1,000,000,000,000$. Some other countries use the long-scale naming system, in which a trillion is expressed as $1,000,000,000,000,000,000,000$. Express each number as a single-digit integer times a power of ten. How many times greater is the long-scale naming system than the short-scale?

$1\,000\,000\,000\,000 = 10^{12}$

$1\,000\,000\,000\,000\,000\,000\,000 = 10^{21}$

$\frac{10^{21}}{10^{12}} = 10^9$. *The long-scale is about 10^9 times greater than the short-scale.*

Problem Set Sample Solutions

Students practice estimating size of quantities and performing operations on numbers written in the form of a single-digit integer times a power of 10.

1. The Atlantic Ocean region contains approximately 2×10^{16} gallons of water. Lake Ontario has approximately $8,000,000,000,000$ gallons of water. How many Lake Ontarios would it take to fill the Atlantic Ocean region in terms of gallons of water?

$$8\,000\,000\,000\,000 = 8 \times 10^{12}$$

$$\begin{aligned}\frac{2 \times 10^{16}}{8 \times 10^{12}} &= \frac{2}{8} \times \frac{10^{16}}{10^{12}}\\ &= \frac{1}{4} \times 10^4\\ &= 0.25 \times 10^4\\ &= 2500\end{aligned}$$

$2,500$ *Lake Ontario's would be needed to fill the Atlantic Ocean region.*

2. U.S. national forests cover approximately $300,000$ square miles. Conservationists want the total square footage of forests to be $300,000^2$ square miles. When Ivanna used her phone to do the calculation, her screen showed the following:

a. What does the answer on her screen mean? Explain how you know.

The answer means 9×10^{10}. This is because:

$$\begin{aligned}(300\,000)^2 &= (3 \times 10^5)^2\\ &= 3^2 \times (10^5)^2\\ &= 9 \times 10^{10}\end{aligned}$$

b. Given that the U.S. has approximately 4 million square miles of land, is this a reasonable goal for conservationists? Explain.

$4\,000\,000 = 4 \times 10^6$. *It is unreasonable for conservationists to think the current square mileage of forests could increase that much because that number is greater than the number that represents the total number of square miles in the U.S,* $9 \times 10^{10} > 4 \times 10^6$.

3. The average American is responsible for about $20,000$ kilograms of carbon emission pollution each year. Express this number as a single-digit integer times a power of 10.

$$20\,000 = 2 \times 10^4$$

4. The United Kingdom is responsible for about 1×10^4 kilograms of carbon emission pollution each year. Which country is responsible for greater carbon emission pollution each year? By how much?

$$2 \times 10^4 > 1 \times 10^4$$

America is responsible for greater carbon emission pollution each year. America produces twice the amount of the U.K. pollution.

Number Correct: ______

Applying Properties of Exponents to Generate Equivalent Expressions—Round 1

Directions: Simplify each expression using the laws of exponents. Use the least number of bases possible and only positive exponents. When appropriate, express answers without parentheses or as equal to 1. All letters denote numbers.

1.	$4^5 \cdot 4^{-4} =$	
2.	$4^5 \cdot 4^{-3} =$	
3.	$4^5 \cdot 4^{-2} =$	
4.	$7^{-4} \cdot 7^{11} =$	
5.	$7^{-4} \cdot 7^{10} =$	
6.	$7^{-4} \cdot 7^{9} =$	
7.	$9^{-4} \cdot 9^{-3} =$	
8.	$9^{-4} \cdot 9^{-2} =$	
9.	$9^{-4} \cdot 9^{-1} =$	
10.	$9^{-4} \cdot 9^{0} =$	
11.	$5^0 \cdot 5^1 =$	
12.	$5^0 \cdot 5^2 =$	
13.	$5^0 \cdot 5^3 =$	
14.	$(12^3)^9 =$	
15.	$(12^3)^{10} =$	
16.	$(12^3)^{11} =$	
17.	$(7^{-3})^{-8} =$	
18.	$(7^{-3})^{-9} =$	
19.	$(7^{-3})^{-10} =$	
20.	$\left(\frac{1}{2}\right)^9 =$	
21.	$\left(\frac{1}{2}\right)^8 =$	
22.	$\left(\frac{1}{2}\right)^7 =$	

23.	$\left(\frac{1}{2}\right)^6 =$	
24.	$(3x)^5 =$	
25.	$(3x)^7 =$	
26.	$(3x)^9 =$	
27.	$(8^{-2})^3 =$	
28.	$(8^{-3})^3 =$	
29.	$(8^{-4})^3 =$	
30.	$(22^0)^{50} =$	
31.	$(22^0)^{55} =$	
32.	$(22^0)^{60} =$	
33.	$\left(\frac{1}{11}\right)^{-5} =$	
34.	$\left(\frac{1}{11}\right)^{-6} =$	
35.	$\left(\frac{1}{11}\right)^{-7} =$	
36.	$\frac{56^{-23}}{56^{-34}} =$	
37.	$\frac{87^{-12}}{87^{-34}} =$	
38.	$\frac{23^{-15}}{23^{-17}} =$	
39.	$(-2)^{-12} \cdot (-2)^1 =$	
40.	$\frac{2y}{y^3} =$	
41.	$\frac{5xy^7}{15x^7y} =$	
42.	$\frac{16x^6y^9}{8x^{-5}y^{-11}} =$	
43.	$(2^3 \cdot 4)^{-5} =$	
44.	$(9^{-8})(27^{-2}) =$	

Applying Properties of Exponents to Generate Equivalent Expressions—Round 1 [KEY]

Directions: Simplify each expression using the laws of exponents. Use the least number of bases possible and only positive exponents. When appropriate, express answers without parentheses or as equal to 1. All letters denote numbers.

1.	$4^5 \cdot 4^{-4} =$	4^1	23.	$\left(\frac{1}{2}\right)^6 =$	$\frac{1}{2^6}$
2.	$4^5 \cdot 4^{-3} =$	4^2	24.	$(3x)^5 =$	3^5x^5
3.	$4^5 \cdot 4^{-2} =$	4^3	25.	$(3x)^7 =$	3^7x^7
4.	$7^{-4} \cdot 7^{11} =$	7^7	26.	$(3x)^9 =$	3^9x^9
5.	$7^{-4} \cdot 7^{10} =$	7^6	27.	$(8^{-2})^3 =$	$\frac{1}{8^6}$
6.	$7^{-4} \cdot 7^9 =$	7^5	28.	$(8^{-3})^3 =$	$\frac{1}{8^9}$
7.	$9^{-4} \cdot 9^{-3} =$	$\frac{1}{9^7}$	29.	$(8^{-4})^3 =$	$\frac{1}{8^{12}}$
8.	$9^{-4} \cdot 9^{-2} =$	$\frac{1}{9^6}$	30.	$(22^0)^{50} =$	1
9.	$9^{-4} \cdot 9^{-1} =$	$\frac{1}{9^5}$	31.	$(22^0)^{55} =$	1
10.	$9^{-4} \cdot 9^0 =$	$\frac{1}{9^4}$	32.	$(22^0)^{60} =$	1
11.	$5^0 \cdot 5^1 =$	5^1	33.	$\left(\frac{1}{11}\right)^{-5} =$	11^5
12.	$5^0 \cdot 5^2 =$	5^2	34.	$\left(\frac{1}{11}\right)^{-6} =$	11^6
13.	$5^0 \cdot 5^3 =$	5^3	35.	$\left(\frac{1}{11}\right)^{-7} =$	11^7
14.	$(12^3)^9 =$	12^{27}	36.	$\frac{56^{-23}}{56^{-34}} =$	56^{11}
15.	$(12^3)^{10} =$	12^{30}	37.	$\frac{87^{-12}}{87^{-34}} =$	87^{22}
16.	$(12^3)^{11} =$	12^{33}	38.	$\frac{23^{-15}}{23^{-17}} =$	23^2
17.	$(7^{-3})^{-8} =$	7^{24}	39.	$(-2)^{-12} \cdot (-2)^1 =$	$\frac{1}{(-2)^{11}}$
18.	$(7^{-3})^{-9} =$	7^{27}	40.	$\frac{2y}{y^3} =$	$\frac{2}{y^2}$
19.	$(7^{-3})^{-10} =$	7^{30}	41.	$\frac{5xy^7}{15x^7y} =$	$\frac{y^6}{3x^6}$
20.	$\left(\frac{1}{2}\right)^9 =$	$\frac{1}{2^9}$	42.	$\frac{16x^6y^9}{8x^{-5}y^{-11}} =$	$2x^{11}y^{20}$
21.	$\left(\frac{1}{2}\right)^8 =$	$\frac{1}{2^8}$	43.	$(2^3 \cdot 4)^{-5} =$	$\frac{1}{2^{25}}$
22.	$\left(\frac{1}{2}\right)^7 =$	$\frac{1}{2^7}$	44.	$(9^{-8})(27^{-2}) =$	$\frac{1}{3^{22}}$

Number Correct: ______

Improvement: ______

Applying Properties of Exponents to Generate Equivalent Expressions—Round 2

Directions: Simplify each expression using the laws of exponents. Use the least number of bases possible and only positive exponents. When appropriate, express answers without parentheses or as equal to 1. All letters denote numbers.

1.	$11^5 \cdot 11^{-4} =$	
2.	$11^5 \cdot 11^{-3} =$	
3.	$11^5 \cdot 11^{-2} =$	
4.	$7^{-7} \cdot 7^9 =$	
5.	$7^{-8} \cdot 7^9 =$	
6.	$7^{-9} \cdot 7^9 =$	
7.	$(-6)^{-4} \cdot (-6)^{-3} =$	
8.	$(-6)^{-4} \cdot (-6)^{-2} =$	
9.	$(-6)^{-4} \cdot (-6)^{-1} =$	
10.	$(-6)^{-4} \cdot (-6)^0 =$	
11.	$x^0 \cdot x^1 =$	
12.	$x^0 \cdot x^2 =$	
13.	$x^0 \cdot x^3 =$	
14.	$(12^5)^9 =$	
15.	$(12^6)^9 =$	
16.	$(12^7)^9 =$	
17.	$(7^{-3})^{-4} =$	
18.	$(7^{-4})^{-4} =$	
19.	$(7^{-5})^{-4} =$	
20.	$\left(\frac{3}{7}\right)^8 =$	
21.	$\left(\frac{3}{7}\right)^7 =$	
22.	$\left(\frac{3}{7}\right)^6 =$	

23.	$\left(\frac{3}{7}\right)^5 =$	
24.	$(18xy)^5 =$	
25.	$(18xy)^7 =$	
26.	$(18xy)^9 =$	
27.	$(5.2^{-2})^3 =$	
28.	$(5.2^{-3})^3 =$	
29.	$(5.2^{-4})^3 =$	
30.	$(22^6)^0 =$	
31.	$(22^{12})^0 =$	
32.	$(22^{18})^0 =$	
33.	$\left(\frac{4}{5}\right)^{-5} =$	
34.	$\left(\frac{4}{5}\right)^{-6} =$	
35.	$\left(\frac{4}{5}\right)^{-7} =$	
36.	$\left(\frac{6^{-2}}{7^5}\right)^{-11} =$	
37.	$\left(\frac{6^{-2}}{7^5}\right)^{-12} =$	
38.	$\left(\frac{6^{-2}}{7^5}\right)^{-13} =$	
39.	$\left(\frac{6^{-2}}{7^5}\right)^{-15} =$	
40.	$\frac{42ab^{10}}{14a^{-9}b} =$	
41.	$\frac{5xy^7}{25x^7y} =$	
42.	$\frac{22a^{15}b^{32}}{121ab^{-5}} =$	
43.	$(7^{-8} \cdot 49)^{-5} =$	
44.	$(36^9)(216^{-2}) =$	

Applying Properties of Exponents to Generate Equivalent Expressions—Round 2 [KEY]

Directions: Simplify each expression using the laws of exponents. Use the least number of bases possible and only positive exponents. When appropriate, express answers without parentheses or as equal to 1. All letters denote numbers.

1.	$11^5 \cdot 11^{-4} =$	11^1
2.	$11^5 \cdot 11^{-3} =$	11^2
3.	$11^5 \cdot 11^{-2} =$	11^3
4.	$7^{-7} \cdot 7^9 =$	7^2
5.	$7^{-8} \cdot 7^9 =$	7^1
6.	$7^{-9} \cdot 7^9 =$	1
7.	$(-6)^{-4} \cdot (-6)^{-3} =$	$\frac{1}{(-6)^7}$
8.	$(-6)^{-4} \cdot (-6)^{-2} =$	$\frac{1}{(-6)^6}$
9.	$(-6)^{-4} \cdot (-6)^{-1} =$	$\frac{1}{(-6)^5}$
10.	$(-6)^{-4} \cdot (-6)^0 =$	$\frac{1}{(-6)^4}$
11.	$x^0 \cdot x^1 =$	x^1
12.	$x^0 \cdot x^2 =$	x^2
13.	$x^0 \cdot x^3 =$	x^3
14.	$(12^5)^9 =$	12^{45}
15.	$(12^6)^9 =$	12^{54}
16.	$(12^7)^9 =$	12^{63}
17.	$(7^{-3})^{-4} =$	7^{12}
18.	$(7^{-4})^{-4} =$	7^{16}
19.	$(7^{-5})^{-4} =$	7^{20}
20.	$\left(\frac{3}{7}\right)^8 =$	$\frac{3^8}{7^8}$
21.	$\left(\frac{3}{7}\right)^7 =$	$\frac{3^7}{7^7}$
22.	$\left(\frac{3}{7}\right)^6 =$	$\frac{3^6}{7^6}$

23.	$\left(\frac{3}{7}\right)^5 =$	$\frac{3^5}{7^5}$
24.	$(18xy)^5 =$	$18^5x^5y^5$
25.	$(18xy)^7 =$	$18^7x^7y^7$
26.	$(18xy)^9 =$	$18^9x^9y^9$
27.	$(5.2^{-2})^3 =$	$\frac{1}{(5.2)^6}$
28.	$(5.2^{-3})^3 =$	$\frac{1}{(5.2)^9}$
29.	$(5.2^{-4})^3 =$	$\frac{1}{(5.2)^{12}}$
30.	$(22^6)^0 =$	1
31.	$(22^{12})^0 =$	1
32.	$(22^{18})^0 =$	1
33.	$\left(\frac{4}{5}\right)^{-5} =$	$\frac{5^5}{4^5}$
34.	$\left(\frac{4}{5}\right)^{-6} =$	$\frac{5^6}{4^6}$
35.	$\left(\frac{4}{5}\right)^{-7} =$	$\frac{5^7}{4^7}$
36.	$\left(\frac{6^{-2}}{7^5}\right)^{-11} =$	$6^{22}7^{55}$
37.	$\left(\frac{6^{-2}}{7^5}\right)^{-12} =$	$6^{24}7^{60}$
38.	$\left(\frac{6^{-2}}{7^5}\right)^{-13} =$	$6^{26}7^{65}$
39.	$\left(\frac{6^{-2}}{7^5}\right)^{-15} =$	$6^{30}7^{75}$
40.	$\frac{42ab^{10}}{14a^{-9}b} =$	$3a^{10}b^9$
41.	$\frac{5xy^7}{25x^7y} =$	$\frac{y^6}{5x^6}$
42.	$\frac{22a^{15}b^{32}}{121ab^{-5}} =$	$\frac{2a^{14}b^{37}}{11}$
43.	$(7^{-8} \cdot 49)^{-5} =$	7^{30}
44.	$(36^9)(216^{-2}) =$	6^{12}

EUREKA MATH™

Lesson 9: Scientific Notation

Student Outcomes

- Students write, add, and subtract numbers in scientific notation and understand what is meant by the term leading digit.

Classwork

Discussion (5 minutes)

Our knowledge of the integer powers of 10 (i.e., Fact 1 and Fact 2 in Lesson 7) enable us to understand the next concept, *scientific* notation. Until now, we have been approximating a number like 6187 as 6×10^3. In this lesson we learn how to write the number exactly, using scientific notation.

Consider the estimated number of stars in the universe: 6×10^{22}. This is a 23-digit *whole number* with the **leading digit** (the leftmost digit) 6 followed by 22 zeros. When it is written in the form 6×10^{22}, it is said to be expressed in *scientific* notation.

A positive, finite decimal[1] s is said to be written in **scientific notation** if it is expressed as a product $d \times 10^n$, where d is a finite decimal ≥ 1 and < 10 (i.e., $1 \leq d < 10$), and n is an integer (i.e., d is a finite decimal with only a single, nonzero digit to the left of the decimal point). The integer n is called the **order of magnitude** of the decimal $d \times 10^n$.[2] (Note that now we present the order of magnitude, building from what was learned about magnitude in Lesson 7.)

> **A positive, finite decimal s is said to be written in scientific notation if it is expressed as a product $d \times 10^n$, where d is a finite decimal so that $1 \leq d < 10$, and n is an integer.**
>
> **The integer n is called the order of magnitude of the decimal $d \times 10^n$.**

Example 1 (2 minutes)

The finite decimal 234.567 is equal to every one of the following:

$2.345\,67 \times 10^2$	$0.234\,567 \times 10^3$	23.4567×10
234.567×10^0	$234\,567 \times 10^{-3}$	$234\,567\,000 \times 10^{-6}$

However, only the first is a representation of 234.567 in scientific notation. Ask students to explain why the first representation of 234.567 is the only example of scientific notation.

---

[1]Recall that every whole number is a finite decimal.
[2]Sometimes the place value, 10^n, of the leading digit of $d \times 10^n$ is called the **order of magnitude**. There is little chance of confusion.

Exercises 1–6 (3 minutes)

Students complete Exercises 1–6 independently.

Are the following numbers written in scientific notation? If not, state the reason.

Exercise 1

1.908×10^{17} *yes*

Exercise 2

0.325×10^{-2} *no, $d < 1$*

Exercise 3

7.99×10^{32} *yes*

Exercise 4

$4.0701 + 10^{7}$ *no, it must be a product*

Exercise 5

18.432×5^{8} *no, $d > 10$ and it is $\times 5$ instead of $\times 10$*

Exercise 6

8×10^{-11} *yes*

Discussion (2 minutes)

Exponent n is called the *order of magnitude* of the positive number $s = d \times 10^{n}$ (in scientific notation) because the following inequalities hold:

$$10^{n} \leq s \quad \text{and} \quad s < 10^{n+1}. \tag{18}$$

Thus, the exponent n serves to give an approximate location of s on the number line. That is, n gives the approximate magnitude of s.

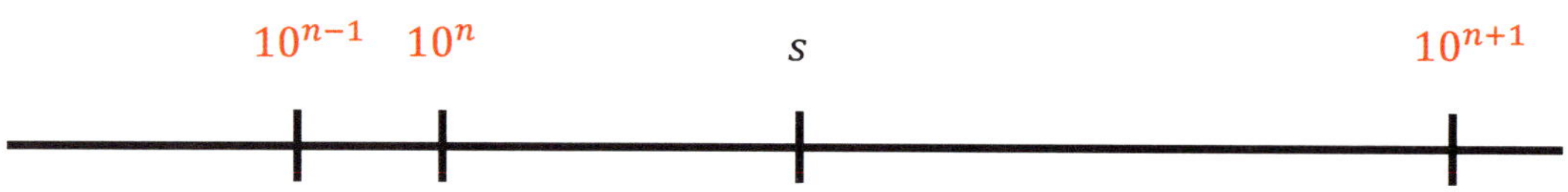

The inequalities in (18) above can be written as $10^{n} \leq s < 10^{n+1}$, and the number line shows that the number s is between 10^{n} and 10^{n+1}.

Examples 2–3 (10 minutes)

In the previous lesson, we approximated numbers by writing them as a single-digit integer times a power of 10. The guidelines provided by scientific notation allow us to continue to approximate numbers but now with more precision. For example, we use a finite decimal times a power of 10 instead of using only a single digit times a power of 10.

This allows us to compute with greater accuracy while still enjoying the benefits of completing basic computations with numbers and using the laws of exponents with powers of 10.

Example 2: Let's say we need to determine the difference in the populations of Texas and North Dakota. In 2012, Texas had a population of about 26 million people, and North Dakota had a population of about 6.9×10^4.

Scaffold:

Based on the needs of your students, for this and subsequent examples, choose to have students practice writing numbers in scientific notation or have them check the accuracy of numbers given in scientific notation.

We begin by writing each number in scientific notation:

- Texas: $26{,}000{,}000 = 2.6 \times 10^7$
- North Dakota: $69{,}000 = 6.9 \times 10^4$.

To find the difference, we subtract:

$$2.6 \times 10^7 - 6.9 \times 10^4.$$

To compute this easily, we need to make the order of magnitude of each number equal. That is, each number must have the same order of magnitude and the same base. When numbers have the same order of magnitude and the same base, we can use the distributive property to perform operations because each number has a common factor. Specifically for this problem, we can rewrite the population of Texas so that it is an integer multiplied by 10^4 and then subtract.

$$\begin{aligned} 2.6 \times 10^7 - 6.9 \times 10^4 &= (2.6 \times 10^3) \times 10^4 - 6.9 \times 10^4 && \text{By the first law of exponents} \\ &= 2600 \times 10^4 - 6.9 \times 10^4 \\ &= (2600 - 6.9) \times 10^4 && \text{By the distributive property} \\ &= 2593.1 \times 10^4 \\ &= 25{,}931{,}000 \end{aligned}$$

Example 3: Let's say that we need to find the combined mass of two hydrogen atoms and one oxygen atom, which is normally written as H_2O or otherwise known as water. To appreciate the value of scientific notation, the mass of each atom will be given in standard notation:

- One hydrogen atom is approximately $0.000\ 000\ 000\ 000\ 000\ 000\ 000\ 000\ 001\ 7$ kilograms.
- One oxygen atom is approximately $0.000\ 000\ 000\ 000\ 000\ 000\ 000\ 000\ 027$ kilograms.

To determine the combined mass of water, we need the mass of 2 hydrogen atoms plus one oxygen atom. First, we should write each atom in scientific notation.

- Hydrogen: 1.7×10^{-27}
- Oxygen: 2.7×10^{-26}
- 2 Hydrogen atoms $= 2(1.7 \times 10^{-27}) = 3.4 \times 10^{-27}$
- 2 Hydrogen atoms $+1$ Oxygen atom $= 3.4 \times 10^{-27} + 2.7 \times 10^{-26}$

As in the previous example, we must have the same order of magnitude for each number. Thus, changing them both to 10^{-26}:

$$\begin{aligned} 3.4 \times 10^{-27} + 2.7 \times 10^{-26} &= (3.4 \times 10^{-1}) \times 10^{-26} + 2.7 \times 10^{-26} && \text{By the first law of exponents} \\ &= 0.34 \times 10^{-26} + 2.7 \times 10^{-26} \\ &= (0.34 + 2.7) \times 10^{-26} && \text{By the distributive property} \\ &= 3.04 \times 10^{-26} \end{aligned}$$

We can also choose to do this problem a different way, by making both numbers have 10^{-27} as the common order of magnitude:

$$\begin{aligned} 3.4 \times 10^{-27} + 2.7 \times 10^{-26} &= 3.4 \times 10^{-27} + (2.7 \times 10) \times 10^{-27} && \text{By the first law of exponents} \\ &= 3.4 \times 10^{-27} + 27 \times 10^{-27} \\ &= (3.4 + 27) \times 10^{-27} && \text{By the distributive property} \\ &= 30.4 \times 10^{-27} \\ &= 3.04 \times 10^{-26}. \end{aligned}$$

Exercises 7–9 (10 minutes)

Have students complete Exercises 7–9 independently.

Use the table below to complete Exercises 7 and 8.

The table below shows the debt of the three most populous states and the three least populous states.

State	Debt (in dollars)	Population (2012)
California	$407,000,000,000$	$38,000,000$
New York	$337,000,000,000$	$19,000,000$
Texas	$276,000,000,000$	$26,000,000$
North Dakota	$4,000,000,000$	$690,000$
Vermont	$4,000,000,000$	$626,000$
Wyoming	$2,000,000,000$	$576,000$

Exercise 7

a. **What is the sum of the debts for the three most populous states? Express your answer in scientific notation.**

$$\begin{aligned} (4.07 \times 10^{11}) + (3.37 \times 10^{11}) + (2.76 \times 10^{11}) &= (4.07 + 3.37 + 2.76) \times 10^{11} \\ &= 10.2 \times 10^{11} \\ &= (1.02 \times 10) \times 10^{11} \\ &= 1.02 \times 10^{12} \end{aligned}$$

b. **What is the sum of the debt for the three least populous states? Express your answer in scientific notation.**

$$\begin{aligned} (4 \times 10^{9}) + (4 \times 10^{9}) + (2 \times 10^{9}) &= (4 + 4 + 2) \times 10^{9} \\ &= 10 \times 10^{9} \\ &= (1 \times 10) \times 10^{9} \\ &= 1 \times 10^{10} \end{aligned}$$

c. **How much larger is the combined debt of the three most populous states than that of the three least populous states? Express your answer in scientific notation.**

$$\begin{aligned} (1.02 \times 10^{12}) - (1 \times 10^{10}) &= (1.02 \times 10^{2} \times 10^{10}) - (1 \times 10^{10}) \\ &= (102 \times 10^{10}) - (1 \times 10^{10}) \\ &= (102 - 1) \times 10^{10} \\ &= 101 \times 10^{10} \\ &= (1.01 \times 10^{2}) \times 10^{10} \\ &= 1.01 \times 10^{12} \end{aligned}$$

Exercise 8

a. **What is the sum of the population of the three most populous states? Express your answer in scientific notation.**

$$(3.8 \times 10^7) + (1.9 \times 10^7) + (2.6 \times 10^7) = (3.8 + 1.9 + 2.6) \times 10^7$$
$$= 8.3 \times 10^7$$

b. **What is the sum of the population of the three least populous states? Express your answer in scientific notation.**

$$(6.9 \times 10^5) + (6.26 \times 10^5) + (5.76 \times 10^5) = (6.9 + 6.26 + 5.76) \times 10^5$$
$$= 18.92 \times 10^5$$
$$= (1.892 \times 10) \times 10^5$$
$$= 1.892 \times 10^6$$

c. **Approximately how many times greater is the total population of California, New York, and Texas compared to the total population of North Dakota, Vermont, and Wyoming?**

$$\frac{8.3 \times 10^7}{1.892 \times 10^6} = \frac{8.3}{1.892} \times \frac{10^7}{10^6}$$
$$\approx 4.39 \times 10$$
$$= 43.9$$

The combined population of California, New York, and Texas is about 43.9 *times greater than the combined population of North Dakota, Vermont, and Wyoming.*

Exercise 9

All planets revolve around the sun in elliptical orbits. Uranus's furthest distance from the sun is approximately 3.004×10^9 km, and its closest distance is approximately 2.749×10^9 km. Using this information, what is the average distance of Uranus from the sun?

$$\textit{average distance} = \frac{(3.004 \times 10^9) + (2.749 \times 10^9)}{2}$$
$$= \frac{(3.004 + 2.749) \times 10^9}{2}$$
$$= \frac{5.753 \times 10^9}{2}$$
$$= 2.8765 \times 10^9$$

On average, Uranus is 2.8765×10^9 km *from the sun.*

Discussion (5 minutes)

- Why are we interested in writing numbers in scientific notation?
  - *It is essential that we express very large and very small numbers in scientific notation. For example, consider once again the estimated number of stars in the universe. The advantage of presenting it as* 6×10^{22}, *rather than as* 60,000,000,000,000,000,000,000 *(in the standard notation), is perhaps too obvious for discussion. In the standard form, we cannot keep track of the number of zeros!*

- There is another deeper advantage of scientific notation. When faced with very large numbers, one's natural first question is *roughly* how big? Is it near a few hundred billion (a number with 11 digits) or even a few trillion (a number with 12 digits)? The exponent 22 in the scientific notation 6×10^{22} lets us know immediately that there is a 23-digit number and, therefore, far larger than *a few trillion*.
- We should elaborate on the last statement. Observe that the number 6234.5×10^{22} does *not* have 23 digits but 26 digits because it is the number $62{,}345{,}000{,}000{,}000{,}000{,}000{,}000{,}000$, which equals 6.2345×10^{25}.

Have students check to see that this number actually has 26 digits.

Ask them to think about why it has 26 digits when the exponent is 22. If they need help, point out what we started with: 6×10^{22} and 6234.5×10^{22}. Ask students what makes these numbers different. They should see that the first number is written in proper scientific notation, so the exponent of 22 tells us that this number will have $(22 + 1)$ digits. The second number has a value of d that is in the thousands (recall: $s = d \times 10^n$ and $1 \leq d < 10$). So, we are confident that 6×10^{22} has only 23 digits because 6 is greater than 1 and less than 10.

- Therefore, by normalizing (i.e., standardizing) the d in $d \times 10^n$ to satisfy $1 \leq d < 10$, we can rely on the exponent n to give us a sense of a number's *order of magnitude* of $d \times 10^n$.

Closing (3 minutes)

Summarize, or have students summarize, the lesson.

- Knowing how to write numbers in scientific notation allows us to determine the order of magnitude of a finite decimal.
- We now know how to compute with numbers expressed in scientific notation.

Exit Ticket (5 minutes)

EUREKA MATH™

G8-M1-B1-v1.3.0_05/04/15

Name ______________________________ Date ____________

Lesson 9: Scientific Notation

Exit Ticket

1. The approximate total surface area of Earth is $5.1 \times 10^8 \text{ km}^2$. All the salt water on Earth has an approximate surface area of $352{,}000{,}000 \text{ km}^2$, and all the freshwater on Earth has an approximate surface area of $9 \times 10^6 \text{ km}^2$. How much of Earth's surface is covered by water, including both salt and fresh water? Write your answer in scientific notation.

2. How much of Earth's surface is covered by land? Write your answer in scientific notation.

3. Approximately how many times greater is the amount of Earth's surface that is covered by water compared to the amount of Earth's surface that is covered by land?

Exit Ticket Sample Solutions

1. The approximate total surface area of Earth is 5.1×10^8 km^2. All the salt water on Earth has an approximate surface area of $352,000,000$ km^2, and all the freshwater on Earth has an approximate surface area of 9×10^6 km^2. How much of Earth's surface is covered by water, including both salt and fresh water? Write your answer in scientific notation.

$$\begin{aligned}(3.52 \times 10^8) + (9 \times 10^6) &= (3.52 \times 10^2 \times 10^6) + (9 \times 10^6)\\ &= (352 \times 10^6) + (9 \times 10^6)\\ &= (352 + 9) \times 10^6\\ &= 361 \times 10^6\\ &= 3.61 \times 10^8\end{aligned}$$

The Earth's surface is covered by 3.61×10^8 km^2 *of water.*

2. How much of Earth's surface is covered by land? Write your answer in scientific notation.

$$\begin{aligned}(5.1 \times 10^8) - (3.61 \times 10^8) &= (5.1 - 3.61) \times 10^8\\ &= 1.49 \times 10^8\end{aligned}$$

The Earth's surface is covered by 1.49×10^8 km^2 *of land.*

3. Approximately how many times greater is the amount of Earth's surface that is covered by water compared to the amount of Earth's surface that is covered by land?

$$\frac{3.61 \times 10^8}{1.49 \times 10^8} \approx 2.4$$

About 2.4 *times more of the Earth's surface is covered by water than by land.*

Problem Set Sample Solutions

Students practice working with numbers written in scientific notation.

1. Write the number $68,127,000,000,000,000$ in scientific notation. Which of the two representations of this number do you prefer? Explain.

$$68\,127\,000\,000\,000\,000 = 6.8127 \times 10^{16}$$

Most likely, students will say that they like the scientific notation better because it allows them to write less. However, they should also take note of the fact that counting the number of zeros in $68,127,000,000,000,000$ *is a nightmare. A strong reason for using scientific notation is to circumvent this difficulty: right away, the exponent* 16 *shows that this is a* 17*-digit number.*

2. Here are the masses of the so-called inner planets of the solar system.

Mercury: 3.3022×10^{23} kg | Earth: 5.9722×10^{24} kg

Venus: 4.8685×10^{24} kg | Mars: 6.4185×10^{23} kg

What is the average mass of all four inner planets? Write your answer in scientific notation.

$$\begin{aligned} \textit{average mass} &= \frac{(3.3022 \times 10^{23}) + (4.8685 \times 10^{24}) + (5.9722 \times 10^{24}) + (6.4185 \times 10^{23})}{4} \\ &= \frac{(3.3022 \times 10^{23}) + (48.685 \times 10^{23}) + (59.722 \times 10^{23}) + (6.4185 \times 10^{23})}{4} \\ &= \frac{(3.3022 + 48.685 + 59.722 + 6.4185) \times 10^{23}}{4} \\ &= \frac{118.1277 \times 10^{23}}{4} \\ &= 29.531925 \times 10^{23} \\ &= 2.9531925 \times 10^{24} \end{aligned}$$

The average mass of the inner planets is 2.9531925×10^{24} kg.

Lesson 10: Operations with Numbers in Scientific Notation

Student Outcomes

- Students practice operations with numbers expressed in scientific notation and standard notation.

Classwork

Examples 1–2 (8 minutes)

Example 1: The world population is about 7 billion. There are 4.6×10^7 ants for every human on the planet. About how many ants are there in the world?

First, write 7 billion in scientific notation: (7×10^9).

To find the number of ants in the world, we need to multiply the world population by the known number of ants for each person: $(7 \times 10^9)(4.6 \times 10^7)$.

$(7 \times 10^9)(4.6 \times 10^7) = (7 \times 4.6)(10^9 \times 10^7)$	By repeated use of the associative and commutative properties
$= 32.2 \times 10^{16}$	By the first law of exponents
$= 3.22 \times 10 \times 10^{16}$	
$= 3.22 \times 10^{17}$	By the first law of exponents

There are about 3.22×10^{17} *ants in the world!*

Example 2: A certain social media company processes about 990 billion *likes* per year. If the company has approximately 8.9×10^8 users of the social media, about how many *likes* is each user responsible for per year? Write your answer in scientific and standard notation.

First, write 990 billion in scientific notation: 9.9×10^{11}.

To find the number of *likes* per person, divide the total number of *likes* by the total number of users: $\frac{9.9 \times 10^{11}}{8.9 \times 10^8}$

$\frac{9.9 \times 10^{11}}{8.9 \times 10^8} = \frac{9.9}{8.9} \times \frac{10^{11}}{10^8}$	By the product formula
$= 1.11235\ldots \times 10^3$	By the first law of exponents
$\approx 1.1 \times 10^3$	
≈ 1100	

Each user is responsible for about 1.1×10^3*, or* 1,100*, likes per year.*

Exercises 1– 2 (10 minutes)

Have students complete Exercises 1 and 2 independently.

Exercise 1

The speed of light is $300,000,000$ meters per second. The sun is approximately 1.5×10^{11} meters from Earth. How many seconds does it take for sunlight to reach Earth?

$$300\,000\,000 = 3 \times 10^8$$

$$\frac{1.5 \times 10^{11}}{3 \times 10^8} = \frac{1.5}{3} \times \frac{10^{11}}{10^8}$$
$$= 0.5 \times 10^3$$
$$= 0.5 \times 10 \times 10^2$$
$$= 5 \times 10^2$$

It takes 500 seconds for sunlight to reach Earth.

Exercise 2

The mass of the moon is about 7.3×10^{22} kg. It would take approximately $26,000,000$ moons to equal the mass of the sun. Determine the mass of the sun.

$$26\,000\,000 = 2.6 \times 10^7$$

$$(2.6 \times 10^7)(7.3 \times 10^{22}) = (2.6 \times 7.3)(10^7 \times 10^{22})$$
$$= 18.98 \times 10^{29}$$
$$= 1.898 \times 10 \times 10^{29}$$
$$= 1.898 \times 10^{30}$$

The mass of the sun is 1.898×10^{30} kg.

Example 3 (8 minutes)

In 2010, Americans generated 2.5×10^8 tons of garbage. There are about 2,000 landfills in the United States. Assuming that each landfill is the same size and that trash is divided equally among them, determine how many tons of garbage were sent to each landfill in 2010.

First, write 2,000 in scientific notation: 2×10^3.

To find the number of tons of garbage sent to each landfill, divide the total weight of the garbage by the number of landfills: $\frac{2.5 \times 10^8}{2 \times 10^3}$.

$$\frac{2.5 \times 10^8}{2 \times 10^3} = \frac{2.5}{2} \times \frac{10^8}{10^3} \quad \text{By the product formula}$$
$$= 1.25 \times 10^5 \quad \text{By the first law of exponents}$$

Each landfill received 1.25×10^5 tons of garbage in 2010.

Actually, not all garbage went to landfills. Some of it was recycled and composted. The amount of recycled and composted material accounted for about 85 million tons of the 2.5×10^8 tons of garbage. Given this new information, how much garbage was actually sent to each landfill?

First, write 85 million in scientific notation: 8.5×10^7.

Next, subtract the amount of recycled and composted material from the garbage: $2.5 \times 10^8 - 8.5 \times 10^7$. To subtract, we must give each number the same order of magnitude and then use the distributive property.

$$\begin{aligned} 2.5 \times 10^8 - 8.5 \times 10^7 &= (2.5 \times 10) \times 10^7 - 8.5 \times 10^7 && \text{By the first law of exponents} \\ &= (2.5 \times 10) - 8.5)) \times 10^7 && \text{By the distributive property} \\ &= (25 - 8.5) \times 10^7 \\ &= 16.5 \times 10^7 \\ &= 1.65 \times 10 \times 10^7 \\ &= 1.65 \times 10^8 && \text{By the first law of exponents} \end{aligned}$$

Now, divide the new amount of garbage by the number of landfills: $\frac{1.65 \times 10^8}{2 \times 10^3}$.

$$\begin{aligned} \frac{1.65 \times 10^8}{2 \times 10^3} &= \frac{1.65}{2} \times \frac{10^8}{10^3} && \text{By the product formula} \\ &= 0.825 \times 10^5 && \text{By the first law of exponents} \\ &= 0.825 \times 10 \times 10^4 && \text{By the first law of exponents} \\ &= 8.25 \times 10^4 \end{aligned}$$

Each landfill actually received 8.25×10^4 *tons of garbage in* 2010.

Exercises 3–5 (10 minutes)

Have students complete Exercises 3–5 independently.

Exercise 3

The mass of Earth is 5.9×10^{24} kg. The mass of Pluto is $13,000,000,000,000,000,000,000$ kg. Compared to Pluto, how much greater is Earth's mass than Pluto's mass?

$$13\,000\,000\,000\,000\,000\,000\,000 = 1.3 \times 10^{22}$$

$$\begin{aligned} 5.9 \times 10^{24} - 1.3 \times 10^{22} &= (5.9 \times 10^2) \times 10^{22} - 1.3 \times 10^{22} \\ &= (590 - 1.3) \times 10^{22} \\ &= 588.7 \times 10^{22} \\ &= 5.887 \times 10^2 \times 10^{22} \\ &= 5.887 \times 10^{24} \end{aligned}$$

The mass of Earth is 5.887×10^{24} kg *greater than the mass of Pluto.*

Exercise 4

Using the information in Exercises 2 and 3, find the combined mass of the moon, Earth, and Pluto.

$$\begin{aligned} 7.3 \times 10^{22} + 1.3 \times 10^{22} + 5.9 \times 10^{24} &= (7.3 \times 10^{22} + 1.3 \times 10^{22}) + 5.9 \times 10^{24} \\ &= 8.6 \times 10^{22} + 5.9 \times 10^{24} \\ &= (8.6 + 590) \times 10^{22} \\ &= 598.6 \times 10^{22} \\ &= 5.986 \times 10^{2} \times 10^{22} \\ &= 5.986 \times 10^{24} \end{aligned}$$

The combined mass of the moon, Earth, and Pluto is 5.986×10^{24} kg.

Exercise 5

How many combined moon, Earth, and Pluto masses (i.e., the answer to Exercise 4) are needed to equal the mass of the sun (i.e., the answer to Exercise 2)?

$$\begin{aligned} \frac{1.898 \times 10^{30}}{5.986 \times 10^{24}} &= \frac{1.898}{5.986} \times \frac{10^{30}}{10^{24}} \\ &= 0.3170\ldots \times 10^{6} \\ &\approx 0.32 \times 10^{6} \\ &= 0.32 \times 10 \times 10^{5} \\ &= 3.2 \times 10^{5} \end{aligned}$$

It would take 3.2×10^{5} combined masses of the moon, Earth, and Pluto to equal the mass of the sun.

Closing (4 minutes)

Summarize, or have students summarize, the lesson.

- We can perform all operations for numbers expressed in scientific notation or standard notation.

Exit Ticket (5 minutes)

Name ______________________________ Date ______________

Lesson 10: Operations with Numbers in Scientific Notation

Exit Ticket

1. The speed of light is 3×10^8 meters per second. The sun is approximately $230,000,000,000$ meters from Mars. How many seconds does it take for sunlight to reach Mars?

2. If the sun is approximately 1.5×10^{11} meters from Earth, what is the approximate distance from Earth to Mars?

EUREKA MATH™

Exit Ticket Sample Solutions

1. The speed of light is 3×10^8 meters per second. The sun is approximately $230,000,000,000$ meters from Mars. How many seconds does it take for sunlight to reach Mars?

$$230\,000\,000\,000 = 2.3 \times 10^{11}$$
$$\frac{2.3 \times 10^{11}}{3 \times 10^8} = \frac{2.3}{3} \times \frac{10^{11}}{10^8}$$
$$= 0.7666\ldots \times 10^3$$
$$\approx 0.77 \times 10 \times 10^2$$
$$\approx 7.7 \times 10^2$$

It takes approximately 770 seconds for sunlight to reach Mars.

2. If the sun is approximately 1.5×10^{11} meters from Earth, what is the approximate distance from Earth to Mars?

$$(2.3 \times 10^{11}) - (1.5 \times 10^{11}) = (2.3 - 1.5) \times 10^{11}$$
$$= 0.8 \times 10^{11}$$
$$= 0.8 \times 10 \times 10^{10}$$
$$= 8 \times 10^{10}$$

The distance from Earth to Mars is 8×10^{10} meters.

Problem Set Sample Solutions

Have students practice operations with numbers written in scientific notation and standard notation.

1. The sun produces 3.8×10^{27} joules of energy per second. How much energy is produced in a year? (Note: a year is approximately $31,000,000$ seconds).

$$31\,000\,000 = 3.1 \times 10^7$$
$$(3.8 \times 10^{27})(3.1 \times 10^7) = (3.8 \times 3.1)(10^{27} \times 10^7)$$
$$= 11.78 \times 10^{34}$$
$$= 1.178 \times 10 \times 10^{34}$$
$$= 1.178 \times 10^{35}$$

The sun produces 1.178×10^{35} joules of energy in a year.

2. On average, Mercury is about $57,000,000$ km from the sun, whereas Neptune is about 4.5×10^9 km from the sun. What is the difference between Mercury's and Neptune's distances from the sun?

$$57\,000\,000 = 5.7 \times 10^7$$
$$\begin{aligned} 4.5 \times 10^9 - 5.7 \times 10^7 &= (4.5 \times 10^2) \times 10^7 - 5.7 \times 10^7 \\ &= 450 \times 10^7 - 5.7 \times 10^7 \\ &= (450 - 5.7) \times 10^7 \\ &= 444.3 \times 10^7 \\ &= 4.443 \times 10^2 \times 10^7 \\ &= 4.443 \times 10^9 \end{aligned}$$

The difference in the distance of Mercury and Neptune from the sun is 4.443×10^9 *km.*

3. The mass of Earth is approximately 5.9×10^{24} kg, and the mass of Venus is approximately 4.9×10^{24} kg.

   a. Find their combined mass.

$$\begin{aligned} 5.9 \times 10^{24} + 4.9 \times 10^{24} &= (5.9 + 4.9) \times 10^{24} \\ &= 10.8 \times 10^{24} \\ &= 1.08 \times 10 \times 10^{24} \\ &= 1.08 \times 10^{25} \end{aligned}$$

   The combined mass of Earth and Venus is 1.08×10^{25} *kg.*

   b. Given that the mass of the sun is approximately 1.9×10^{30} kg, how many Venuses and Earths would it take to equal the mass of the sun?

$$\begin{aligned} \frac{1.9 \times 10^{30}}{1.08 \times 10^{25}} &= \frac{1.9}{1.08} \times \frac{10^{30}}{10^{25}} \\ &= 1.75925\ldots \times 10^5 \\ &\approx 1.8 \times 10^5 \end{aligned}$$

   It would take approximately 1.8×10^5 *Venuses and Earths to equal the mass of the sun.*

EUREKA MATH

Lesson 11: Efficacy of Scientific Notation

Student Outcomes

- Students continue to practice working with very small and very large numbers expressed in scientific notation.
- Students read, write, and perform operations on numbers expressed in scientific notation.

Lesson Notes

Powers of Ten, a video that demonstrates positive and negative powers of 10, is available online. The video should pique students' interest in why exponential notation is necessary. A link to the video is provided below (9 minutes).

http://www.youtube.com/watch?v=0fKBhvDjuy0

Classwork

Exercises 1–2 (3 minutes)

Exercise 1

The mass of a proton is

$$\mathbf{0.000\,000\,000\,000\,000\,000\,000\,000\,001\,672\,622\ kg.}$$

In scientific notation it is

$\mathbf{1.672622 \times 10^{-27}\ kg.}$

Exercise 2

The mass of an electron is

$$\mathbf{0.000\,000\,000\,000\,000\,000\,000\,000\,000\,000\,910\,938\,291\ kg.}$$

In scientific notation it is

$\mathbf{9.10938291 \times 10^{-31}\ kg.}$

Discussion (3 minutes)

MP.6

We continue to discuss why it is important to express numbers using scientific notation.

Consider the mass of the proton

$$0.000\,000\,000\,000\,000\,000\,000\,000\,001\,672\,622\ \text{kg}.$$

It is more informative to write it in scientific notation

$$1.672622 \times 10^{-27}\ \text{kg}.$$

The exponent -27 is used because the first nonzero digit (i.e., 1) of this number occurs in the 27th digit after the decimal point.

MP.6

Similarly, the mass of the electron is

$$0.000\,000\,000\,000\,000\,000\,000\,000\,000\,000\,910\,938\,291 \text{ kg}.$$

It is much easier to read this number in scientific notation

$$9.10938291 \times 10^{-31} \text{ kg}.$$

In this case, the exponent -31 is used because the first nonzero digit (i.e., 9) of this number occurs in the 31st digit to the right of the decimal point.

> *Scaffolding:*
>
> Some time should be spent making sense of these statements with students. This can be accomplished by, for example, providing a series of simpler numbers (e.g., $0.1, 0.01, 0.001$) and demonstrating that when they are expressed in exponential notation, $10^{-1}, 10^{-2}, 10^{-3}$, that the exponent tells that the first nonzero digit occurs in the first, second, and third digit to the right of the decimal point, respectively.

Exercise 3 (3 minutes)

Before students write the ratio that compares the mass of a proton to the mass of an electron, ask them whether they would rather write the ratio in standard (i.e., decimal) or scientific notation. If necessary, help them understand why scientific notation is more advantageous.

> **Exercise 3**
>
> **Write the ratio that compares the mass of a proton to the mass of an electron.**
>
> ***Ratio:*** $(1.672622 \times 10^{-27}) : (9.10938291 \times 10^{-31})$

Discussion (20 minutes)

This discussion includes Example 1, Exercise 4, and Example 2.

Example 1

The advantage of the scientific notation becomes even more pronounced when we have to compute how many times heavier a proton is than an electron. Instead of writing the value of the ratio, r, as

$$r = \frac{0.000\,000\,000\,000\,000\,000\,000\,000\,001\,672\,622}{0.000\,000\,000\,000\,000\,000\,000\,000\,000\,000\,910\,938\,291}$$

we express it as

$$r = \frac{1.672622 \times 10^{-27}}{9.10938291 \times 10^{-31}}.$$

MP.2

- Should we eliminate the power of 10 in the numerator or denominator? Explain.
  - *Using the theorem on generalized equivalent fractions, we can eliminate the negative power of* 10 *in the numerator and denominator to see what we are doing more clearly. Anticipating that* $10^{-31} \times 10^{31} = 1$, *we can multiply the numerator and denominator of the (complex) fraction by* 10^{31}

> *Scaffolding:*
>
> When students ask why we eliminated the negative power of 10 in the denominator (instead of the numerator), explain that positive powers of 10 are easier to interpret.

$$r = \frac{1.672622 \times 10^{-27}}{9.10938291 \times 10^{-31}} \times \frac{10^{31}}{10^{31}}.$$

Using the first law of exponents (10) presented in Lesson 5, we get

$$r = \frac{1.672622 \times 10^{4}}{9.10938291} = \frac{1.672622}{9.10938291} \times 10^{4}.$$

MP.2 Note that since we are using scientific notation, we can interpret an approximate value of r right away. For example, we see

$$\frac{1.672622}{9.10938291} \approx \frac{1.7}{9.1} = \frac{17}{91} \approx \frac{1}{5}$$

so that r is approximately $\frac{1}{5} \times 10{,}000$, which is 2,000. Thus, we expect a proton to be about two thousand times heavier than an electron.

Exercise 4

Students find a more precise answer for Example 1. Allow students to use a calculator to divide 1.672 622 by 9.109 382 91. When they finish, have students compare the approximate answer (2,000) to their more precise answer (1,836).

Exercise 4

Compute how many times heavier a proton is than an electron (i.e., find the value of the ratio). Round your final answer to the nearest one.

Let r = the value of the ratio, then:

$$\begin{aligned} r &= \frac{1.672622 \times 10^{-27}}{9.10938291 \times 10^{-31}} \\ &= \frac{1.672622 \times 10^{-27} \times 10^{31}}{9.10938291 \times 10^{-31} \times 10^{31}} \\ &= \frac{1.672622 \times 10^{4}}{9.10938291} \\ &= \frac{1.672622}{9.10938291} \times 10^{4} \\ &= \frac{1.672622 \times 10^{8}}{9.10938291 \times 10^{8}} \times 10^{4} \\ &= \frac{167{,}262{,}200}{910{,}938{,}291} \times 10^{4} \\ &= 0.183615291675 \times 10^{4} \\ &= 1836.15291675 \\ &\approx 1836 \end{aligned}$$

Example 2

As of March 23, 2013, the U.S. national debt was $\$16{,}755{,}133{,}009{,}522$ (rounded to the nearest dollar). According to the 2012 U.S. census, there are about $313{,}914{,}040$ American citizens. What is each citizen's approximate share of the debt?

- How precise should we make our answer? Do we want to know the exact amount, to the nearest dollar, or is a less precise answer alright?
    - *The most precise answer uses the exact numbers listed in the problem. The more the numbers are rounded, the precision of the answer decreases. We should aim for the most precise answer when necessary, but the following problem does not require it since we are finding the "approximate share of the debt."*

Let's round off the debt to the nearest *billion* (10^9). It is $\$16{,}755{,}000{,}000{,}000$, which is 1.6755×10^{13} dollars. Let's also round off the population to the nearest *million* (10^6), making it $314{,}000{,}000$, which is 3.14×10^8. Therefore, using the product formula and equation (13) from Lesson 5, we see that each citizen's share of the debt, in dollars, is

$$\begin{aligned}\frac{1.6755 \times 10^{13}}{3.14 \times 10^8} &= \frac{1.6755}{3.14} \times \frac{10^{13}}{10^8} \\ &= \frac{1.6755}{3.14} \times 10^5.\end{aligned}$$

Once again, we note the advantages of computing numbers expressed in scientific notation. Immediately, we can approximate the answer, about half of 10^5, or a hundred thousand dollars, (i.e., about $\$50{,}000$), because

$$\frac{1.6755}{3.14} \approx \frac{1.7}{3.1} = \frac{17}{31} \approx \frac{1}{2}.$$

More precisely, with the help of a calculator,

$$\frac{1.6755}{3.14} = \frac{16755}{31410} = 0.533598\ldots \approx 0.5336.$$

Therefore, each citizen's share of the U.S. national debt is about $\$53{,}360$.

Example 2

The U.S. national debt as of March 23, 2013, rounded to the nearest dollar, is $\$16,755,133,009,522$. According to the 2012 U.S. census, there are about $313,914,040$ U.S. citizens. What is each citizen's approximate share of the debt?

$$\begin{aligned}\frac{1.6755 \times 10^{13}}{3.14 \times 10^8} &= \frac{1.6755}{3.14} \times \frac{10^{13}}{10^8} \\ &= \frac{1.6755}{3.14} \times 10^5 \\ &= 0.533598\ldots \times 10^5 \\ &\approx 0.5336 \times 10^5 \\ &= 53360\end{aligned}$$

***Each U.S. citizen's share of the national debt is about* $\$53,360$.**

Exercises 5–6 (8 minutes)

Students work on Exercises 5 and 6 independently.

Exercise 5

The geographic area of California is $163,696$ sq. mi., and the geographic area of the U.S. is $3,794,101$ sq. mi. Let's round off these figures to 1.637×10^5 and 3.794×10^6. In terms of area, roughly estimate how many Californias would make up one U.S. Then compute the answer to the nearest ones.

$$\begin{aligned}\frac{3.794 \times 10^6}{1.637 \times 10^5} &= \frac{3.794}{1.637} \times \frac{10^6}{10^5} \\ &= \frac{3.794}{1.637} \times 10 \\ &= 2.3176\ldots \times 10 \\ &\approx 2.318 \times 10 \\ &= 23.18\end{aligned}$$

It would take about 23 Californias to make up one U.S.

Exercise 6

The average distance from Earth to the moon is about 3.84×10^5 km, and the distance from Earth to Mars is approximately 9.24×10^7 km in year 2014. On this simplistic level, how much farther is traveling from Earth to Mars than from Earth to the moon?

$$\begin{aligned}9.24 \times 10^7 - 3.84 \times 10^5 &= 924 \times 10^5 - 3.84 \times 10^5 \\ &= (924 - 3.84) \times 10^5 \\ &= 920.16 \times 10^5 \\ &= 92\,016\,000\end{aligned}$$

It is $92,016,000$ km further to travel from Earth to Mars than from Earth to the moon.

Closing (3 minutes)

Summarize, or have students summarize, the lesson.

- We can read, write, and operate with numbers expressed in scientific notation.

Exit Ticket (5 minutes)

Name ____ Date ____________________

Lesson 11: Efficacy of the Scientific Notation

Exit Ticket

1. Two of the largest mammals on earth are the blue whale and the African elephant. An adult male blue whale weighs about 170 tonnes or long tons. (1 tonne $= 1000$ kg)

   Show that the weight of an adult blue whale is 1.7×10^5 kg.

2. An adult male African elephant weighs about 9.07×10^3 kg.

   Compute how many times heavier an adult male blue whale is than an adult male African elephant (i.e., find the value of the ratio). Round your final answer to the nearest one.

Exit Ticket Sample Solutions

1. Two of the largest mammals on earth are the blue whale and the elephant. An adult male blue whale weighs about 170 tonnes or long tons. (1 tonne $= 1000$ kg)

   Show that the weight of an adult blue whale is 1.7×10^5 kg.

   Let x (or any other symbol) represent the number of kilograms an adult blue whale weighs.

$$170 \times 1000 = x$$
$$1.7 \times 10^5 = x$$

2. An adult male elephant weighs about 9.07×10^3 kg.

   Compute how many times heavier an adult male blue whale is than an adult male elephant (i.e., find the value of the ratio). Round your final answer to the nearest one.

   Let r be the value of the ratio.

$$\begin{aligned} r &= \frac{1.7 \times 10^5}{9.07 \times 10^3} \\ &= \frac{1.7}{9.07} \times 10^2 \\ &= 0.18743 \times 10^2 \\ &= 18.743 \\ &\approx 19 \end{aligned}$$

   The blue whale is 19 times heavier than the elephant.

Problem Set Sample Solutions

1. There are approximately 7.5×10^{18} grains of sand on Earth. There are approximately 7×10^{27} atoms in an average human body. Are there more grains of sand on Earth or atoms in an average human body? How do you know?

   There are more atoms in the average human body. When comparing the order of magnitude of each number, $27 > 18$; therefore, $7 \times 10^{27} > 7.5 \times 10^{18}$.

2. About how many times more atoms are in a human body compared to grains of sand on Earth?

$$\begin{aligned} \frac{7 \times 10^{27}}{7.5 \times 10^{18}} &= \frac{7}{7.5} \times \frac{10^{27}}{10^{18}} \\ &\approx 1 \times 10^{27-18} \\ &\approx 1 \times 10^9 \\ &\approx 10^9 \end{aligned}$$

   There are about $1,000,000,000$ times more atoms in the human body compared to grains of sand on Earth.

3. Suppose the geographic areas of California and the U.S. are 1.637×10^5 and 3.794×10^6 sq. mi., respectively. California's population (as of 2012) is approximately 3.804×10^7 people. If population were proportional to area, what would be the U.S. population?

We already know from Exercise 5 that it would take about 23 Californias to make up one U.S. Then the population of the U.S. would be 23 times the population of California, which is

$$\begin{aligned} 23 \times 3.804 \times 10^7 &= 87.492 \times 10^7 \\ &= 8.7492 \times 10^8 \\ &= 874{,}920{,}000. \end{aligned}$$

4. The actual population of the U.S. (as of 2012) is approximately 3.14×10^8. How does the population density of California (i.e., the number of people per square mile) compare with the population density of the U.S.?

Population density of California per square mile:

$$\begin{aligned} \frac{3.804 \times 10^7}{1.637 \times 10^5} &= \frac{3.804}{1.637} \times \frac{10^7}{10^5} \\ &= 2.32376\ldots \times 10^2 \\ &\approx 2.32 \times 10^2 \\ &= 232 \end{aligned}$$

Population density of the U.S. per square mile:

$$\begin{aligned} \frac{3.14 \times 10^8}{3.794 \times 10^6} &= \frac{3.14}{3.794} \times \frac{10^8}{10^6} \\ &= 0.8276\ldots \times 10^2 \\ &\approx 0.83 \times 10^2 \\ &= 83 \end{aligned}$$

Population density of California compared to the population density of the U.S.:

$$\begin{aligned} \frac{232}{83} &= 2.7951\ldots \\ &\approx 2.8 \end{aligned}$$

California is about 3 times as dense as the U.S. in terms of population.

Lesson 12: Choice of Unit

Student Outcomes

- Students understand how choice of unit determines how easy or difficult it is to understand an expression of measurement.
- Students determine appropriate units for various measurements and rewrite measurements based on new units.

Lesson Notes

This lesson focuses on choosing appropriate units. It is important for students to see the simple example (i.e., dining table measurements), as well as more sophisticated examples from physics and astronomy. We want students to understand the *necessity of learning to read, write, and operate in scientific notation*. For this very reason, we provide real data and explanations for why scientific notation is important and necessary in advanced sciences. This is a challenging, but crucial, lesson and should not be skipped.

Classwork

MP.4

Concept Development (2 minutes): The main reason for using scientific notation is to sensibly and efficiently record and convey the results of measurements. When we use scientific notation, the question of what unit to use naturally arises. In everyday context, this issue is easy to understand. For example, suppose we want to measure the horizontal dimensions of a dining table. In this case, measurements of 42×60 sq. in., or for that matter, $3\frac{1}{2} \times 5$ sq. ft. are commonly accepted. However, when the same measurement is presented as

$$\frac{0.7}{1056} \times \frac{1}{1056} \text{ sq. mi.},$$

it is confusing because we cannot relate a unit as long as a mile to a space as small as a dining room (*recall:* 1 mile is 5,280 feet), not to mention that the numbers $\frac{0.7}{1056}$ and $\frac{1}{1056}$ are unmanageable.

Exercises 1–3 (5 minutes)

Have students complete Exercises 1–3 in small groups.

Exercise 1

A certain brand of MP3 player will display how long it will take to play through its entire music library. If the maximum number of songs the MP3 player can hold is $1,000$ (and the average song length is 4 minutes), would you want the time displayed in terms of seconds-, days-, or years-worth of music? Explain.

It makes the most sense to have the time displayed in days because numbers such as $240,000$ seconds-worth of music and $\frac{5}{657}$ of a year are more difficult to understand than about 2.8 days.

Exercise 2

You have been asked to make frosted cupcakes to sell at a school fundraiser. Each frosted cupcake contains about 20 grams of sugar. Bake sale coordinators expect 500 people will attend the event. Assume everyone who attends will buy a cupcake; does it make sense to buy sugar in grams, pounds, or tons? Explain.

Because each cupcake contains about 20 grams of sugar, we will need 500×20 grams of sugar. Therefore, grams are too small of a measurement, while tons are too large. Therefore, the sugar should be purchased in pounds.

Exercise 3

The seafloor spreads at a rate of approximately 10 cm per year. If you were to collect data on the spread of the seafloor each week, which unit should you use to record your data? Explain.

The seafloor spreads 10 cm per year, which is less than 1 cm per month. Data will be collected each week, so it makes the most sense to measure the spread with a unit like millimeters.

Example 1 (3 minutes)

Now let's look at the field of particle physics or the study of subatomic particles, such as protons, electrons, neutrons, and mesons. In the previous lesson, we worked with the masses of protons and electrons, which are

$$1.672622 \times 10^{-27} \text{ and } 9.10938291 \times 10^{-31} \text{ kg, respectively.}$$

The factors 10^{-27} and 10^{-31} suggest that we are dealing with very small quantities; therefore, the use of a unit other than kilograms may be necessary. Should we use gram instead of kilogram? At first glance, yes, but when we do, we get the numbers 1.672622×10^{-24} g and $9.10938291 \times 10^{-28}$ g. One cannot claim that these are much easier to deal with.

> *Scaffolding:*
>
> Remind students that a kilogram is 1,000 grams, so we can quickly write the masses in the new unit using our knowledge of powers of 10.

- Is it easier to visualize something that is 10^{-24} compared to 10^{-27}?
  - *Of course not. That is why a better unit, the gigaelectronvolt, is used.*

MP.2

For this and other reasons, particle physicists use the **gigaelectronvolt**, $\frac{\text{GeV}}{c^2}$ as a unit of mass:

$$1\frac{\text{GeV}}{c^2} = 1.783 \times 10^{-27} \text{ kg.}$$

The gigaelectronvolt, $\frac{\text{GeV}}{c^2}$, is what particle physicists use for a unit of mass.

1 gigaelectronvolt = 1.783×10^{-27} kg

Mass of 1 proton = $1.672\,622 \times 10^{-27}$ kg

MP.2 The very name of the unit gives a hint that it was created for a purpose, but we do not need to explore that at this time. The important piece of information is to understand that 1.783×10^{-27} kg is a unit, and it represents 1 gigaelectronvolt. Thus, the mass of a proton is $0.938 \frac{\text{GeV}}{c^2}$ rounded to the nearest 10^{-3}, and the mass of an electron is $0.000511 \frac{\text{GeV}}{c^2}$ rounded to the nearest 10^{-6}. A justification[1] for this new unit is that the masses of a large class of subatomic particles have the same order of magnitude as 1 gigaelectronvolt.

Exercise 4 (3 minutes)

Have students complete Exercise 4 independently.

Exercise 4

Show that the mass of a proton is $0.938 \frac{\text{GeV}}{c^2}$.

Let x represent the number of gigaelectronvolts equal to the mass of a proton.

$$x\left(\frac{\text{GeV}}{c^2}\right) = \textit{mass of proton}$$
$$x(1.783 \times 10^{-27}) = 1.672622 \times 10^{-27}$$
$$x = \frac{1.672622 \times 10^{-27}}{1.783 \times 10^{-27}}$$
$$= \frac{1.672622}{1.783}$$
$$\approx 0.938$$

Example 2 (4 minutes)

Choosing proper units is also essential for work with very large numbers, such as those involved in astronomy (e.g., astronomical distances). The distance from the sun to the nearest star (Proxima Centauri) is approximately

$$4.013\,336\,473 \times 10^{13} \text{ km}.$$

In 1838, F.W. Bessel[2] was the first to measure the distance to a star, 61 Cygni, and its distance from the sun was

$$1.078\,807 \times 10^{14} \text{ km}.$$

MP.2 For numbers of this magnitude, we need to use a unit other than kilometers.

In popular science writing, a commonly used unit is the light-year, or the <u>distance</u> light travels in one year (note: one year is defined as 365.25 days).

$$\textbf{1 light-year} = 9{,}460{,}730{,}472{,}580.800 \text{ km} \approx 9.46073 \times 10^{12} \text{ km}$$

[1]There are other reasons coming from considerations within physics.
[2]Students will discover *Bessel functions* if they pursue STEM subjects at universities. We now know that 61 Cygni is actually a binary system consisting of *two* stars orbiting each other around a point called their *center of gravity*, but Bessel did not have a sufficiently powerful telescope to resolve the binary system.

One light-year is approximately 9.46073×10^{12} km. Currently, the distance of Proxima Centauri to the sun is approximately 4.2421 light-years, and the distance of 61 Cygni to the sun is approximately 11.403 light-years. When you ignore that 10^{12} is an enormous number, it is easier to think of the distances as 4.2 light-years and 11.4 light-years for these stars. For example, we immediately see that 61 Cygni is almost 3 times further from the sun than Proxima Centauri.

To measure the distance of stars in our galaxy, the light-year is a logical unit to use. Since launching the powerful Hubble Space Telescope in 1990, galaxies billions of light-years from the sun have been discovered. For these galaxies, the **gigalight-year** (or 10^9 light-years) is often used.

Exercise 5 (3 minutes)

Have students work on Exercise 5 independently.

Exercise 5

The distance of the nearest star (Proxima Centauri) to the sun is approximately $4.013\,336\,473 \times 10^{13}$ km. Show that Proxima Centauri is 4.2421 light-years from the sun.

Let x represent the number of light-years Proxima Centauri is from the sun.

$$\begin{aligned} x(9.46073 \times 10^{12}) &= 4.013336473 \times 10^{13} \\ x &= \frac{4.013336473 \times 10^{13}}{9.46073 \times 10^{12}} \\ &= \frac{4.013336473}{9.46073} \times 10 \\ &= 0.424210021 \times 10 \\ &\approx 4.2421 \end{aligned}$$

Exploratory Challenge 1 (8 minutes)

Finally, let us look at an example involving the masses of the planets in our solar system. They range from Mercury's 3.3022×10^{23} kg to Jupiter's 1.8986×10^{27} kg. However, Earth's mass is the fourth heaviest among the eight planets[3], and it seems reasonable to use it as the point of reference for discussions among planets. Therefore, a new unit is M_E, the mass of the Earth, or 5.97219×10^{24} kg.

Suggested white-board activity: Show students the table below, leaving the masses for Mercury and Jupiter blank. Demonstrate how to rewrite the mass of Mercury in terms of the new unit, M_E. Then, have students rewrite the mass of Jupiter using the new unit. Finally, complete the chart with the rewritten masses.

[3]Since 2006, Pluto is no longer classified as a planet. If Pluto was still considered a planet, then Earth would be the fifth-heaviest planet, right in the middle, which would further boost Earth's claim to be the point of reference.

Mercury: Let x represent the mass of Mercury in the unit M_E. We want to determine what number times the new unit is equal to the mass of Mercury in kilograms. Since $M_E = 5.97219 \times 10^{24}$, then:

$$(5.97219 \times 10^{24})x = 3.3022 \times 10^{23}$$
$$x = \frac{3.3022 \times 10^{23}}{5.97219 \times 10^{24}}$$
$$= \frac{3.3022}{5.97219} \times \frac{10^{23}}{10^{24}}$$
$$\approx 0.553 \times 10^{-1}$$
$$= 0.0553.$$

Mercury's mass is $0.0553\ M_E$.

Jupiter: Let x represent the mass of Jupiter in the unit M_E. We want to determine what number times the new unit is equal to the mass of Jupiter in kilograms. Since $M_E = 5.97219 \times 10^{24}$, then:

$$(5.97219 \times 10^{24})x = 1.8986 \times 10^{27}$$
$$x = \frac{1.8986 \times 10^{27}}{5.97219 \times 10^{24}}$$
$$= \frac{1.8986}{5.97219} \times \frac{10^{27}}{10^{24}}$$
$$\approx 0.318 \times 10^{3}$$
$$= 318.$$

Jupiter's mass is $318\ M_E$.

Mercury	$\mathbf{0.0553\ M_E}$	Jupiter	$\mathbf{318\ M_E}$
Venus	$0.815\ M_E$	Saturn	$95.2\ M_E$
Earth	$1\ M_E$	Uranus	$14.5\ M_E$
Mars	$0.107\ M_E$	Neptune	$17.2\ M_E$

Exploratory Challenge 2/Exercises 6–8 (10 minutes)

Have students complete Exercises 6–8 independently or in small groups. Allow time for groups to discuss their choice of unit and the reasoning for choosing it.

Exploratory Challenge 2

Suppose you are researching atomic diameters and find that credible sources provided the diameters of five different atoms as shown in the table below. All measurements are in centimeters.

1×10^{-8}	1×10^{-12}	5×10^{-8}	5×10^{-10}	5.29×10^{-11}

Exercise 6

What new unit might you introduce in order to discuss the differences in diameter measurements?

There are several answers that students could give for their choice of unit. Accept any reasonable answer, provided the explanation is clear and correct. Some students may choose 10^{-12} as their unit because all measurements could then be expressed without exponential notation. Other students may decide that 10^{-8} should be the unit because two measurements are already of that order of magnitude. Still, other students may choose 10^{-10} because that is the average of the powers.

Exercise 7

Name your unit, and explain why you chose it.

Students can name their unit anything reasonable, as long as they clearly state what their unit is and how it will be written. For example, if a student chooses a unit of 10^{-10}, then he or she should state that the unit will be represented with a letter. For example, Y, then $Y = 10^{-10}$.

Exercise 8

Using the unit you have defined, rewrite the five diameter measurements.

Using the unit $Y = 10^{-10}$, then:

$1 \times 10^{-8} = 100\ Y$	$1 \times 10^{-12} = 0.01\ Y$	$5 \times 10^{-8} = 500\ Y$	$5 \times 10^{-10} = 5\ Y$	$5.29 \times 10^{-11} = 0.529\ Y$

Closing (2 minutes)

Summarize the lesson:

- Choosing an appropriate unit allows us to determine the size of the numbers we are dealing with.

  For example, the dining table measurement:

$$42 \times 60 \text{ sq. in.} = 3\frac{1}{2} \times 5 \text{ sq. ft.} = \frac{0.7}{1056} \times \frac{1}{1056} \text{ sq. mi.}$$

- We have reinforced their ability to read, write, and operate with numbers in scientific notation.

Exit Ticket (5 minutes)

Name ______________________________ Date ________________

Lesson 12: Choice of Unit

Exit Ticket

1. The table below shows an approximation of the national debt at the beginning of each decade over the last century. Choose a unit that would make a discussion about the growth of the national debt easier. Name your unit, and explain your choice.

Year	Debt in Dollars
1900	2.1×10^9
1910	2.7×10^9
1920	2.6×10^{10}
1930	1.6×10^{10}
1940	4.3×10^{10}
1950	2.6×10^{11}
1960	2.9×10^{11}
1970	3.7×10^{11}
1980	9.1×10^{11}
1990	3.2×10^{12}
2000	5.7×10^{12}

2. Using the new unit you have defined, rewrite the debt for years 1900, 1930, 1960, and 2000.

Exit Ticket Sample Solutions

1. The table below shows an approximation of the national debt at the beginning of each decade over the last century. Choose a unit that would make a discussion about the growth of the national debt easier. Name your unit, and explain your choice.

Students will likely choose 10^{11} as their unit because the majority of the data is of that magnitude. Accept any reasonable answer that students provide. Verify that they have named their unit.

Year	Debt in Dollars
1900	2.1×10^{9}
1910	2.7×10^{9}
1920	2.6×10^{10}
1930	1.6×10^{10}
1940	4.3×10^{10}
1950	2.6×10^{11}
1960	2.9×10^{11}
1970	3.7×10^{11}
1980	9.1×10^{11}
1990	3.2×10^{12}
2000	5.7×10^{12}

2. Using the new unit you have defined, rewrite the debt for years 1900, 1930, 1960, and 2000.

Let D represent the unit 10^{11}. Then, the debt in 1900 is $0.021D$, in 1930 it is $0.16D$, in 1960 it is $2.9D$, and $57D$ in 2000.

Problem Set Sample Solution

1. Verify the claim that, in terms of gigaelectronvolts, the mass of an electron is 0.000511.

Let x represent the number of gigaelectronvolts equal to the mass of an electron.

$$\begin{aligned} x\left(\frac{\text{GeV}}{c^2}\right) &= \text{Mass of electron} \\ x(1.783 \times 10^{-27}) &= 9.109\,382\,91 \times 10^{-31} \\ x &= \frac{9.109\,382\,91 \times 10^{-31} \times 10^{31}}{1.783 \times 10^{-27} \times 10^{31}} \\ &= \frac{9.109\,382\,91}{1.783 \times 10^{4}} \\ &= \frac{9.109\,382\,91}{17\,830} \\ &\approx 0.000\,511 \end{aligned}$$

2. The maximum distance between Earth and the sun is 1.52098232×10^8 km, and the minimum distance is 1.47098290×10^8 km.[4] What is the average distance between Earth and the sun in scientific notation?

$$\textit{average distance} = \frac{1.52098232 \times 10^8 + 1.47098290 \times 10^8}{2}$$
$$= \frac{(1.52098232 + 1.47098290) \times 10^8}{2}$$
$$= \frac{2.99196522 \times 10^8}{2}$$
$$= \frac{2.99196522}{2} \times 10^8$$
$$= 1.49598261 \times 10^8 \text{km}$$

3. Suppose you measure the following masses in terms of kilograms:

2.6×10^{21}	9.04×10^{23}
8.82×10^{23}	2.3×10^{18}
1.8×10^{12}	2.103×10^{22}
8.1×10^{20}	6.23×10^{18}
6.723×10^{19}	1.15×10^{20}
7.07×10^{21}	7.210×10^{29}
5.11×10^{25}	7.35×10^{24}
7.8×10^{19}	5.82×10^{26}

What new unit might you introduce in order to aid discussion of the masses in this problem? Name your unit, and express it using some power of 10. Rewrite each number using your newly defined unit.

A very motivated student may search the Internet and find that units exist that convert large masses to reasonable numbers, such as petagrams (10^{12} kg), exagrams (10^{15} kg), or zetagrams (10^{18} kg). More likely, students will decide that something near 10^{20} should be used as a unit because many of the numbers are near that magnitude. There is one value, 1.8×10^{12}, that serves as an outlier and should be ignored because it is much smaller than the majority of the data. Students can name their unit anything reasonable. The answers provided are suggestions, but any reasonable answers should be accepted.

Let U be defined as the unit 10^{20}.

$2.6 \times 10^{21} = 26U$	$9.04 \times 10^{23} = 9040U$
$8.82 \times 10^{23} = 8820U$	$2.3 \times 10^{18} = 0.023U$
$1.8 \times 10^{12} = 0.000\,000\,018U$	$2.103 \times 10^{22} = 210.3U$
$8.1 \times 10^{0} = 8.1U$	$6.23 \times 10^{18} = 0.0623U$
$6.723 \times 10^{19} = 0.6723U$	$1.15 \times 10^{20} = 1.15U$
$7.07 \times 10^{21} = 70.7U$	$7.210 \times 10^{29} = 7\,210\,000\,000U$
$5.11 \times 10^{25} = 511\,000U$	$7.35 \times 10^{24} = 73\,500U$
$7.8 \times 10^{19} = 0.78U$	$5.82 \times 10^{26} = 5\,820\,000U$

[4]Note: Earth's orbit is elliptical, not circular.

Lesson 13: Comparison of Numbers Written in Scientific Notation and Interpreting Scientific Notation Using Technology

Student Outcomes

- Students compare numbers expressed in scientific notation.
- Students apply the laws of exponents to interpret data and use technology to compute with very large numbers.

Classwork

Examples 1–2/ Exercises 1–2 (10 minutes)

Concept Development: We have learned why scientific notation is indispensable in science. This means that we have to learn how to compute and compare numbers in scientific notation. We have already done some computations, so we are ready to take a closer look at comparing the size of different numbers.

> **There is a general principle that underlies the comparison of two numbers in scientific notation: *Reduce everything to whole numbers if possible.* To this end, we recall two basic facts.**
>
> 1. **Inequality (A): Let x and y be numbers and let $z > 0$. Then $x < y$ if and only if $xz < yz$.**
> 2. **Comparison of whole numbers:**
>   a. **If two whole numbers have different numbers of digits, then the one with more digits is greater.**
>   b. **Suppose two whole numbers p and q have the same number of digits and, moreover, they agree digit-by-digit (starting from the left) until the nth place. If the digit of p in the $(n+1)^{\text{th}}$ place is greater than the corresponding digit in q, then $p > q$.**

Example 1

MP.8

Among the galaxies closest to Earth, M82 is about 1.15×10^7 light-years away, and Leo I Dwarf is about 8.2×10^5 light-years away. Which is closer?

- *First solution:* This is the down-to-earth, quick, and direct solution. The number 8.2×10^5 equals the 6-digit number $820{,}000$. On the other hand, 1.15×10^7 equals the 8-digit number $11{,}500{,}000$. By (2a), above, $8.2 \times 10^5 < 1.15 \times 10^7$. Therefore, Leo I Dwarf is closer.

MP.8

- *Second Solution:* This solution is for the long haul, that is, the solution that works every time no matter how large (or small) the numbers become. First, we express both numbers as a product with the same power of 10. Since $10^7 = 10^2 \times 10^5$, we see that the distance to M82 is

$$1.15 \times 10^2 \times 10^5 = 115 \times 10^5.$$

The distance to Leo I Dwarf is 8.2×10^5. By (1) above, comparing 1.15×10^7 and 8.2×10^5 is equivalent to comparing 115 and 8.2. Since $8.2 < 115$, we see that $8.2 \times 10^5 < 1.15 \times 10^7$. Thus, Leo I Dwarf is closer.

> *Scaffolding:*
> - Display the second solution.
> - Guide students through the solution.

Exercise 1

Have students complete Exercise 1 independently, using the logic modeled in the second solution.

Exercise 1

The Fornax Dwarf galaxy is 4.6×10^5 light-years away from Earth, while Andromeda I is 2.430×10^6 light-years away from Earth. Which is closer to Earth?

$$2.430 \times 10^6 = 2.430 \times 10 \times 10^5 = 24.30 \times 10^5$$

Because $4.6 < 24.30$, then $4.6 \times 10^5 < 24.30 \times 10^5$, and since $24.30 \times 10^5 = 2.430 \times 10^6$, we know that $4.6 \times 10^5 < 2.430 \times 10^6$. Therefore, Fornax Dwarf is closer to Earth.

Example 2

Background information for the teacher: The next example brings us back to the world of subatomic particles. In the early 20th century, the picture of elementary particles was straightforward: Electrons, protons, neutrons, and photons were the fundamental constituents of matter. But in the 1930s, positrons, mesons, and neutrinos were discovered, and subsequent developments rapidly increased the number of subatomic particle types observed. Many of these newly observed particle types are extremely short-lived (see Example 2 below and Exercise 2). The so-called Standard Model developed during the latter part of the last century finally restored some order, and it is now theorized that different kinds of quarks and leptons are the basic constituents of matter.

Many subatomic particles are unstable: charged pions have an average lifetime of 2.603×10^{-8} seconds, while muons have an average lifetime of 2.197×10^{-6} seconds. Which has a longer average lifetime?

We follow the same method as the second solution in Example 1. We have

$$2.197 \times 10^{-6} = 2.197 \times 10^2 \times 10^{-8} = 219.7 \times 10^{-8}.$$

Therefore, comparing 2.603×10^{-8} with 2.197×10^{-6} is equivalent to comparing 2.603 with 219.7 (by (1) above). Since $2.603 < 219.7$, we see that $2.603 \times 10^{-8} < 2.197 \times 10^{-6}$. Thus, muons have a longer lifetime.

Exercise 2 (3 minutes)

Have students complete Exercise 2 independently.

Exercise 2

The average lifetime of the tau lepton is 2.906×10^{-13} seconds, and the average lifetime of the neutral pion is 8.4×10^{-17} seconds. Explain which subatomic particle has a longer average lifetime.

$$2.906 \times 10^{-13} = 2.906 \times 10^{4} \times 10^{-17} = 29,060 \times 10^{-17}$$

Since $8.4 < 29,060$, then $8.4 \times 10^{-17} < 29,060 \times 10^{-17}$, and since $29,060 \times 10^{-17} = 2.906 \times 10^{-13}$, we know that $8.4 \times 10^{-17} < 2.906 \times 10^{-13}$. Therefore, tau lepton has a longer average lifetime.

This problem, as well as others, can be solved using an alternate method. Our goal is to make the magnitude of the numbers we are comparing the same, which will allow us to reduce the comparison to that of whole numbers.

Here is an alternate solution:

$$8.4 \times 10^{-17} = 8.4 \times 10^{-4} \times 10^{-13} = 0.000\,84 \times 10^{-13}.$$

Since $0.00084 < 2.906$, then $0.000\,84 \times 10^{-13} < 2.906 \times 10^{-13}$, and since $0.000\,84 \times 10^{-13} = 8.4 \times 10^{-17}$, we know that $8.4 \times 10^{-17} < 2.906 \times 10^{-13}$. Therefore, tau lepton has a longer average lifetime.

Exploratory Challenge 1/Exercise 3 (8 minutes)

Examples 1 and 2 illustrate the following general fact:

MP.8 THEOREM: *Given two positive numbers in scientific notation, $a \times 10^m$ and $b \times 10^n$, if $m < n$, then $a \times 10^m < b \times 10^n$.* Allow time for students to discuss, in small groups, how to prove the theorem.

Scaffolding:

- Use the suggestions below, as needed, for the work related to the theorem.
- Remind students about order of magnitude.
- Remind them that if $m < n$, then there is a positive integer k so that $n = k + m$. Therefore, by the first law of exponents (10),

$$b \times 10^n = b \times 10^k \times 10^m = (b \times 10^k) \times 10^m.$$

- Point out that we just spent time on forcing numbers that were expressed in scientific notation to have the same power of 10, which allowed us to easily compare the numbers. This proof is no different. We just wrote an equivalent expression $(b \times 10^k) \times 10^m$ for $b \times 10^n$, so that we could look at and compare two numbers that both have a magnitude of m.

Exploratory Challenge 1/Exercise 3

THEOREM: Given two positive numbers in scientific notation, $a \times 10^m$ and $b \times 10^n$, if $m < n$, then $a \times 10^m < b \times 10^n$.

Prove the theorem.

If $m < n$, then there is a positive integer k so that $n = k + m$.

By the first law of exponents (10) in Lesson 5, $b \times 10^n = b \times 10^k \times 10^m = (b \times 10^k) \times 10^m$. Because we are comparing with $a \times 10^m$, we know by (1) that we only need to prove $a < (b \times 10^k)$. By the definition of scientific notation, $a < 10$ and also $(b \times 10^k) \geq 10$ because $k \geq 1$ and $b \geq 1$, so that $(b \times 10^k) \geq 1 \times 10 = 10$. This proves $a < (b \times 10^k)$, and therefore, $a \times 10^m < b \times 10^n$.

Explain to students that we know that $a < 10$ because of the statement given that $a \times 10^m$ is a number expressed in scientific notation. That is not enough information to convince students that $a < b \times 10^k$; therefore, we need to say something about the right side of the inequality. We know that $k \geq 1$ because k is a positive integer so that $n = k + m$. We also know that $b \geq 1$ because of the definition of scientific notation. That means that the minimum possible value of $b \times 10^k$ is 10 because $1 \times 10^1 = 10$. Therefore, we can be certain that $a < b \times 10^k$.

Therefore, by (1), $a \times 10^m < (b \times 10^k) \times 10^m$. Since $n = k + m$, we can rewrite the right side of the inequality as $b \times 10^n$, and finally $a \times 10^m < b \times 10^n$.

Example 3 (2 minutes)

Compare 1.815×10^{14} with 1.82×10^{14}.

By (1), we only have to compare 1.815 with 1.82, and for the same reason, we only need to compare 1.815×10^3 with 1.82×10^3.

Thus, we compare 1,815 and 1,820: Clearly $1{,}815 < 1{,}820$ (use (2b) if you like). Therefore, using (1) repeatedly,

$$1{,}815 < 1{,}820 \to 1.815 < 1.82 \to 1.815 \times 10^{14} < 1.82 \times 10^{14}.$$

Scaffolding:

Remind students that it is easier to compare whole numbers; that's why each number is multiplied by 10^3. However, if students can accurately compare 1.815 to 1.82, it is not necessary that they multiply each number by 10^3 to make them whole numbers.

Exercises 4–5 (2 minutes)

Have students complete Exercises 4 and 5 independently.

Exercise 4

Compare 9.3×10^{28} and 9.2879×10^{28}.

We only need to compare 9.3 and 9.2879. $9.3 \times 10^4 = 93,000$ and $9.2879 \times 10^4 = 92,879$, so we see that $93,000 > 92,879$. Therefore, $9.3 \times 10^{28} > 9.2879 \times 10^{28}$.

Exercise 5

Chris said that $5.3 \times 10^{41} < 5.301 \times 10^{41}$ because 5.3 has fewer digits than 5.301. Show that even though his answer is correct, his reasoning is flawed. Show him an example to illustrate that his reasoning would result in an incorrect answer. Explain.

Chris is correct that $5.3 \times 10^{41} < 5.301 \times 10^{41}$, but that is because when we compare 5.3 and 5.301, we only need to compare 5.3×10^3 and 5.301×10^3 (by (1) above). But, $5.3 \times 10^3 < 5.301 \times 10^3$ or rather $5,300 < 5,301$, and this is the reason that $5.3 \times 10^{41} < 5.301 \times 10^{41}$. However, Chris's reasoning would lead to an incorrect answer for a problem that compares 5.9×10^{41} and 5.199×10^{41}. His reasoning would lead him to conclude that $5.9 \times 10^{41} < 5.199 \times 10^{41}$, but $5,900 > 5,199$, which is equivalent to $5.9 \times 10^3 > 5.199 \times 10^3$. By (1) again, $5.9 > 5.199$, meaning that $5.9 \times 10^{41} > 5.199 \times 10^{41}$.

Exploratory Challenge 2/Exercise 6 (10 minutes)

Students use snapshots of technology displays to determine the exact product of two numbers.

Exploratory Challenge 2/Exercise 6

You have been asked to determine the exact number of Google searches that are made each year. The only information you are provided is that there are $35,939,938,877$ searches performed each week. Assuming the exact same number of searches are performed each week for the 52 weeks in a year, how many total searches will have been performed in one year? Your calculator does not display enough digits to get the exact answer. Therefore, you must break down the problem into smaller parts. Remember, you cannot approximate an answer because you need to find an exact answer. Use the screen shots below to help you reach your answer.

First, I need to rewrite the number of searches for each week using numbers that can be computed using my calculator.

$$\begin{aligned} 35\,939\,938\,877 &= 35\,939\,000\,000 + 938\,877 \\ &= 35\,939 \times 10^6 + 938\,877 \end{aligned}$$

Next, I need to multiply each term of the sum by 52***, using the distributive law.***

$$(35\,939 \times 10^6 + 938\,877) \times 52 = (35\,939 \times 10^6) \times 52 + (938\,877 \times 52)$$

By repeated use of the commutative and associative properties, I can rewrite the problem as

$$(35\,939 \times 52) \times 10^6 + (938\,877 \times 52).$$

According to the screen shots, I get

$$\begin{aligned} 1\,868\,828 \times 10^6 + 48\,821\,604 &= 1\,868\,828\,000\,000 + 48\,821\,604 \\ &= 1\,868\,876\,821\,604. \end{aligned}$$

Therefore, $1,868,876,821,604$ ***Google searches are performed each year.***

Yahoo! is another popular search engine. Yahoo! receives requests for $1,792,671,335$ searches each month. Assuming the same number of searches are performed each month, how many searches are performed on Yahoo! each year? Use the screen shots below to help determine the answer.

First, I need to rewrite the number of searches for each month using numbers that can be computed using my calculator.

$$\begin{aligned} 1\,792\,671\,335 &= 1\,792\,000\,000 + 671\,335 \\ &= 1\,792 \times 10^6 + 671\,335. \end{aligned}$$

Next, I need to multiply each term of the sum by 12***, using the distributive law.***

$$(1\,792 \times 10^6 + 671\,335) \times 12 = (1\,792 \times 10^6) \times 12 + (671\,335 \times 12).$$

By repeated use of the commutative and associative properties, I can rewrite the problem as

$$(1\,792 \times 12) \times 10^6 + (671\,335 \times 12)$$

According to the screen shots, I get

$$\begin{aligned} 21\,504 \times 10^6 + 8\,056\,020 &= 21\,504\,000\,000 + 8\,056\,020 \\ &= 21\,512\,056\,020 \end{aligned}$$

Therefore, $21,512,056,020$ ***Yahoo! searches are performed each year.***

Closing (2 minutes)

Summarize the lesson and Module 1:

- We have completed the lessons on exponential notation, the properties of integer exponents, magnitude, and scientific notation.
- We can read, write, and operate with numbers expressed in scientific notation, which is the language of many sciences. Additionally, they can interpret data using technology.

Exit Ticket (3 minutes)

Fluency Exercise (5 minutes)

Rapid White Board Exchange: Have students respond to your prompts for practice with operations with numbers expressed in scientific notation using white boards (or other display options as available). This exercise can be conducted at any point throughout the lesson. The prompts are listed at the end of the lesson. Refer to the Rapid White Board Exchanges section in the Module Overview for directions to administer a Rapid White Board Exchange.

Name ______________________________ Date ______________

Lesson 13: Comparison of Numbers Written in Scientific Notation and Interpreting Scientific Notation Using Technology

Exit Ticket

1. Compare 2.01×10^{15} and 2.8×10^{13}. Which number is larger?

2. The wavelength of the color red is about 6.5×10^{-9} m. The wavelength of the color blue is about 4.75×10^{-9} m. Show that the wavelength of red is longer than the wavelength of blue.

Exit Ticket Sample Solutions

1. Compare 2.01×10^{15} and 2.8×10^{13}. Which number is larger?

   $2.01 \times 10^{15} = 2.01 \times 10^{2} \times 10^{13} = 201 \times 10^{13}$

   Since $201 > 2.8$, we have $201 \times 10^{13} > 2.8 \times 10^{13}$, and since $201 \times 10^{13} = 2.01 \times 10^{15}$, we conclude $2.01 \times 10^{15} > 2.8 \times 10^{13}$.

2. The wavelength of the color red is about 6.5×10^{-9} m. The wavelength of the color blue is about 4.75×10^{-9} m. Show that the wavelength of red is longer than the wavelength of blue.

   We only need to compare 6.5 and 4.75:

   $6.5 \times 10^{-9} = 650 \times 10^{-7}$ and $4.75 \times 10^{-9} = 475 \times 10^{-7}$, so we see that $650 > 475$. Therefore, $6.5 \times 10^{-9} > 4.75 \times 10^{-9}$.

Problem Set Sample Solutions

1. Write out a detailed proof of the fact that, given two numbers in scientific notation, $a \times 10^{n}$ and $b \times 10^{n}$, $a < b$, if and only if $a \times 10^{n} < b \times 10^{n}$.

   Because $10^{n} > 0$, we can use inequality (A) (i.e., (1) above) twice to draw the necessary conclusions. First, if $a < b$, then by inequality (A), $a \times 10^{n} < b \times 10^{n}$. Second, given $a \times 10^{n} < b \times 10^{n}$, we can use inequality (A) again to show $a < b$ by multiplying each side of $a \times 10^{n} < b \times 10^{n}$ by 10^{-n}.

   a. Let A and B be two positive numbers, with no restrictions on their size. Is it true that $A \times 10^{-5} < B \times 10^{5}$?

      No, it is not true that $A \times 10^{-5} < B \times 10^{5}$. Using inequality (A), we can write $A \times 10^{-5} \times 10^{5} < B \times 10^{5} \times 10^{5}$, which is the same as $A < B \times 10^{10}$. To disprove the statement, all we would need to do is find a value of A that exceeds $B \times 10^{10}$.

   b. Now, if $A \times 10^{-5}$ and $B \times 10^{5}$ are written in scientific notation, is it true that $A \times 10^{-5} < B \times 10^{5}$? Explain.

      Yes, since the numbers are written in scientific notation, we know that the restrictions for A and B are $1 \leq A < 10$ and $1 \leq B < 10$. The maximum value for A, when multiplied by 10^{-5}, will still be less than 1. The minimum value of B will produce a number at least 10^{5} in size.

2. The mass of a neutron is approximately 1.674927×10^{-27} kg. Recall that the mass of a proton is 1.672622×10^{-27} kg. Explain which is heavier.

   Since both numbers have a factor of 10^{-27}, we only need to look at 1.674927 and 1.672622. When we multiply each number by 10^{6}, we get

   $$1.674927 \times 10^{6} \text{ and } 1.672622 \times 10^{6},$$

   which is the same as

   $$1,674,927 \text{ and } 1,672,622.$$

   Now that we are looking at whole numbers, we can see that $1,674,927 > 1,672,622$ (by (2b) above), which means that $1.674927 \times 10^{-27} > 1.672622 \times 10^{-27}$. Therefore, the mass of a neutron is heavier.

3. The average lifetime of the Z boson is approximately 3×10^{-25} seconds, and the average lifetime of a neutral rho meson is approximately 4.5×10^{-24} seconds.

   a. Without using the theorem from today's lesson, explain why the neutral rho meson has a longer average lifetime.

      Since $3 \times 10^{-25} = 3 \times 10^{-1} \times 10^{-24}$, we can compare $3 \times 10^{-1} \times 10^{-24}$ and 4.5×10^{-24}. Based on Example 3 or by use of (1) above, we only need to compare 3×10^{-1} and 4.5, which is the same as 0.3 and 4.5. If we multiply each number by 10, we get whole numbers 3 and 45. Since $3 < 45$, then $3 \times 10^{-25} < 4.5 \times 10^{-24}$. Therefore, the neutral rho meson has a longer average lifetime.

   b. Approximately how much longer is the lifetime of a neutral rho meson than a Z boson?

      $45:3$ or 15 times longer

Rapid White Board Exchange: Operations with Numbers Expressed in Scientific Notation

1. $(5 \times 10^4)^2 =$

   2.5×10^9

2. $(2 \times 10^9)^4 =$

   1.6×10^{37}

3. $\frac{(1.2\times10^4)+(2\times10^4)+(2.8\times10^4)}{3} =$

   2×10^4

4. $\frac{7\times10^{15}}{14\times10^9} =$

   5×10^5

5. $\frac{4\times10^2}{2\times10^8} =$

   2×10^{-6}

6. $\frac{(7\times10^9)+(6\times10^9)}{2} =$

   6.5×10^9

7. $(9 \times 10^{-4})^2 =$

   8.1×10^{-7}

8. $(9.3 \times 10^{10}) - (9 \times 10^{10}) =$

   3×10^9

Name ______________________________ Date ______________

1. You have been hired by a company to write a report on Internet companies' Wi-Fi ranges. They have requested that all values be reported in feet using scientific notation.

   Ivan's Internet Company boasts that its wireless access points have the greatest range. The company claims that you can access its signal up to $2{,}640$ feet from its device. A competing company, Winnie's Wi-Fi, has devices that extend to up to $2\frac{1}{2}$ miles.

   a. Rewrite the range of each company's wireless access devices in feet using scientific notation, and state which company actually has the greater range ($5{,}280$ feet $= 1$ mile).

   b. You can determine how many times greater the range of one Internet company is than the other by writing their ranges as a ratio. Write and find the value of the ratio that compares the range of Winnie's wireless access devices to the range of Ivan's wireless access devices. Write a complete sentence describing how many times greater Winnie's Wi-Fi range is than Ivan's Wi-Fi range.

EUREKA MATH

c. UC Berkeley uses Wi-Fi over Long Distances (WiLD) to create long-distance, point-to-point links. UC Berkeley claims that connections can be made up to 10 miles away from its device. Write and find the value of the ratio that compares the range of Ivan's wireless access devices to the range of Berkeley's WiLD devices. Write your answer in a complete sentence.

2. There is still controversy about whether or not Pluto should be considered a planet. Although planets are mainly defined by their orbital path (the condition that prevented Pluto from remaining a planet), the issue of size is something to consider. The table below lists the planets, including Pluto, and their approximate diameters in meters.

Planet	*Approximate Diameter* (m)
Mercury	4.88×10^6
Venus	1.21×10^7
Earth	1.28×10^7
Mars	6.79×10^6
Jupiter	1.43×10^8
Saturn	1.2×10^8
Uranus	5.12×10^7
Neptune	4.96×10^7
Pluto	2.3×10^6

a. Name the planets (including Pluto) in order from smallest to largest.

b. Comparing only diameters, about how many times larger is Jupiter than Pluto?

c. Again, comparing only diameters, find out about how many times larger Jupiter is compared to Mercury.

d. Assume you are a voting member of the International Astronomical Union (IAU) and the classification of Pluto is based entirely on the length of the diameter. Would you vote to keep Pluto a planet or reclassify it? Why or why not?

e. Just for fun, Scott wondered how big a planet would be if its diameter was the square of Pluto's diameter. If the diameter of Pluto in terms of meters were squared, what would the diameter of the new planet be? (Write the answer in scientific notation.) Do you think it would meet any size requirement to remain a planet? Would it be larger or smaller than Jupiter?

3. Your friend Pat bought a fish tank that has a volume of 175 liters. The brochure for Pat's tank lists a "fun fact" that it would take 7.43×10^{18} tanks of that size to fill all the oceans in the world. Pat thinks the both of you can quickly calculate the volume of all the oceans in the world using the fun fact and the size of her tank.

a. Given that 1 liter $= 1.0 \times 10^{-12}$ cubic kilometers, rewrite the size of the tank in cubic kilometers using scientific notation.

b. Determine the volume of all the oceans in the world in cubic kilometers using the "fun fact."

c. You liked Pat's fish so much you bought a fish tank of your own that holds an additional 75 liters. Pat asked you to figure out a different "fun fact" for your fish tank. Pat wants to know how many tanks of this new size would be needed to fill the Atlantic Ocean. The Atlantic Ocean has a volume of $323,600,000$ cubic kilometers.

EUREKA MATH™

A Progression Toward Mastery					
Assessment Task Item		**STEP 1** Missing or incorrect answer and little evidence of reasoning or application of mathematics to solve the problem.	**STEP 2** Missing or incorrect answer but evidence of some reasoning or application of mathematics to solve the problem.	**STEP 3** A correct answer with some evidence of reasoning or application of mathematics to solve the problem, <u>OR</u> An incorrect answer with substantial evidence of solid reasoning or application of mathematics to solve the problem.	**STEP 4** A correct answer supported by substantial evidence of solid reasoning or application of mathematics to solve the problem.
1	**a–c** 8.EE.A.3 8.EE.A.4	Student completes part (a) correctly by writing each company's Wi-Fi range in scientific notation and determines which is greater. Student is unable to write ratios in parts (b)–(c). OR Student is unable to perform operations with numbers written in scientific notation and does not complete parts (b)–(c). OR Student is able to write the ratios in parts (b)–(c) but is unable to find the value of the ratios.	Student completes part (a) correctly. Student is able to write ratios in parts (b)–(c). Student is able to perform operations with numbers written in scientific notation in parts (b)–(c) but makes computational errors leading to incorrect answers. Student does not interpret calculations to answer questions.	Student answers at least two parts of (a)–(c) correctly. Student makes a computational error that leads to an incorrect answer. Student interprets calculations correctly and justifies the answers. Student uses a complete sentence to answer part (b) or (c).	Student answers all parts of (a)–(c) correctly. Ratios written are correct and values are calculated accurately. Calculations are interpreted correctly and answers are justified. Student uses a complete sentence to answer parts (b) and (c).
2	**a–c** 8.EE.A.3 8.EE.A.4	Student correctly orders the planets in part (a). Student is unable to perform operations with numbers written in scientific notation.	Student completes two or three parts of (a)–(c) correctly. Calculations have minor errors. Student provides partial justifications for conclusions made.	Student completes two or three parts of (a)–(c) correctly. Calculations are precise. Student provides justifications for conclusions made.	Student completes all three parts of (a)–(c) correctly. Calculations are precise. Student responses demonstrate mathematical reasoning leading to strong justifications for conclusions made.

	d 8.EE.A.3 8.EE.A.4	Student states a position but provides *no* explanation to defend it.	Student states a position and provides *weak* arguments to defend it.	Student states a position and provides a *reasonable* explanation to defend it.	Student states a position and provides a *compelling* explanation to defend it.
	e 8.EE.A.3 8.EE.A.4	Student is unable to perform the calculation or answer questions.	Student performs the calculation but does not write answer in scientific notation. Student provides an explanation for why the new planet would remain a planet by stating it would be the largest.	Student correctly performs the calculation. Student provides an explanation for why the new planet would remain a planet without reference to the calculation. Student correctly states that the new planet would be the largest planet.	Student correctly performs the calculation. Student provides an explanation for why the new planet would remain a planet, including reference to the calculation performed. Student correctly states that the new planet would be the largest planet.
3	a–c 8.EE.A.3 8.EE.A.4	Student completes all parts of the problem incorrectly. Evidence that student has some understanding of scientific notation but cannot integrate use of properties of exponents to perform operations. Student makes gross errors in computation.	Student completes one part of (a)–(c) correctly. Student makes several minor errors in computation. Student performs operations on numbers written in scientific notation but does not rewrite answers in scientific notation.	Student completes two parts of (a)–(c) correctly. Student makes a minor error in computation. Evidence shown that student understands scientific notation and can use properties of exponents with numbers in this form.	Student completes all parts of (a)–(c) correctly. Student has precise calculations. Evidence shown of mastery with respect to scientific notation usage and performing operations on numbers in this form using properties of exponents.

Name ______________________________ Date ______________

1. You have been hired by a company to write a report on Internet companies' Wi-Fi ranges. They have requested that all values be reported in feet using scientific notation.

   Ivan's Internet Company boasts that its wireless access points have the greatest range. The company claims that you can access its signal up to 2,640 feet from its device. A competing company, Winnie's Wi-Fi, has devices that extend to up to $2\frac{1}{2}$ miles.

   a. Rewrite the range of each company's wireless access devices in feet using scientific notation, and state which company actually has the greater range (5,280 feet = 1 mile).

   IVAN'S RANGE: $2{,}640 = 2.64 \times 10^3$ ft

   WINNIE'S RANGE: $(2.5)5280 = 13200 = 1.32 \times 10^4$ ft.

   WINNIE'S WI-FI HAS THE GREATER RANGE.

   b. You can determine how many times greater the range of one Internet company is than the other by writing their ranges as a ratio. Write and find the value of the ratio that compares the range of Winnie's wireless access devices to the range of Ivan's wireless access devices. Write a complete sentence describing how many times greater Winnie's Wi-Fi range is than Ivan's Wi-Fi range.

   WINNIE TO IVAN'S RATIO – $(1.32 \times 10^4) : (2.64 \times 10^3)$

   VALUE OF RATIO – $\frac{1.32 \times 10^4}{2.64 \times 10^3} = \frac{1.32}{2.64} \times \frac{10^4}{10^3} = \frac{1}{2} \times 10 = 5$

   WINNIE'S WI-FI IS 5 TIMES GREATER IN RANGE THAN IVAN'S INTERNET COMPANY.

c. UC Berkeley uses Wi-Fi over Long Distances (WiLD) to create long-distance, point-to-point links. UC Berkeley claims that connections can be made up to 10 miles away from its device. Write and find the value of the ratio that compares the range of Ivan's wireless access devices to the range of Berkeley's WiLD devices. Write your answer in a complete sentence.

$(10)5280 = 52800 = 5.28 \times 10^4$

IVAN'S TO BERKELEY RATIO: $(2.64 \times 10^3) : (5.28 \times 10^4)$

VALUE OF RATIO – $\frac{2.64 \times 10^3}{5.28 \times 10^4} = \frac{2.64}{5.28} \times \frac{10^3}{10^4} = \frac{1}{2} \times \frac{1}{10} = \frac{1}{20}$

IVAN'S INTERNET DEVICES HAVE A RANGE $\frac{1}{20}$ THE RANGE OF UC BERKELEY'S WiLD DEVICES.

2. There is still controversy about whether or not Pluto should be considered a planet. Although planets are mainly defined by their orbital path (the condition that prevented Pluto from remaining a planet), the issue of size is something to consider. The table below lists the planets, including Pluto, and their approximate diameters in meters.

Planet	*Approximate Diameter* (m)
Mercury	4.88×10^6
Venus	1.21×10^7
Earth	1.28×10^7
Mars	6.79×10^6
Jupiter	1.43×10^8
Saturn	1.2×10^8
Uranus	5.12×10^7
Neptune	4.96×10^7
Pluto	2.3×10^6

a. Name the planets (including Pluto) in order from smallest to largest.

PLUTO, MERCURY, MARS, VENUS, EARTH, NEPTUNE, URANUS, SATURN, JUPITER

EUREKA MATH™

b. Comparing only diameters, about how many times larger is Jupiter than Pluto?

$$\frac{1.43 \times 10^8}{2.3 \times 10^6} = \frac{1.43}{2.3} \times \frac{10^8}{10^6}$$

$$\approx 0.622 \times 10^2$$

$$\approx 62.2$$

THE DIAMETER OF JUPITER IS ABOUT 62 TIMES LARGER THAN PLUTO.

c. Again, comparing only diameters, find out about how many times larger Jupiter is compared to Mercury.

$$\frac{1.43 \times 10^8}{4.88 \times 10^6} = \frac{1.43}{4.88} \times \frac{10^8}{10^6}$$

$$\approx 0.293 \times 10^2$$

$$\approx 29.3$$

THE DIAMETER OF JUPITER IS ABOUT 29 TIMES LARGER THAN MERCURY.

d. Assume you are a voting member of the International Astronomical Union (IAU) and the classification of Pluto is based entirely on the length of the diameter. Would you vote to keep Pluto a planet or reclassify it? Why or why not?

I WOULD VOTE TO RECLASSIFY IT. KNOWING THAT JUPITER IS 29 TIMES LARGER THAN MERCURY MEANS MERCURY IS PRETTY SMALL. JUPITER IS 62 TIMES LARGER THAN PLUTO, WHICH MEANS PLUTO IS EVEN SMALLER THAN MERCURY. FOR THAT REASON I'D VOTE THAT THE LENGTH OF THE DIAMETER OF PLUTO IS TOO SMALL COMPARED TO OTHER PLANETS (EVEN THE SMALL ONE).

e. Just for fun, Scott wondered how big a planet would be if its diameter was the square of Pluto's diameter. If the diameter of Pluto in terms of meters were squared, what would the diameter of the new planet be? (Write answer in scientific notation.) Do you think it would meet any size requirement to remain a planet? Would it be larger or smaller than Jupiter?

$$(2.3 \times 10^6)^2 = 2.3^2 \times (10^6)^2$$
$$= 5.29 \times 10^{12}$$

YES, 5.29×10^{12} WOULD LIKELY MEET ANY SIZE REQUIREMENT FOR PLANETS. IT WOULD BE LARGER THAN JUPITER.

3. Your friend Pat bought a fish tank that has a volume of 175 liters. The brochure for Pat's tank lists a "fun fact" that it would take 7.43×10^{18} tanks of that size to fill all the oceans in the world. Pat thinks the both of you can quickly calculate the volume of all the oceans in the world using the fun fact and the size of her tank.

a. Given that 1 liter $= 1.0 \times 10^{-12}$ cubic kilometers, rewrite the size of the tank in cubic kilometers using scientific notation.

$$175 \text{ LITERS} = 175(1.0 \times 10^{-12}) \text{ CUBIC KILOMETERS}$$
$$= 175 \times 10^{-12} \text{ KM}^3$$
$$= 1.75 \times 10^{-10} \text{ KM}^3$$

b. Determine the volume of all the oceans in the world in cubic kilometers using the "fun fact."

$$(1.75 \times 10^{-10})(7.43 \times 10^{18}) = (1.75 \times 7.43)(10^{-10} \times 10^{18})$$
$$= 13.0025 \times 10^8$$
$$= 1.30025 \times 10^9$$

THE VOLUME OF ALL THE OCEANS IN THE WORLD IS (1.30025×10^9) KM^3.

c. You liked Pat's fish so much you bought a fish tank of your own that holds an additional 75 liters. Pat asked you to figure out a different "fun fact" for your fish tank. Pat wants to know how many tanks of this new size would be needed to fill the Atlantic Ocean. The Atlantic Ocean has a volume of 323,600,000 cubic kilometers.

TANK : 175 + 75 = 250 LITERS

250 LITERS $= 250(1.0 \times 10^{-12})$ KM^3
$= 250 \times 10^{-12}$
$= 2.5 \times 10^{-10}$

ATLANTIC OCEAN : 323,600,000
$= 3.236 \times 10^8$ KM^3

$$\frac{3.236 \times 10^8}{2.5 \times 10^{-10}} = \frac{3.236}{2.5} \times \frac{10^8}{10^{-10}}$$
$$= 1.2944 \times 10^{18}$$

IT WOULD TAKE 1.2944×10^{18} TANKS (OF SIZE 250 LITERS) TO FILL THE ATLANTIC OCEAN.

This page intentionally left blank

Teacher Edition

Eureka Math
Grade 8
Module 2

Special thanks go to the Gordan A. Cain Center and to the Department of Mathematics at Louisiana State University for their support in the development of *Eureka Math*.

Published by Great Minds

Printed in the U.S.A.
This book may be purchased from the publisher at eureka-math.org
10 9 8 7 6 5 4 3

ISBN 978-1-63255-620-2

Table of Contents[1]

The Concept of Congruence

---

[1]Each lesson is ONE day, and ONE day is considered a 45-minute period.

Grade 8 • Module 2

The Concept of Congruence

OVERVIEW

In this module, students learn about translations, reflections, and rotations in the plane and, more importantly, how to use them to precisely define the concept of *congruence*. Up to this point, *congruence* has been taken to mean, intuitively, *same size and same shape*. Because this module begins with a serious study of geometry, this intuitive definition must be replaced by a precise definition. This module is a first step; its goal is to provide the needed intuitive background for the precise definitions that are introduced in this module for the first time.

Translations, reflections, and rotations are examples of *rigid motions*, which are, intuitively, rules of moving points in the plane in such a way that preserves distance. For the sake of brevity, these three rigid motions are referred to exclusively as the *basic rigid motions*. Initially, the exploration of these basic rigid motions is done via hands-on activities using an overhead projector transparency, but with the availability of geometry software, the use of technology in this learning environment is inevitable, and some general guidelines for this usage are laid out at the end of Lesson 2. What needs to be emphasized is that the importance of these basic rigid motions lies not in the fun activities they bring but in the *mathematical* purpose they serve in clarifying the meaning of congruence.

Throughout Topic A, on the definitions and properties of the basic rigid motions, students verify experimentally their basic properties and, when feasible, deepen their understanding of these properties using reasoning. In particular, what students learned in Grade 4 about angles and angle measurement (**4.MD.C.5**) is put to good use here. They learn that the basic rigid motions preserve angle measurements as well as segment lengths.

Topic B is a critical foundation to the understanding of congruence. All the lessons of Topic B demonstrate to students the ability to sequence various combinations of rigid motions while maintaining the basic properties of individual rigid motions. Lesson 7 begins this work with a sequence of translations. Students verify experimentally that a sequence of translations has the same properties as a single translation. Lessons 8 and 9 demonstrate sequences of reflections and translations and sequences of rotations. The concept of sequencing a combination of all three rigid motions is introduced in Lesson 10; this paves the way for the study of congruence in the next topic.

In Topic C, which introduces the definition and properties of congruence, students learn that congruence is just a sequence of basic rigid motions. The fundamental properties shared by all the basic rigid motions are then inherited by congruence: Congruence moves lines to lines and angles to angles, and it is both distance- and angle-preserving (Lesson 11). In Grade 7, students used facts about supplementary, complementary, vertical, and adjacent angles to find the measures of unknown angles (**7.G.B.5**). This module extends that knowledge to angle relationships that are formed when two parallel lines are cut by a transversal. In Topic C, on angle relationships related to parallel lines, students learn that pairs of angles are congruent because they are angles that have been translated along a transversal, rotated around a point, or reflected across a line.

EUREKA MATH™

Students use this knowledge of angle relationships in Lessons 13 and 14 to show why a triangle has a sum of interior angles equal to $180°$ and why the measure of each exterior angle of a triangle is the sum of the measures of the two remote interior angles of the triangle.

Optional Topic D introduces the Pythagorean theorem. Students are shown the "square within a square" proof of the Pythagorean theorem. The proof uses concepts learned in previous topics of the module, that is, the concept of congruence and concepts related to degrees of angles. Students begin the work of finding the length of a leg or hypotenuse of a right triangle using $a^2 + b^2 = c^2$. Note that this topic is not assessed until Module 7.

Focus Standards

Understand congruence and similarity using physical models, transparencies, or geometry software.

8.G.A.1 Verify experimentally the properties of rotations, reflections, and translations:

a. Lines are taken to lines, and line segments to line segments of the same length.

b. Angles are taken to angles of the same measure.

c. Parallel lines are taken to parallel lines.

8.G.A.2 Understand that a two-dimensional figure is congruent to another if the second can be obtained from the first by a sequence of rotations, reflections, and translations; given two congruent figures, describe a sequence that exhibits the congruence between them.

8.G.A.5 Use informal arguments to establish facts about the angle sum and exterior angle of triangles, about the angles created when parallel lines are cut by a transversal, and the angle-angle criterion for similarity of triangles. *For example, arrange three copies of the same triangle so that the sum of the three angles appears to form a line, and give an argument in terms of transversals why this is so.*

Understand and apply the Pythagorean Theorem.

8.G.B.6 Explain a proof of the Pythagorean Theorem and its converse.

8.G.B.7 Apply the Pythagorean Theorem to determine unknown side lengths in right triangles in real-world and mathematical problems in two and three dimensions.

Foundational Standards

Geometric measurement: understand concepts of angle and measure angles.

4.MD.C.5 Recognize angles as geometric shapes that are formed wherever two rays share a common endpoint, and understand concepts of angle measurement:

a. An angle is measured with reference to a circle with its center at the common endpoint of the rays, by considering the fraction of the circular arc between the points where the two rays intersect the circle. An angle that turns through 1/360 of a circle is called a "one-degree angle," and can be used to measure angles.

b. An angle that turns through n one-degree angles is said to have an angle measure of n degrees.

Draw and identify lines and angles, and classify shapes by properties of their lines and angles.

4.G.A.1 Draw points, lines, line segments, rays, angles (right, acute, obtuse), and perpendicular and parallel lines. Identify these in two-dimensional figures.

4.G.A.2 Classify two-dimensional figures based on the presence or absence of parallel or perpendicular lines, or the presence or absence of angles of a specified size. Recognize right triangles as a category, and identify right triangles.

4.G.A.3 Recognize a line of symmetry for a two-dimensional figure as a line across the figure such that the figure can be folded along the line into matching parts. Identify line-symmetric figures and draw lines of symmetry.

Solve real-life and mathematical problems involving angle measure, area, surface area, and volume.

7.G.B.5 Use facts about supplementary, complementary, vertical, and adjacent angles in a multi-step problem to write and solve simple equations for an unknown angle in a figure.

Focus Standards for Mathematical Practice

MP.2 **Reason abstractly and quantitatively.** This module is rich with notation that requires students to decontextualize and contextualize throughout. Students work with figures and their transformed images using symbolic representations and need to attend to the meaning of the symbolic notation to contextualize problems. Students use facts learned about rigid motions in order to make sense of problems involving congruence.

MP.3 **Construct viable arguments and critique the reasoning of others.** Throughout this module, students construct arguments around the properties of rigid motions. Students make assumptions about parallel and perpendicular lines and use properties of rigid motions to directly or indirectly prove their assumptions. Students use definitions to describe a sequence of rigid motions to prove or disprove congruence. Students build a logical progression of statements to show relationships between angles of parallel lines cut by a transversal, the angle sum of triangles, and properties of polygons like rectangles and parallelograms.

MP.5 **Use appropriate tools strategically**. This module relies on students' fundamental understanding of rigid motions. As a means to this end, students use a variety of tools but none as important as an overhead transparency. Students verify experimentally the properties of rigid motions using physical models and transparencies. Students use transparencies when learning about translation, rotation, reflection, and congruence in general. Students determine when they need to use the transparency as a tool to justify conjectures or when critiquing the reasoning of others.

MP.6 **Attend to precision**. This module begins with precise definitions related to transformations and statements about transformations being distance- and angle-preserving. Students are expected to attend to the precision of these definitions and statements consistently and appropriately as they communicate with others. Students describe sequences of motions precisely and carefully label diagrams so that there is clarity about figures and their transformed images. Students attend to precision in their verbal and written descriptions of rays, segments, points, angles, and transformations in general.

Terminology

New or Recently Introduced Terms

- **Angle Preserving** (A transformation is *angle preserving* if (1) the image of any angle is again an angle, and (2) for any given angle, the angle measure of the image of that angle is equal to the measure of the angle.)
- **Basic Rigid Motion** (A basic rigid motion is a rotation, reflection, or translation of the plane. Basic rigid motions are the basic examples of transformations that preserve distances and angle measures.)
- **Between** (A point B is *between* A and C if (1) A, B, and C are different points on the same line, and (2) $AB + BC = AC$.)
- **Congruence** (A *congruence* is a finite composition of basic rigid motions (rotation, reflections, translations) of the plane.)
- **Congruent** (Two figures in a plane are *congruent* if there exists a congruence that maps one figure onto the other figure.)
- **Directed Line Segment**[8.G.A.1] (A *directed line segment* $\overrightarrow{AB}$ is the line segment AB together with a direction given by connecting an initial point A to a terminal point B.)

- **Distance Preserving** (A transformation is *distance preserving* if the distance between the images of two points is always equal to the distance between the original two points.

  Since the length of a segment is (by definition) equal to the distance between its endpoints, such a transformation is also said to be *length preserving* or said to *preserve lengths of segments.*)
- **Exterior Angle** (Let $\angle ABC$ be an angle of a triangle, $\triangle ABC$, and let D be a point on $\overleftrightarrow{AB}$ such that B is between A and D. $\angle CBD$ is an *exterior angle* of $\triangle ABC$.)
- **Reflection (description)** (Given a line ℓ in the plane, a *reflection across* ℓ is the transformation of the plane that maps each point on the line ℓ to itself, and maps each remaining point P of the plane to its image P' such that ℓ is the perpendicular bisector of the segment $\overline{PP'}$.)
- **Rotation (description)** (For a number d between 0 and 180, the *rotation of* d *degrees around center* O is the transformation of the plane that maps the point O to itself, and maps each remaining point P of the plane to its image P' in the counterclockwise half-plane of ray $\overrightarrow{OP}$ so that P and P' are the same distance away from O and the measurement of $\angle P'OP$ is d degrees.

  The *counterclockwise half-plane* should be described as the half-plane that lies to the left of $\overrightarrow{OP}$ as you move along $\overrightarrow{OP}$ in the direction from O to P.)
- **Sequence (Composition) of Transformations** (A *composition of transformations* is a transformation that is a sequence of two or more transformations given by applying the first transformation followed by applying the next transformation in the sequence to the image of the first transformation, and so on.

  Given transformations G and F, $G \circ F$ is called the *composition* of F and G, and represents first applying the transformation F and then applying G to the image of F.)
- **Transformation** (A *transformation* F of the plane is a function that assigns to each point P of the plane a point $F(P)$ in the plane.

  Given a transformation F, the *image of a point* A is the point $F(A)$ the transformation maps the point A to in the plane, and is often denoted simply by A'. Similarly, the image of triangle T is often denoted by T'.)
- **Translation (description)** (For vector $\overrightarrow{AB}$, a *translation along* $\overrightarrow{AB}$ is the transformation of the plane that maps each point P of the plane to its image P' so that the line $\overleftrightarrow{PP'}$ is parallel to the vector (or contains it), and the directed line segment $\overrightarrow{PP'}$ points in the same direction and is the same length as the vector.)
- **Transversal** (Given a pair of lines L and M in a plane, a third line T is a *transversal* if it intersects L at a single point and intersects M at a single but different point.)
- **Vector** (A *(bound) vector* is a directed line segment, i.e., an "arrow." A vector's initial point is often called its *tail* and the terminal point is often called its *tip*.

  The *length* of a directed line segment (vector) $\overrightarrow{AB}$ is the length of the segment, i.e., AB.)

Familiar Terms and Symbols[2]

- Area and perimeter
- Parallel and perpendicular lines
- Ray, line, line segment, angle
- Supplementary, complementary, vertical, and adjacent angles
- Triangle, quadrilateral

Suggested Tools and Representations

- Transparency or patty paper
- Wet or dry erase markers for use with transparency
- Optional: geometry software
- Composition of Rigid Motions[3]: http://youtu.be/O2XPy3ZLU7Y

Assessment Summary

Assessment Type	Administered	Format	Standards Addressed
Mid-Module Assessment Task	After Topic B	Constructed response with rubric	8.G.A.1
End-of-Module Assessment Task	After Topic C	Constructed response with rubric	8.G.A.2, 8.G.A.5

[2]These are terms and symbols students have seen previously.
[3]The video was developed by Larry Francis.

Topic A

Definitions and Properties of the Basic Rigid Motions

8.G.A.1

Focus Standard:	8.G.A.1	Verify experimentally the properties of rotations, reflections, and translations: a. Lines are taken to lines, and line segments to line segments of the same length. b. Angles are taken to angles of the same measure. c. Parallel lines are taken to parallel lines.
Instructional Days:	6	
Lesson 1:	Why Move Things Around? (E)[1]	
Lesson 2:	Definition of Translation and Three Basic Properties (P)	
Lesson 3:	Translating Lines (S)	
Lesson 4:	Definition of Reflection and Basic Properties (P)	
Lesson 5:	Definition of Rotation and Basic Properties (S)	
Lesson 6:	Rotations of 180 Degrees (P)	

In Topic A, students learn about the mathematical needs for rigid motions and begin by exploring the possible effects of rigid motions in Lesson 1. In particular, the study of rigid motions in this module is not just about moving geometric figures around by the use of reflections, translations, and rotations. Rather, students explore the geometric implications of having an abundance of these basic rigid motions in the plane. Lessons on translation, reflection, and rotation show students that lines are taken to lines, line segments are taken to line segments, and parallel lines are taken to parallel lines. In addition to the intuitive notion of figures retaining the same shape under such motions, students learn to express precisely the fact that lengths of segments and sizes of angles are preserved.

Lessons 2 and 3 focus on translation but also set up precise definitions and statements related to transformations that are used throughout the remainder of the module. In Lesson 2, students learn the basics of translation by translating points, lines, and figures along a vector, and students verify experimentally that translations map lines to lines, segments to segments, rays to rays, and angles to angles. Students also

[1]Lesson Structure Key: **P**-Problem Set Lesson, **M**-Modeling Cycle Lesson, **E**-Exploration Lesson, **S**-Socratic Lesson

verify experimentally that translations preserve length and angle measure. Lesson 3 focuses on the translation of lines, specifically the idea that a translation maps a line either to itself or to a parallel line.

In Lesson 4, students verify experimentally that reflections are distance- and angle-preserving. In Lesson 5, rotation around a point is investigated in a similar manner as the other rigid motions. Students verify experimentally that rotations take lines to lines, etc., and are distance- and angle-preserving. In Lesson 6, students are provided proof that 180-degree rotations map a line to a parallel line and use that knowledge to prove that vertical angles are equal.

Lesson 1: Why Move Things Around?

Student Outcomes

- Students are introduced to vocabulary and notation related to rigid motions (e.g., *transformation*, *image*, and *map*).
- Students are introduced to *transformations of the plane* and learn that a *rigid motion* is a transformation that is *distance-preserving*.
- Students use transparencies to imitate a rigid motion that moves or maps one figure to another figure in the plane.

Materials (needed for this and subsequent lessons)

- Overhead projector transparencies (one per student)
- Fine point dry erase markers (one per student)
- Felt cloth or other eraser (one per student or per pair)

Lesson Notes

The goal of this module is to arrive at a clear understanding of the concept of *congruence* (i.e., What does it mean for two geometric figures to have the same size and shape?). We introduce the *basic rigid motions* (i.e., *translations*, *reflections*, and *rotations*) using overhead projector transparencies. We then explain congruence in terms of a sequence of these basic rigid motions. It may be worth pointing out that we are studying the basic rigid motions for a definite mathematical purpose, not for artistic reasons or for the purpose of studying transformational geometry, per se.

The traditional way of dealing with congruence in Euclidian geometry is to write a set of axioms that abstractly guarantees that two figures are the same (i.e., congruent). This method can be confusing to students. Today, we take a more direct approach so that the concept of congruence ceases to be abstract and intangible, becoming instead susceptible to concrete realizations through hands-on activities using an overhead projector transparency. It is important not only that teachers use transparencies for demonstration purposes, but also that students have access to them for hands-on experience.

The two basic references for this module are "Teaching Geometry According to the Common Core Standards" and "Pre-Algebra," both by Hung-Hsi Wu. Incidentally, the latter is identical to the document cited on page 92 of the *Common Core State Standards for Mathematics*, which is "Lecture Notes for the 2009 Pre-Algebra Institute" by Hung-Hsi Wu, September 15, 2009.

Classwork

Concept Development (20 minutes)

- Given two segments AB and CD, which could be very far apart, how can we find out if they have the same length without measuring them individually? Do you think they have the same length? How do you check? (This question is revisited later.)

- For example, given a quadrilateral $ABCD$ where all four angles at A, B, C, D are right angles, are the opposite sides AD and BC of equal length?

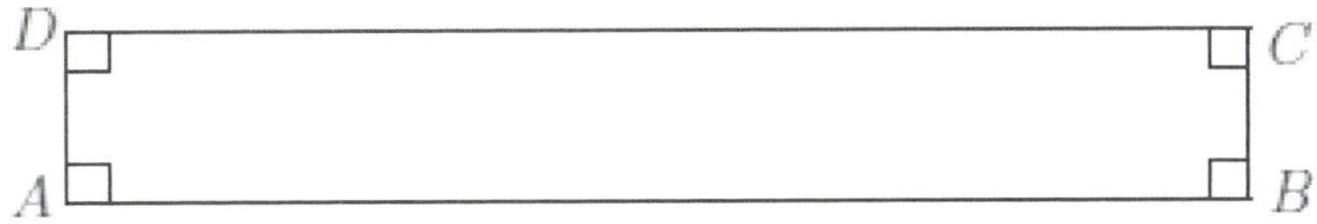

  Later, we *prove* that they have the same length.

- Similarly, given angles $\angle AOB$ and $\angle A'O'B'$, how can we tell whether they have the same degree without having to measure each angle individually?

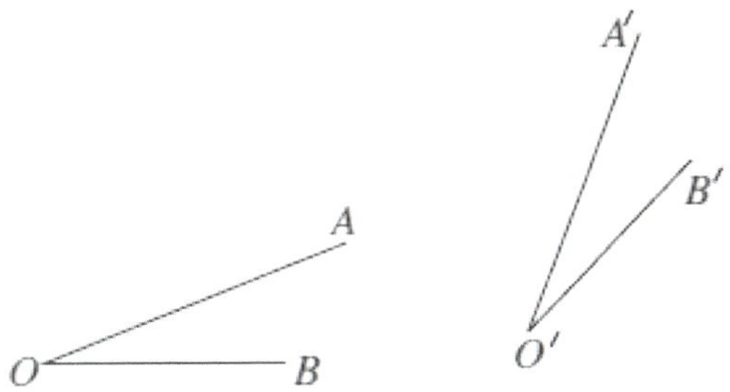

- For example, if two lines L and L' are parallel, and they are intersected by another line, how can we tell if the angles $\angle a$ and $\angle b$ (as shown) have the same degree when measured?

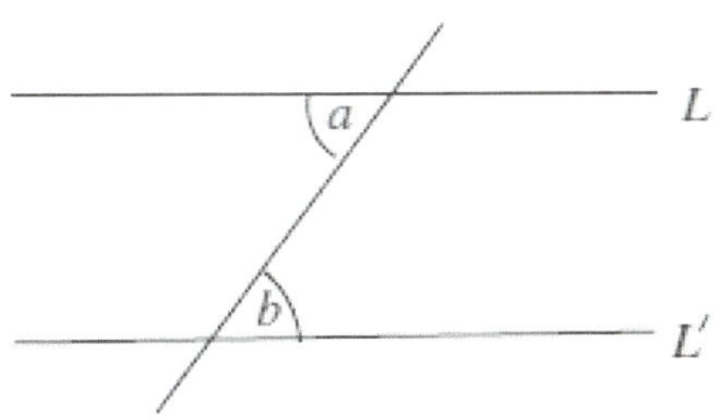

- We are, therefore, confronted with having two geometric figures (two segments, two angles, two triangles, etc.) in different parts of the plane, and we have to find out if they are, in some sense, the same (e.g., same length, same degree, same shape).
- To this end, there are three standard moves we can use to bring one figure on top of another to see if they coincide.
- So, the key question is how do we move things around in a plane, keeping in mind that lines are still lines after being moved and that the lengths of segments and degrees of the measures of angles remain unchanged in the process.
- Moving things around in a plane is exactly where the concept of *transformation* comes in.

MP.2 & MP.4

  - A transformation F of the plane is a function that assigns to each point P of the plane a point $F(P)$.
    - By definition, the symbol $F(P)$ denotes a single point, unambiguously.
    - The point $F(P)$ will be called the image of P by F. Sometimes the image of P by F is denoted simply as P' (read "P prime").
    - The transformation F is sometimes said to "move" the point P to the point $F(P)$.
    - We also say F maps P to $F(P)$.
- In this module, we will mostly be interested in transformations that are given by rules, that is, a set of step-by-step instructions that can be applied to any point P in the plane to get its image.
- The reason for the *image* terminology is that one can, intuitively, think of the plane as a sheet of overhead projector transparency or as a sheet of paper. A transformation F of the plane is a projection (literally, using a light source) from one sheet of the transparency to a sheet of paper with the two sheets identified as being the same plane. Then, the point $F(P)$ is the image on the sheet of paper when the light source projects the point P from the transparency.

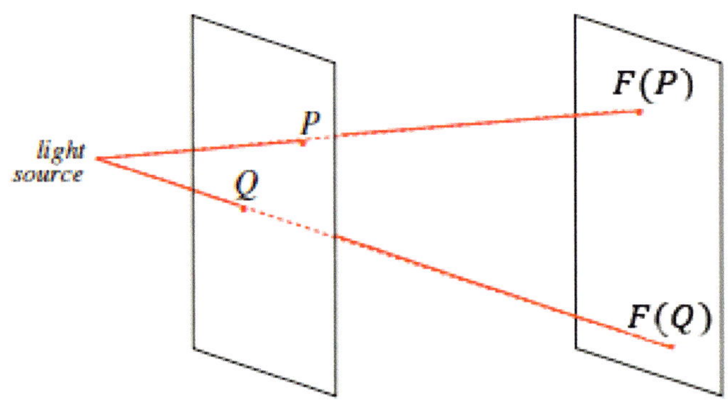

- As to the *map* terminology, think of how you would draw a street map. Drawing a map is a complicated process, but the most mathematically accurate description for the purpose of school mathematics may be that one starts with an aerial view of a particular portion of a city. In the picture below, we look at the area surrounding the Empire State Building (E.S.B.) in New York City. The picture reduces the three-dimensional information into two dimensions and then maps each point on the street to a point on your paper (the map)[1].
  - A point on the street becomes a point on your paper (the map). So, you are mapping each point on the street to your paper.

- Transformations can be defined on spaces of any dimension, but for now we are only concerned with transformations in the plane, in the sense that transformations are those that assign a point of the plane to another point of the plane.
- Transformations can be complicated (i.e., the rule in question can be quite convoluted), but for now we are concentrating on only the simplest transformations, namely those that preserve distance.
- MP.6 A transformation F preserves distance, or is *distance-preserving*, if given any two points P and Q, the distance between the images $F(P)$ and $F(Q)$ is the same as the distance between the original points P and Q.
- An obvious example of this kind of transformation is the *identity transformation*, which assigns each point P of the plane to P itself.
- A main purpose of this module is to introduce many other distance-preserving transformations and show why they are important in geometry.
- A distance-preserving transformation is called a *rigid motion* (or an *isometry*) that, as the name suggests, moves the points of the plane around in a rigid fashion.

[1]This is done in a way that intuitively *preserves the shape*. The correct terminology here is *similar*, as shown in Module 3.

Exploratory Challenge (15 minutes)

Have students complete part (a) independently and part (b) in small groups. Have students share their responses.

Exploratory Challenge

a. **Describe, intuitively, what kind of transformation is required to move the figure on the left to each of the figures (1)–(3) on the right. To help with this exercise, use a transparency to copy the figure on the left. Note: Begin by moving the left figure to each of the locations in (1), (2), and (3).**

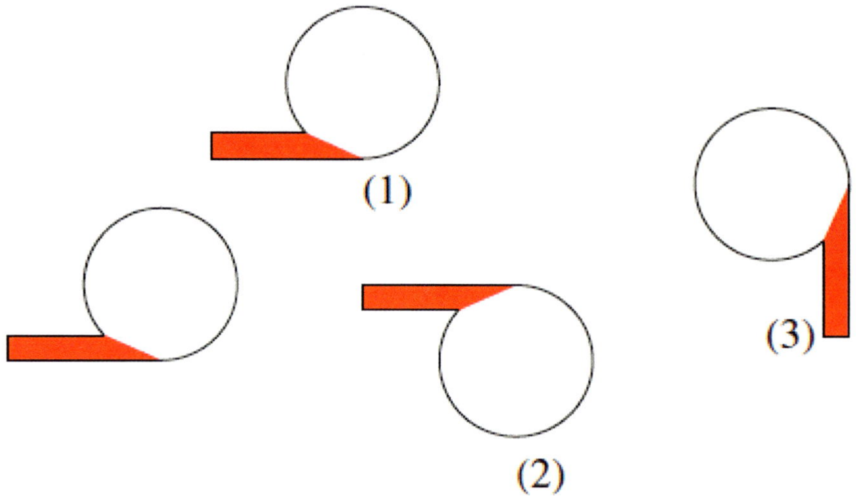

Slide the original figure to the image (1) until they coincide. Slide the original figure to (2), and then flip it so they coincide. Slide the original figure to (3), and then turn it until they coincide.

b. **Given two segments AB and CD, which could be very far apart, how can we find out if they have the same length without measuring them individually? Do you think they have the same length? How do you check? In other words, why do you think we need to move things around on the plane?**

We can trace one of the segments on the transparency and slide it to see if it coincides with the other segment. We move things around in the plane to see if they are exactly the same. This way, we don't have to do any measuring.

Closing (5 minutes)

Summarize, or have students summarize, the lesson.

- We can use a transparency to represent the plane and move figures around.
- We can check to see if one figure is the same as another by mapping one figure onto another and checking to see if they coincide.
- A transformation that preserves distance is known as a rigid motion (the distance between any two corresponding points is the same after the transformation is performed).

Lesson Summary

A *transformation* F of the plane is a function that assigns to each point P of the plane a point $F(P)$ in the plane.

- **By definition, the symbol $F(P)$ denotes a specific single point, unambiguously.**
- **The point $F(P)$ will be called the image of P by F. Sometimes the image of P by F is denoted simply as P' (read "P prime").**
- **The transformation F is sometimes said to "move" the point P to the point $F(P)$.**
- **We also say F maps P to $F(P)$.**

In this module, we will mostly be interested in transformations that are given by rules, that is, a set of step-by-step instructions that can be applied to any point P in the plane to get its image.

If given any two points P and Q, the distance between the images $F(P)$ and $F(Q)$ is the same as the distance between the original points P and Q, and then the transformation F preserves distance, or is distance-preserving.

- **A distance-preserving transformation is called a *rigid motion* (or an *isometry*), and the name suggests that it moves the points of the plane around in a rigid fashion.**

Exit Ticket (5 minutes)

Name ______________________________ Date ______________

Lesson 1: Why Move Things Around?

Exit Ticket

First, draw a simple figure and name it Figure W. Next, draw its image under some transformation (i.e., trace your Figure W on the transparency), and then move it. Finally, draw its image somewhere else on the paper.

Describe, intuitively, how you moved the figure. Use complete sentences.

Exit Ticket Sample Solutions

First, draw a simple figure and name it Figure W. Next, draw its image under some transformation (i.e., trace your Figure W on the transparency), and then move it. Finally, draw its image somewhere else on the paper.

Describe, intuitively, how you moved the figure. Use complete sentences.

Accept any figure and transformation that is correct. Check for the same size and shape. Students should describe the movement of the figure as sliding to the left or right, turning to the left or right, or flipping, similar to how they described the movement of figures in the exercises of the lesson.

Problem Set Sample Solutions

1. Using as much of the new vocabulary as you can, try to describe what you see in the diagram below.

There was a transformation, F, that moved point A to its image $F(A)$ and point B to its image $F(B)$. Since a transformation preserves distance, the distance between points A and B is the same as the distance between points $F(A)$ and $F(B)$.

2. Describe, intuitively, what kind of transformation is required to move Figure A on the left to its image on the right.

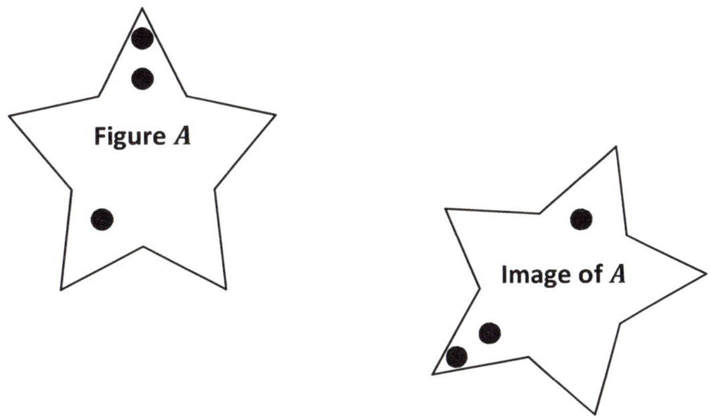

First, I have to slide Figure A so that the point containing two dots maps onto the Image of A in the same location; next, I have to turn (rotate) it so that Figure A maps onto Image of A; finally, I have to flip the figure over so the part of the star with the single dot maps onto the image.

Lesson 2: Definition of Translation and Three Basic Properties

Student Outcomes

- Students perform translations of figures along a specific vector. Students label the image of the figure using appropriate notation.
- Students learn that a translation maps lines to lines, rays to rays, segments to segments, and angles to angles. Students learn that translations preserve lengths of segments and degrees of angles.

Lesson Notes

In this lesson, and those that follow, we emphasize learning about translations, reflections, and rotations, by moving a transparency over a piece of paper. At this initial stage of a student's encounter with these basic rigid motions, such an emphasis provides a tactile familiarity with these transformations, without the worry of technological malfunctions. We do, however, expect students to employ geometry software for further geometric explorations once they have gone beyond this initial phase of learning. Many versions of such software are available, some of which are open source (e.g., GeoGebra and Live Geometry), and others that are not (e.g., Cabri and The Geometer's Sketchpad). Teachers should use discretion about how much technology to incorporate into the lesson, but here are some general guidelines:

1. Students' initial exposure to the basic rigid motions should be made through hands-on activities such as the use of transparencies.
2. Technological tools are just *tools*, and their role in the classroom should be to facilitate learning but not as an end in itself.
3. Students should be made aware of such software because these tools could be of later use, outside of the classroom.

Classwork

Discussion (2 minutes)

- What is the simplest transformation that would map one of the following figures to the other?

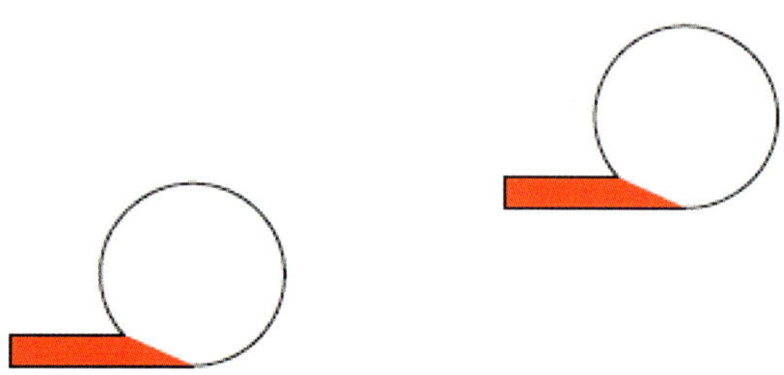

> *Scaffolding:*
> Post new vocabulary words, definitions, and related symbolic notation in a prominent location in the room.

  - *Students will likely answer "a slide," but we want a more precise answer, so we need more information about the topic of transformations to answer this question.*

- In the next few lessons, we learn about three kinds of simple rigid motions: translation, reflection, and rotation.
- We call these the *basic rigid motions*.
- We use the term *basic* because students see that every rigid motion can be obtained by a suitable sequence (see: Topic B) of translations, reflections, and rotations.[1]
- In the following, we describe how to move a transparency over a piece of paper to demonstrate the effect each of these basic rigid motions has on the points in the plane.
- We begin with translations or, more precisely, a *translation along a given vector*.

Example 1 (3 minutes)

Before explaining explicitly what translation along a vector is, review the question that started the discussion. Then, draw a vector, and demonstrate how to translate along the chosen vector to map the lower left figure to the upper right figure.

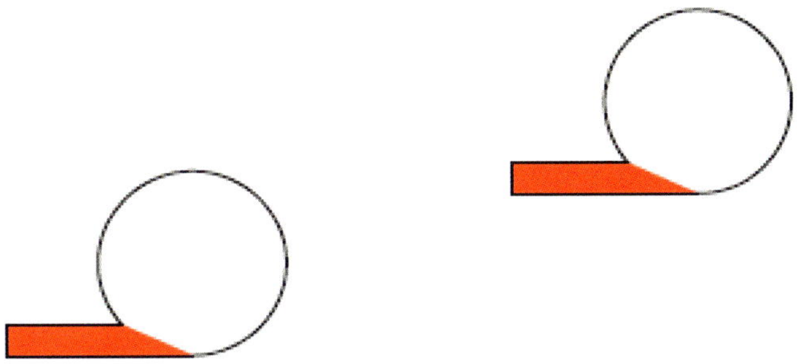

> *Scaffolding:*
> Consider showing several examples of vectors that can be used instead of just one.

- A *vector* is a directed line segment, that is, it is a segment with a direction given by connecting one of its endpoint (called the *initial point or starting point*) to the other endpoint (called the *terminal point* or simply the *endpoint*). It is often represented as an "arrow" with a "tail" and a "tip."
  - The length of a vector is, by definition, the length of its underlying segment.
  - Visually, we distinguish a vector from its underlying segment by adding an arrow above the symbol. Thus, if the segment is AB (A and B being its endpoints), then the vector with starting point A and endpoint B is denoted by $\overrightarrow{AB}$. Likewise, the vector with starting point B and endpoint A is denoted by $\overrightarrow{BA}$.
  - Note that the arrowhead on the endpoint of a vector distinguishes it from the starting point. Here, vector $\overrightarrow{AB}$ is on the left, and vector $\overrightarrow{BA}$ is on the right.

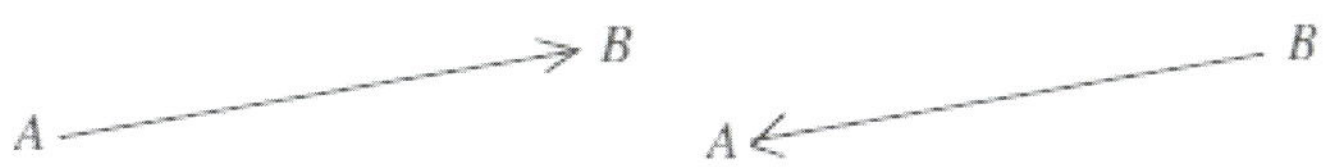

[1]Strictly speaking, all we need are reflections because rotations and translations can be shown to be compositions of reflections. However, for the purpose of fostering geometric intuition, we *should* employ all three.

Example 2 (4 minutes)

MP.4

- We are going to describe how to define a translation T along a vector $\overrightarrow{AB}$ by the use of an overhead projector transparency. Let the plane be represented by a piece of paper on which the vector $\overrightarrow{AB}$ has been drawn. Let P be an arbitrary point in the plane (i.e., the paper), as shown. Note that the line containing the vector $\overrightarrow{AB}$, to be denoted by L_{AB}, is represented by the dotted line segment in the following picture. Note, also, that we are using a finite rectangle to represent the plane, which is infinite in all directions, and a finite segment to represent a line, which is infinite in both directions. The rectangle in the picture below represents the border of the piece of paper.

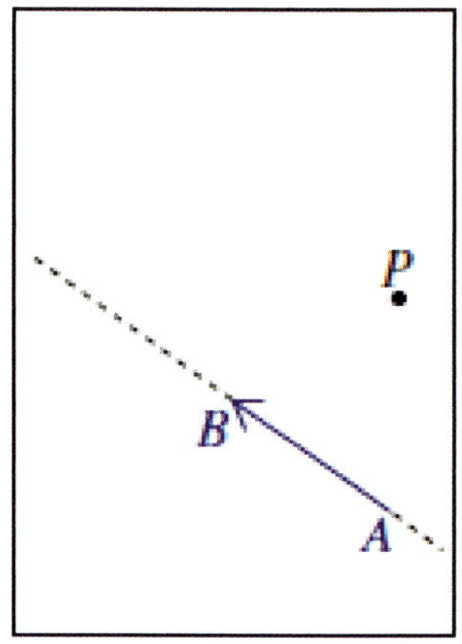

- Now trace $\overrightarrow{AB}$, the line L_{AB}, and the point P exactly on an overhead projector transparency (of exactly the same size as the paper) using a different color, say red. Then, P becomes the red dot on the transparency, and $\overrightarrow{AB}$ becomes a red vector on the transparency; we shall refer to them as the red dot and red vector, respectively, in this example. Keeping the paper fixed, we slide the transparency along $\overrightarrow{AB}$, moving the transparency in the direction from A to B, so that the red vector on the transparency stays on the line L_{AB}, until the starting point of the red vector rests on the endpoint B of the vector $\overrightarrow{AB}$, as shown in the picture. In other words, we slide the transparency along the line AB, in the direction from A to B, for a distance equal to the length of the vector $\overrightarrow{AB}$. The picture shows the transparency after it has been slid along $\overrightarrow{AB}$, and the red rectangle represents the border of the transparency. The point of the plane at the red dot is, by definition, the image $Translation(P)$ of P by the translation T.

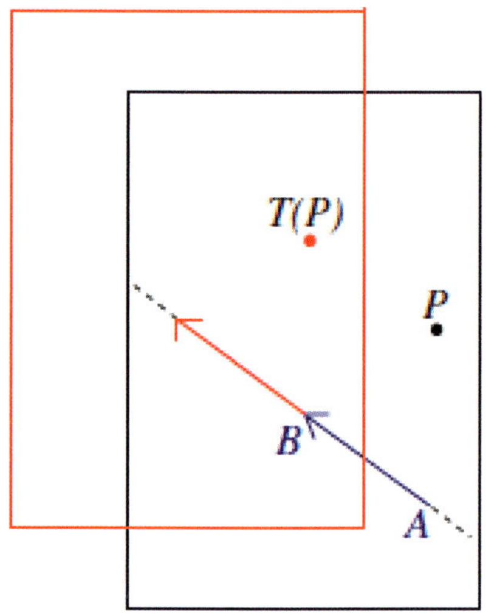

- If we need to be precise, we denote the translation along $\overrightarrow{AB}$ by $Translation_{\overrightarrow{AB}}$. There is some subtlety in this notation: the vector $\overrightarrow{AB}$ has the starting point A and endpoint B, but the vector $\overrightarrow{BA}$ has the starting point B and endpoint A. Thus, $Translation_{\overrightarrow{AB}}$ is *different* from $Translation_{\overrightarrow{BA}}$. Precisely:

$$Translation_{\overrightarrow{AB}}(A) = B \text{ but } Translation_{\overrightarrow{BA}}(B) = A. \quad (1)$$

Note to Teacher:

In reading the descriptions about what translation does, please bear in mind that, in the classroom, a face-to-face demonstration with transparency and paper is far easier to understand than the verbal descriptions given in this lesson.

Video Presentation (2 minutes)

The following animation[2] of a translation would be helpful to a beginner: http://www.harpercollege.edu/~skoswatt/RigidMotions/translation.html

[2]Video developed by Sunil Koswatta.

Example 3 (4 minutes)

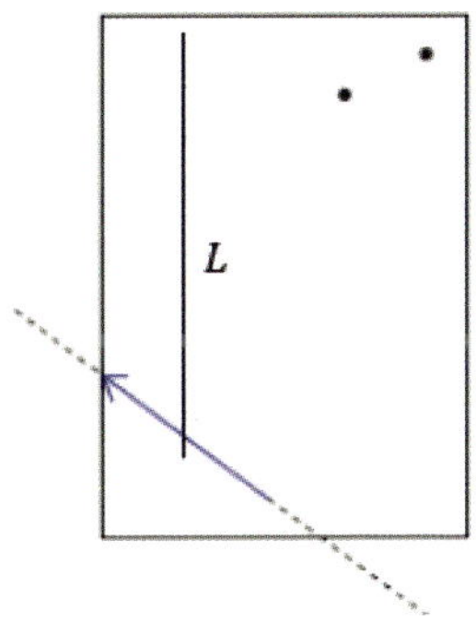

- Suppose we are given a geometric figure consisting of a vertical line L and two points, as shown. We are going to describe the effect on this figure when we translate the plane along the blue vector.

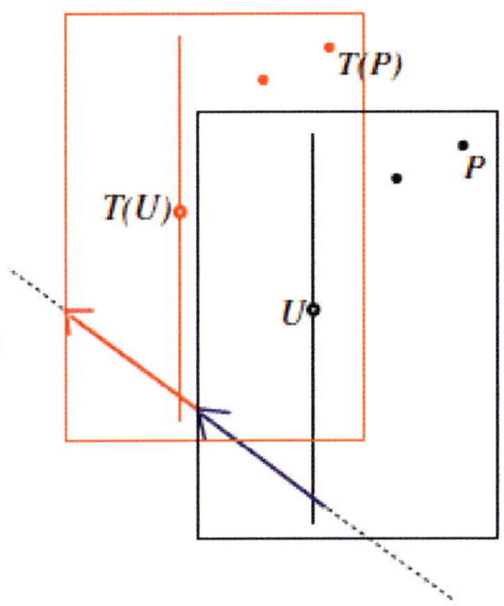

- Again, we copy the whole figure on a transparency in red and then slide the transparency along the blue vector. The whole figure is translated so that the red vertical line and the red dots are in the picture on the right. By definition, the translation T maps the black dots to the red dots, and similarly, T maps a typical point U on the vertical black line (represented by a tiny circle) to the point $T(U)$.

- If we draw the translated figure by itself without reference to the original, it is visually indistinguishable from the original.

Scaffolding:

In this lesson, we try to be clear and, therefore, must define and use terminology precisely. However, in the classroom, it can sometimes be easier to point to the picture without using any definition, at least for a while. Introduce the terminology gradually, repeat often, and remind students of the meaning of the new words.

- So, we put back the original black figure as background information to show where the red figure *used to be*. The dashed arrows should suggest the assignment by the translation T.

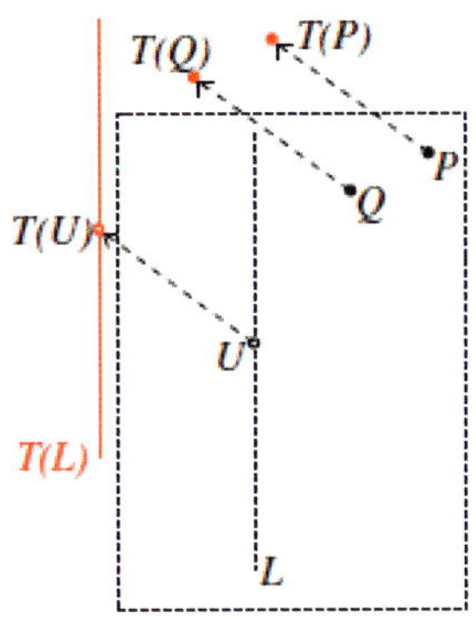

Example 4 (4 minutes)

MP.6

- We now make some observations about the basic properties of translations. We have covered how a translation T along a vector maps some point P to a point $T(P)$. Now, we examine what T does to *the total collection of points in a given figure*. We have actually done that implicitly because the red vertical line in Example 2 *is* the totality of all the points $T(U)$ where U is a point on the black vertical line L. For obvious reasons, we denote the red line by $T(L)$. More formally, if G is a given figure in the plane, then we denote by $T(G)$ the collection of all the points $T(P)$, where P is a point in G. We call $T(G)$ the image *of* G *by* T, and (as in the case of a point) we also say that T *maps* G *to* $T(G)$.

- The diagram to the right shows a translation of a figure along the blue vector $\overrightarrow{AB}$. If G is the black figure, then $T(G)$ is the red figure, as shown.

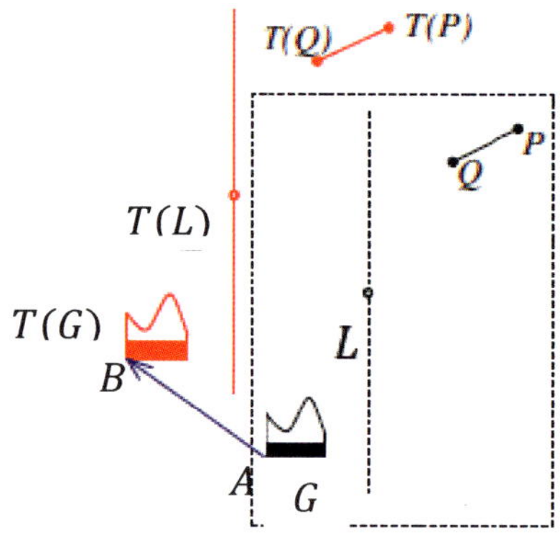

Prime notation is introduced to students in Grade 7, Module 6, Topic B, Lesson 8. Briefly remind students of this notation using the diagram below.

- The notation to represent the image of a point can become cumbersome. However, there are ways to denote the image of the translated points using a more simplified notation. For example, point O after the translation can be denoted by O', said as "O prime"; point P is denoted by P', said as "P prime"; and point Q is denoted by Q', said as "Q prime."

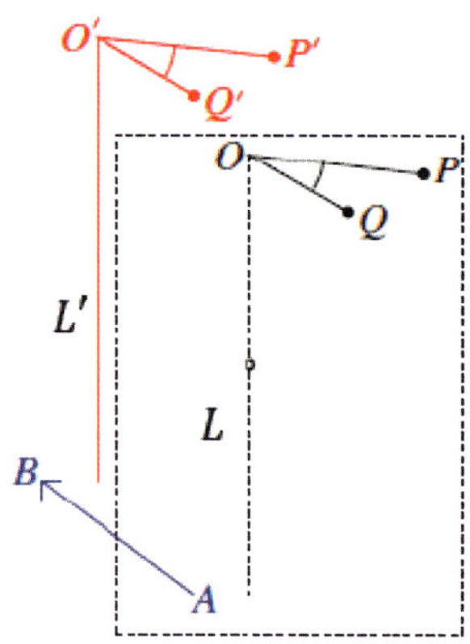

Exercise 1 (5 minutes)

Students complete Exercise 1 in pairs to practice translating a figure along various vectors. Students should describe to their partners what is happening in the translation in order to practice using the vocabulary and notation related to translations. (Note: Students are translating a curved figure because figures that consist entirely of line segments can be reproduced elsewhere on the plane by using a ruler and measuring lengths and degrees of angles, without the use of translation. This defeats the purpose of teaching the concept of translation to move figures around the plane.) Circulate to check student work. Also, consider calling on students to share their work with the class.

Exercise 1

Draw at least three different vectors, and show what a translation of the plane along each vector looks like. Describe what happens to the following figures under each translation using appropriate vocabulary and notation as needed.

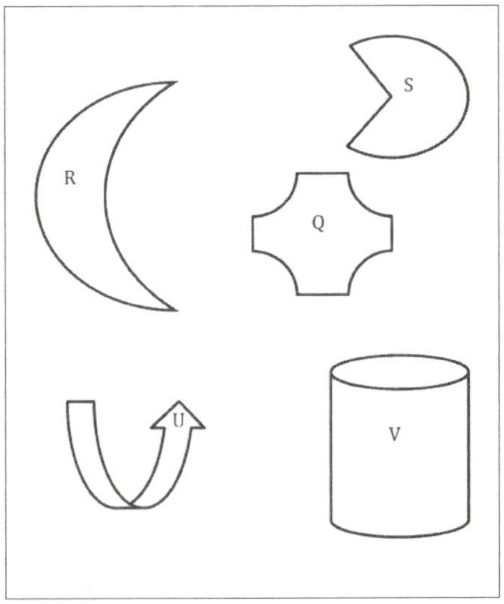

Answers will vary.

Exercise 2 (4 minutes)

Now, students translate specific geometric figures (i.e., lines, angles, segments, and points). They should use prime notation and record their observations as to the lengths of segments and sizes of angles as they complete Exercise 2 independently.

Exercise 2

The diagram below shows figures and their images under a translation along $\overrightarrow{HI}$. Use the original figures and the translated images to fill in missing labels for points and measures.

Solutions are in red, below.

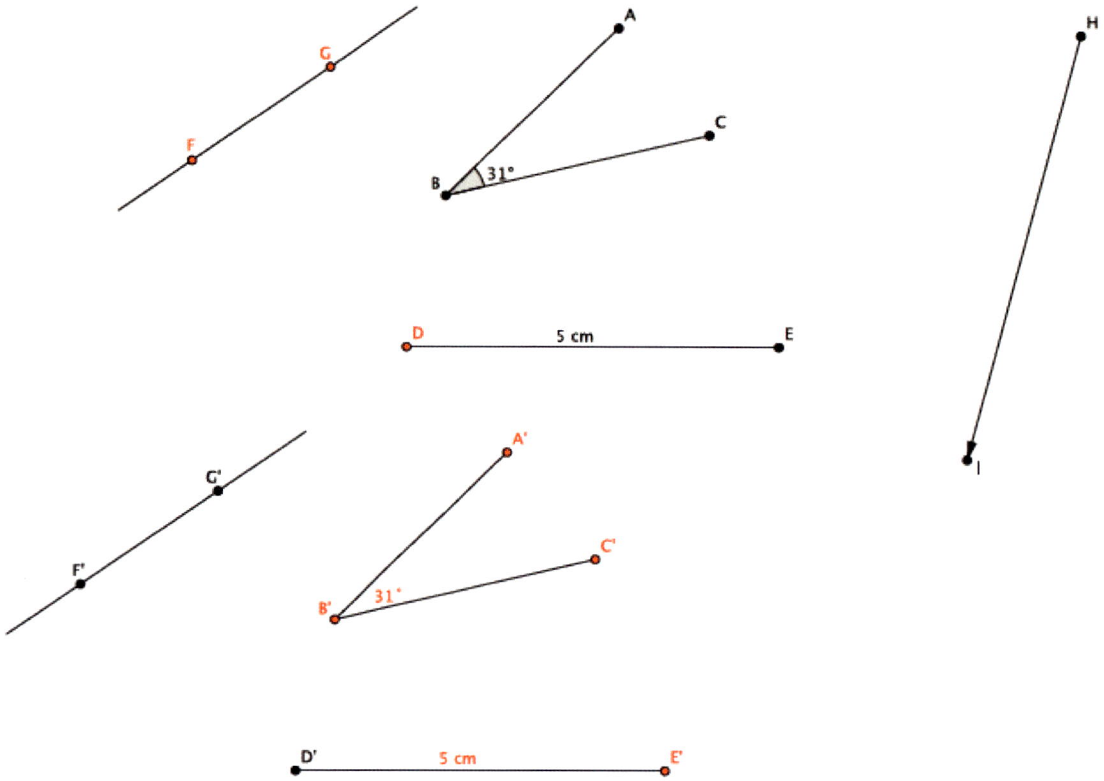

Discussion (10 minutes)

- A translation maps lines to lines, segments to segments, and points to points. Do you believe it?
  - *Yes. After the translation, line FG was still a line, segment DE was still a segment, and the points were still points.*
- A translation preserves lengths of segments. Do you believe it?
  - *Yes. The segment DE in Exercise 2 was the same length after the translation as it was originally.*
- In general, if A and B are any two points in the plane, then the *distance between A and B* is, by definition, the length of the segment joining $Translation(A)$ and $Translation(B)$. Therefore, translations are *distance-preserving* as shown in Lesson 1, and translations are examples of rigid motions.

MP.2 & MP.6

- Translations map angles to angles. Do you believe it?
  - *Yes. After the translation, angle ABC in Exercise 2 was still an angle. Its shape did not change.*
- A translation preserves the degree of an angle. Do you believe it?
  - *Yes. After the translation, angle ABC in Exercise 2 was still 31°. Its size did not change.*

G8-M2-B1-v1.3.0_05/04/15

- The following are some basic properties of translations:

  (Translation 1) A translation maps a line to a line, a ray to a ray, a segment to a segment, and an angle to an angle.

  (Translation 2) A translation preserves lengths of segments.

  (Translation 3) A translation preserves degrees of angles.

- These basic properties of translations are taken for granted in all subsequent discussions of geometry. There are two more points to make about the effect a translation can have on points and lines.
- How can we describe the image of an arbitrary point C by a translation relative to C and the translation itself? In other words, is there a relationship between C, its image C', and the vector C is translated along?
- Let T be the translation along a given vector $\overrightarrow{AB}$. If we slide a transparency along the vector $\overrightarrow{AB}$, it is plausible that C moves to a point C' so that the vector $\overrightarrow{CC'}$ points in the same direction as the vector $\overrightarrow{AB}$, and the length of the segment CC' is the same length as the segment AB. Let's accept these conclusions at this point.

  Here is a pictorial representation of this situation for the case where C does not lie on the line L_{AB}:

  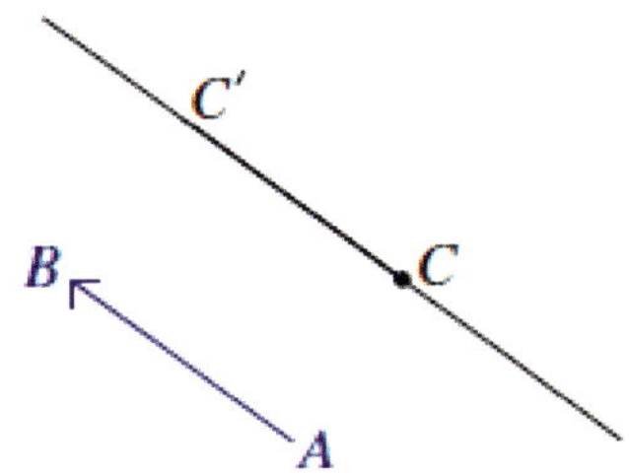

  To clarify, the statement "vector $\overrightarrow{CC'}$ points in the same direction as the vector $\overrightarrow{AB}$" means that the point C' is moved as shown in the picture to the left below, rather than as shown in the picture to the right.

  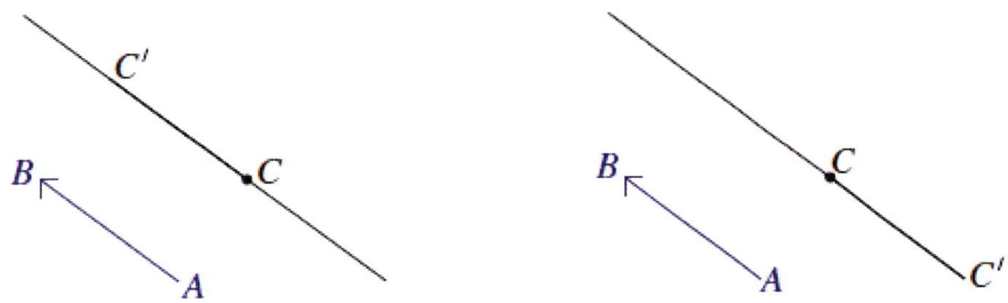

  Still assuming that C is not a point on line L_{AB}, observe from the definition of a translation (in terms of sliding a transparency) that the line $L_{CC'}$ is parallel to the line L_{AB}. This is because the point C on the transparency is not on the blue line L_{AB} on the transparency. Therefore, as we slide the line L_{AB} on the transparency along the original L_{AB}, the point C stays away from the original L_{AB}. For this reason, the point C traces out the line $L_{CC'}$ and has no point in common with the original L_{AB}. In other words, $L_{CC'} \parallel L_{AB}$.

  If C is a point on the line L_{AB}, then C' is also a point on the line L_{AB}, and in this case, $L_{CC'}$ coincides with the line L_{AB} rather than being parallel to the line L_{AB}.

- Our preliminary findings, summarized below, are the basis for the descriptive definition of translation (see the terminology below the Lesson Summary):

  Let T be the translation along $\overrightarrow{AB}$. Let C be a point in the plane, and let the image $T(C)$ be denoted by C'. Then, the vector $\overrightarrow{CC'}$ points in the same direction as $\overrightarrow{AB}$, and the vectors have the same length. If C lies on line L_{AB}, then so does C'. If C does not lie on L_{AB}, then $L_{CC'} \parallel L_{AB}$.

Closing (3 minutes)

Summarize, or have students summarize, the lesson.

- We now know what a translation of a plane along a vector is.
- We know where a point, a line, or a figure on a plane moves to by a translation along a given vector.
- We know that translations have three basic properties:
  - Translations map lines to lines, segments to segments, rays to rays, and angles to angles.
  - Lengths of segments are preserved.
  - Degrees of measures of angles are preserved.
- We can now use a simplified notation, for example, P' to represent the translated point P.

Lesson Summary

Translation occurs along a given vector:

- **A *vector* is a directed line segment, that is, it is a segment with a direction given by connecting one of its endpoint (called the *initial point* or *starting point*) to the other endpoint (called the *terminal point* or simply the *endpoint*). It is often represented as an "arrow" with a "tail" and a "tip."**
- **The *length of a vector* is, by definition, the length of its underlying segment.**
- **Pictorially note the starting and endpoints:**

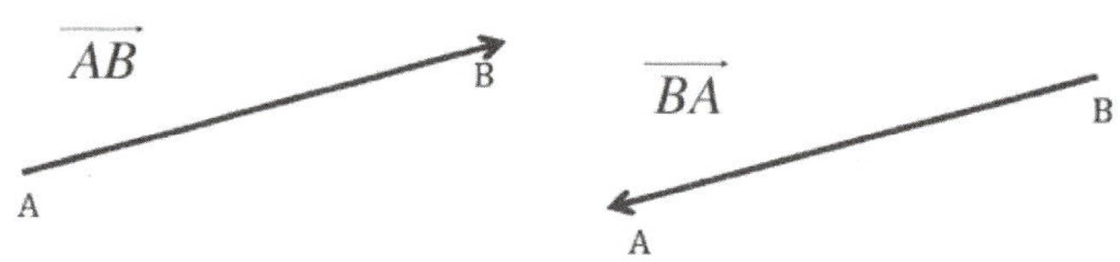

A translation of a plane along a given vector is a basic rigid motion of a plane.

The three basic properties of translation are as follows:

(Translation 1) A translation maps a line to a line, a ray to a ray, a segment to a segment, and an angle to an angle.
(Translation 2) A translation preserves lengths of segments.
(Translation 3) A translation preserves measures of angles.

Terminology

TRANSLATION (description): For vector $\overrightarrow{AB}$, a *translation* along $\overrightarrow{AB}$ is the transformation of the plane that maps each point C of the plane to its image C' so that the line $\overleftrightarrow{CC'}$ is parallel to the vector (or contains it), and the vector $\overrightarrow{CC'}$ points in the same direction and is the same length as the vector $\overrightarrow{AB}$.

Exit Ticket (4 minutes)

Name ___________________________ Date ______________

Lesson 2: Definition of Translation and Three Basic Properties

Exit Ticket

1. Name the vector in the picture below.

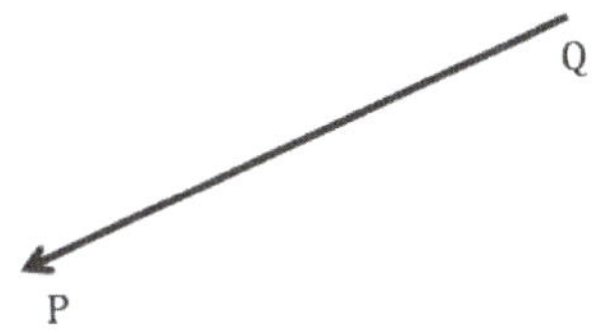

2. Name the vector along which a translation of a plane would map point A to its image $T(A)$.

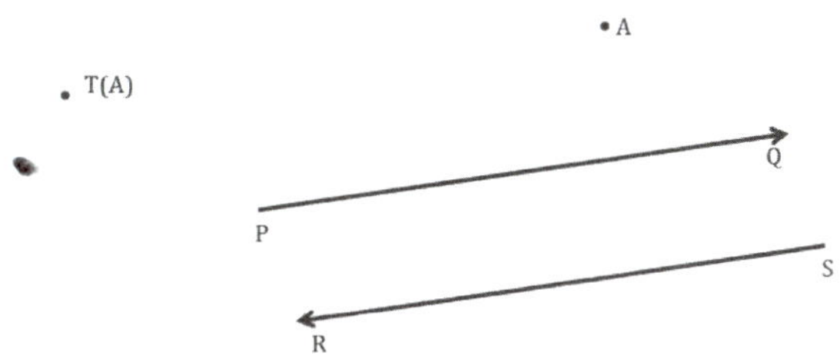

3. Is Maria correct when she says that there is a translation along a vector that maps segment AB to segment CD? If so, draw the vector. If not, explain why not.

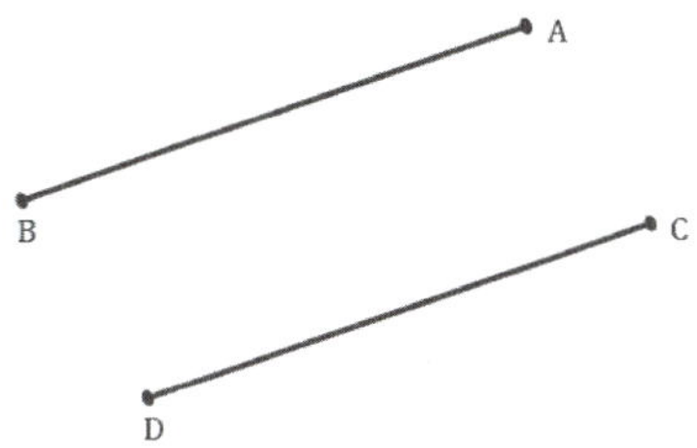

4. Assume there is a translation that maps segment AB to segment CD shown above. If the length of segment CD is 8 units, what is the length of segment AB? How do you know?

Exit Ticket Sample Solutions

1. **Name the vector in the picture below.**

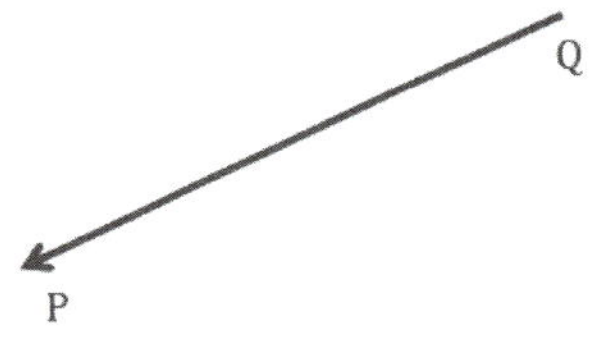

$\overrightarrow{QP}$

2. **Name the vector along which a translation of a plane would map point A to its image $T(A)$.**

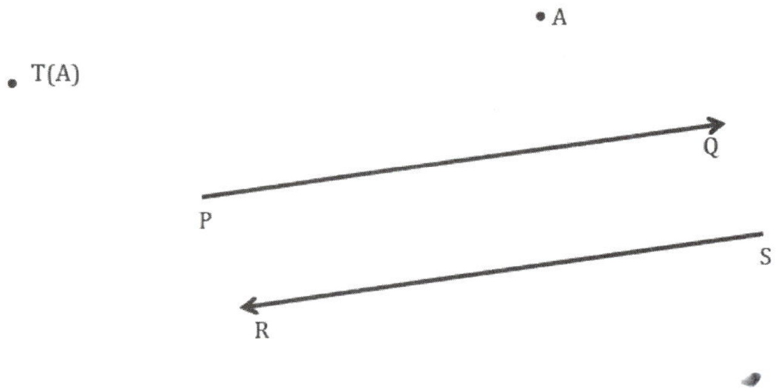

$\overrightarrow{SR}$

3. **Is Maria correct when she says that there is a translation along a vector that maps segment AB to segment CD? If so, draw the vector. If not, explain why not.**

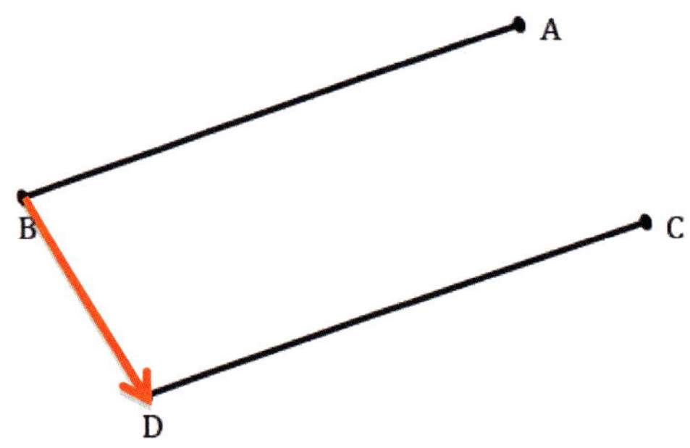

Yes. Accept any vector that would translate the segment AB to segment CD. A possible vector is shown in red, above.

4. **Assume there is a translation that maps segment AB to segment CD shown above. If the length of segment CD is 8 units, what is the length of segment AB? How do you know?**

The length of CD must be 8 units in length because translations preserve the lengths of segments.

Problem Set Sample Solutions

1. Translate the plane containing Figure A along $\overrightarrow{AB}$. Use your transparency to sketch the image of Figure A by this translation. Mark points on Figure A, and label the image of Figure A accordingly.

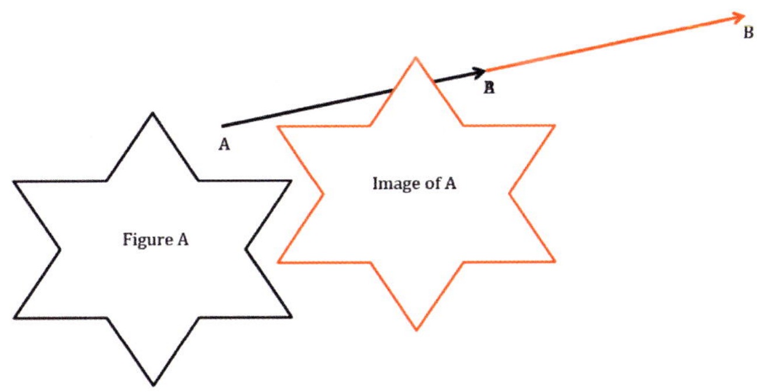

Marked points will vary. Verify that students have labeled their points and images appropriately.

2. Translate the plane containing Figure B along $\overrightarrow{BA}$. Use your transparency to sketch the image of Figure B by this translation. Mark points on Figure B, and label the image of Figure B accordingly.

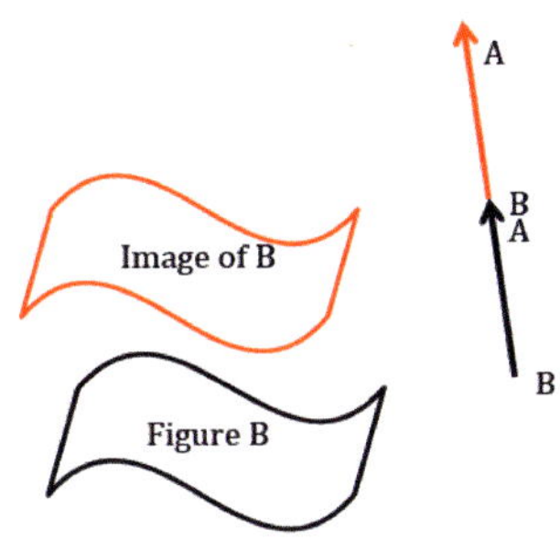

Marked points will vary. Verify that students have labeled their points and images appropriately.

3. Draw an acute angle (your choice of degree), a segment with length 3 cm, a point, a circle with radius 1 in., and a vector (your choice of length, i.e., starting point and ending point). Label points and measures (measurements do not need to be precise, but your figure must be labeled correctly). Use your transparency to translate all of the figures you have drawn along the vector. Sketch the images of the translated figures and label them.

Drawings will vary. Note: Drawing is not to scale.

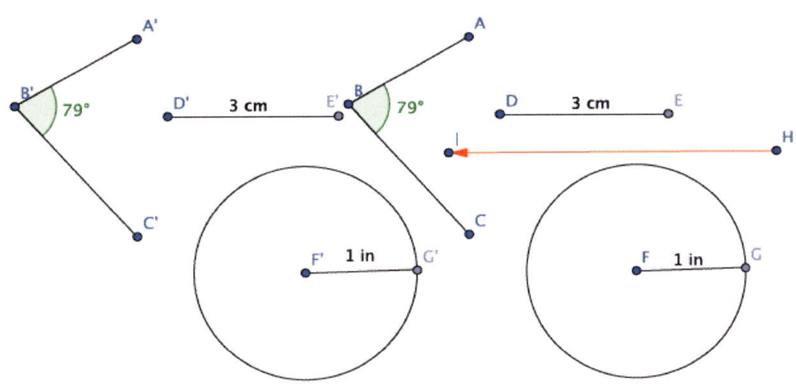

4. **What is the length of the translated segment? How does this length compare to the length of the original segment? Explain.**

The length is 3 cm. *The length is the same as the original because translations preserve the lengths of segments.*

5. **What is the length of the radius in the translated circle? How does this radius length compare to the radius of the original circle? Explain.**

The length is 1 in. *The length is the same as the original because translations preserve lengths of segments.*

6. **What is the degree of the translated angle? How does this degree compare to the degree of the original angle? Explain.**

Answers will vary based on the original size of the angle drawn. The angles will have the same measure because translations preserve degrees of angles.

7. **Translate point D along vector $\overrightarrow{AB}$, and label the image D'. What do you notice about the line containing vector $\overrightarrow{AB}$ and the line containing points D and D'? (Hint: Will the lines ever intersect?)**

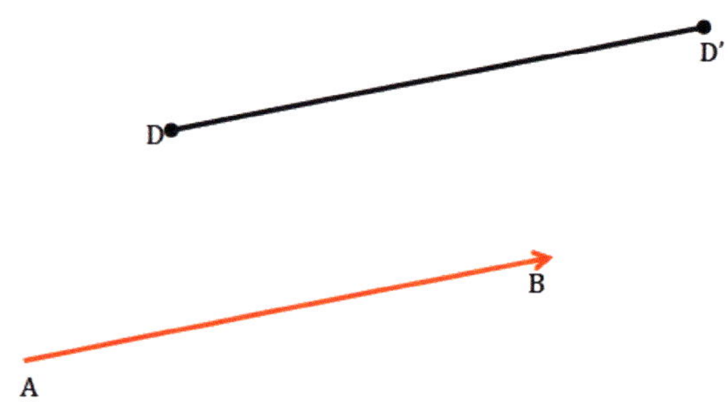

The lines will be parallel.

8. **Translate point E along vector $\overrightarrow{AB}$, and label the image E'. What do you notice about the line containing vector $\overrightarrow{AB}$ and the line containing points E and E'?**

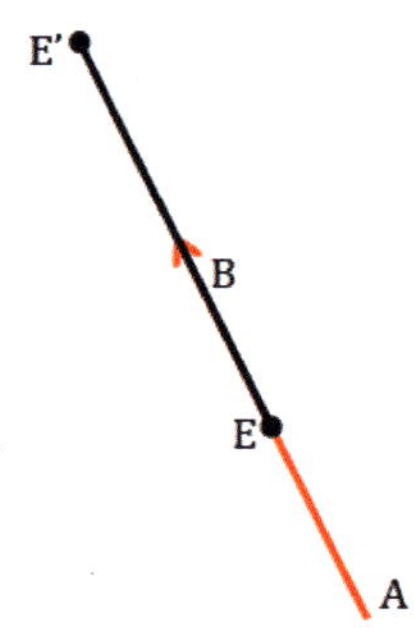

The lines will coincide.

Lesson 3: Translating Lines

Student Outcomes

- Students learn that when lines are translated, they are either parallel to the given line or they coincide.
- Students learn that translations map parallel lines to parallel lines.

Classwork

Exercise 1 (3 minutes)

Students complete Exercise 1 independently in preparation for the discussion that follows.

Exercises

1. **Draw a line passing through point P that is parallel to line L. Draw a second line passing through point P that is parallel to line L and that is distinct (i.e., different) from the first one. What do you notice?**

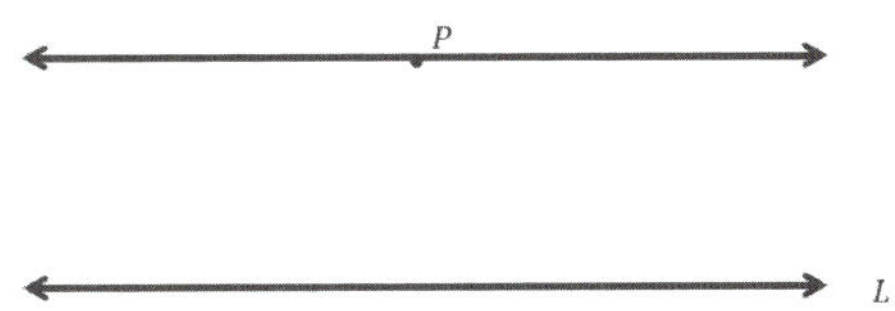

Students should realize that they can only draw one line through point P that is parallel to L.

Discussion (3 minutes)

Bring out a fundamental assumption about the plane (as observed in Exercise 1):

- Given a line L and a point P not lying on L, there is at most one line passing through P and parallel to L.
  - Based on what we have learned up to now, we cannot prove or explain this, so we have to simply agree that this is one of the starting points in the study of the plane.
  - This idea plays a key role in everything we do in the plane. A first consequence is that given a line L and a point P not lying on L, we can now refer to the line (because we agree there is only one) passing through P and parallel to L.

Exercises 2–4 (9 minutes)

Students complete Exercises 2–4 independently in preparation for the discussion that follows.

2. Translate line L along the vector $\overrightarrow{AB}$. What do you notice about L and its image, L'?

L and L' coincide. $L = L'$.

> *Scaffolding:*
>
> Refer to Exercises 2–4 throughout the discussion and in the summary of findings about translating lines.

3. Line L is parallel to vector $\overrightarrow{AB}$. Translate line L along vector $\overrightarrow{AB}$. What do you notice about L and its image, L'?

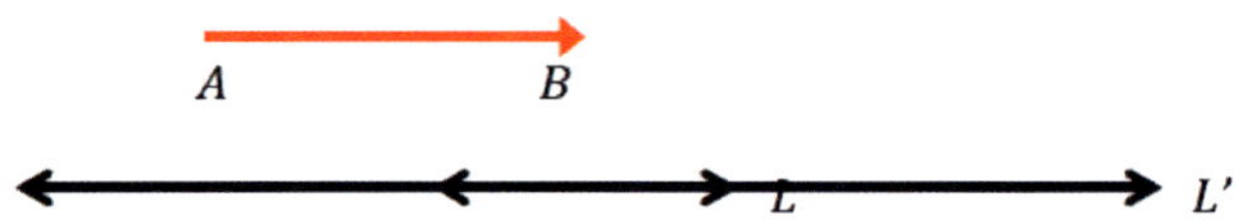

L and L' coincide, again. $L = L'$.

4. Translate line L along the vector $\overrightarrow{AB}$. What do you notice about L and its image, L'?

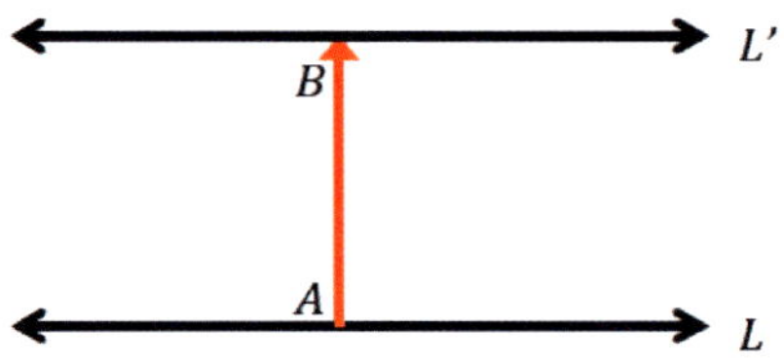

$L \parallel L'$

Discussion (15 minutes)

- Now we examine the effect of a translation on a line. Thus, let line L be given. Again, let the translation be along a given $\overrightarrow{AB}$, and let L' denote the image line of the translated L. We want to know what L' is relative to AB and line L.
- If $L = L_{AB}$, or $L \parallel L_{AB}$, then $L' = L$.
  - If $L = L_{AB}$, then this conclusion follows directly from the work in Lesson 2, which says if C is on L_{AB}, then so is C'; therefore, $L' = L_{AB}$ and $L = L'$ (Exercise 2).
  - If $L \parallel L_{AB}$ and C is on L, then it follows from the work in Lesson 2, which says that C' lies on the line l passing through C and parallel to L_{AB}. However, L is given as a line passing through C and parallel to L_{AB}, so the fundamental assumption that there is just one line passing through a point, parallel to a line (Exercise 1), implies $l = L$. Therefore, C' lies on L after all, and the translation maps every point of L to a point of L'. Therefore, $L = L'$ again (Exercise 3).

> *Note to Teacher:*
>
> The notation $Translation(L)$ is used as a precursor to the notation students encounter in high school Geometry (i.e., $T(L)$). We want to make clear the basic rigid motion that is being performed, so the notation $Translation(L)$ is written to mean *the translation of L along the specified vector.*

- Caution: One must not over-interpret the equality $Translation(L) = L$ (which is the same as $L = L'$).
- All the equality says is that the two lines L and L' coincide completely. It is easy (but wrong) to infer from the equality $Translation(L) = L$ that for any point P on L, $Translation(P) = P$. Suppose the vector $\overrightarrow{AB}$ lying on L is not the zero vector (i.e., assume $A \neq B$). Trace the line L on a transparency to obtain a red line L, and now slide the transparency along $\overrightarrow{AB}$. Then, the red line, as a line, coincides with the original L, but clearly every point on L has been moved by the slide (the translation). Indeed, as we saw in Example 2 of Lesson 2, $Translation(A) = B \neq A$. Therefore, the equality $L' = L$ only says that for any point C on L, $Translation(C)$ is also a point on L, but as long as $\overrightarrow{AB}$ is not a zero vector, $Translation(C) \neq C$.

MP.6

Strictly speaking, we have not completely proved $L = L'$ in either case. To explain this, let us define what it means for two geometric figures F and G to be equal, that is, $F = G$: it means each point of F is also a point of G; conversely, each point of G is also a point of F. In this light, all we have shown above is that if every point C' of L' belongs to L, then Q is also a point of L'. To show the latter, we have to show that this Q is equal to $Translation(P)$ for some P on L. This then completes the reasoning.

However, at this point of students' education in geometry, it may be prudent not to bring up such a sticky point because they are already challenged with all of the new ideas and definitions. Simply allow the preceding reasoning to stand for now, and clarify later in the school year when students are more comfortable with the geometric environment.

- Next, if L is neither L_{AB} nor parallel to L_{AB}, then $L' \parallel L$.
- If we use a transparency to see this translational image of L by the stated translation, then the pictorial evidence is clear: the line L moves in a parallel manner along $\overrightarrow{AB}$, and a typical point C of L is translated to a point C' of L'. The fact that $L' \parallel L$ is unmistakable, as shown. In the classroom, students should be convinced by the pictorial evidence. If so, leave it at that (Exercise 4).

MP.2 & MP.7

Here is a simple proof, but to present it in class, begin by asking students *how* they would prove that two lines are parallel. Ensure students understand that they have no tools in their possession to accomplish this goal. It is only then that they see the need for invoking a proof by contradiction (see discussion above). *If there are no obvious ways to do something, then you just have to do the best you can by trying to see what happens if you assume the opposite is true.* Thus, if L' is not parallel to L, then they intersect at a point C'. Since C' lies on L', it follows from the definition of L' (as the image of L under the translation T) that there is a point C on L so that $Translation(C) = C'$.

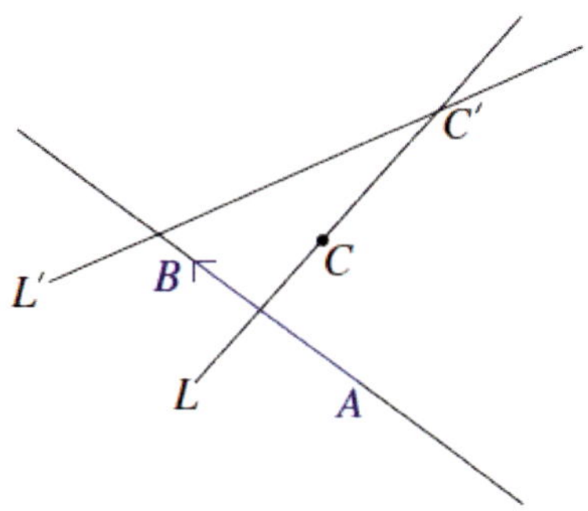

It follows from Lesson 2 that $L_{CC'} \parallel L_{AB}$. However, both C and C' lie on L, so $L_{CC'} \parallel L$, and we get $L \parallel L_{AB}$. This contradicts the assumption that L is not parallel to L_{AB}, so L could not possibly intersect L'. Therefore, $L' \parallel L$.

- Note that a translation maps parallel lines to parallel lines. More precisely, consider a translation T along a vector $\overrightarrow{AB}$. Then:

  If L_1 and L_2 are parallel lines, so are $Translation(L_1)$ and $Translation(L_2)$.

- The reasoning is the same as before: Copy L_1 and L_2 onto a transparency, and then translate the transparency along $\overrightarrow{AB}$. If L_1 and L_2 do not intersect, then their red replicas on the transparency will not intersect either, no matter what $\overrightarrow{AB}$ is used. So, $Translation(L_1)$ and $Translation(L_2)$ are parallel.
- These findings are summarized as follows:

MP.6

Given a translation T along a vector $\overrightarrow{AB}$, let L be a line, and let L' denote the image of L by T.

- If $L \parallel L_{AB}$ or $L = L_{AB}$, then $L' \parallel L$.
- If L is neither parallel to L_{AB} nor equal to L_{AB}, then $L' \parallel L$.

Exercises 5–6 (5 minutes)

Students complete Exercises 5 and 6 in pairs or small groups.

5. Line L has been translated along vector $\overrightarrow{AB}$, resulting in L'. What do you know about lines L and L'?

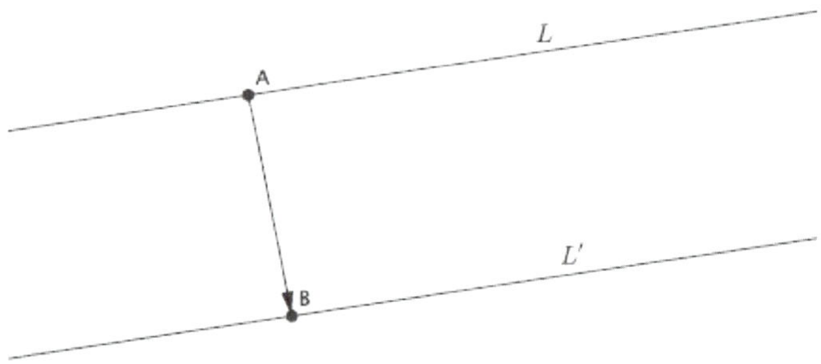

$L \parallel T(L)$

6. Translate L_1 and L_2 along vector $\overrightarrow{DE}$. Label the images of the lines. If lines L_1 and L_2 are parallel, what do you know about their translated images?

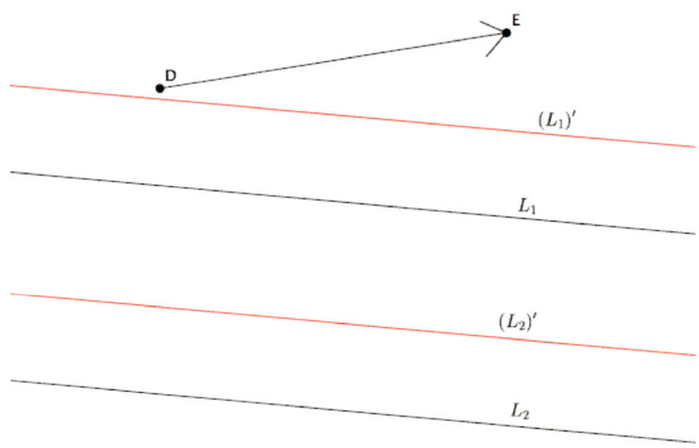

Since $L_1 \parallel L_2$, then $(L_1)' \parallel (L_2)'$.

Closing (5 minutes)

Summarize, or have students summarize, the lesson.

- We know that there exists just one line, parallel to a given line and through a given point not on the line.
- We know that translations map parallel lines to parallel lines.
- We know that when lines are translated, they are either parallel to the given line or they coincide.

> **Lesson Summary**
>
> - **Two lines in the plane are parallel if they do not intersect.**
> - **Translations map parallel lines to parallel lines.**
> - **Given a line L and a point P not lying on L, there is at most one line passing through P and parallel to L.**

Exit Ticket (5 minutes)

Name ____ Date __________________

Lesson 3: Translating Lines

Exit Ticket

1. Translate point Z along vector $\overrightarrow{AB}$. What do you know about the line containing vector $\overrightarrow{AB}$ and the line formed when you connect Z to its image Z'?

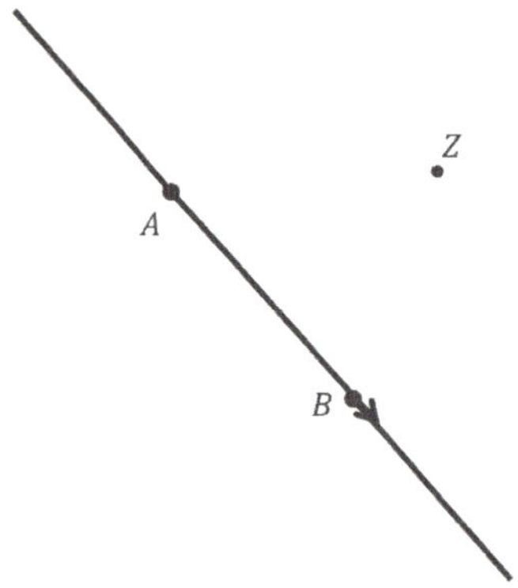

2. Using the above diagram, what do you know about the lengths of segments ZZ' and AB?

3. Let points A and B be on line L and the vector $\overrightarrow{AC}$ be given, as shown below. Translate line L along vector $\overrightarrow{AC}$. What do you know about line L and its image, L'? How many other lines can you draw through point C that have the same relationship as L and L'? How do you know?

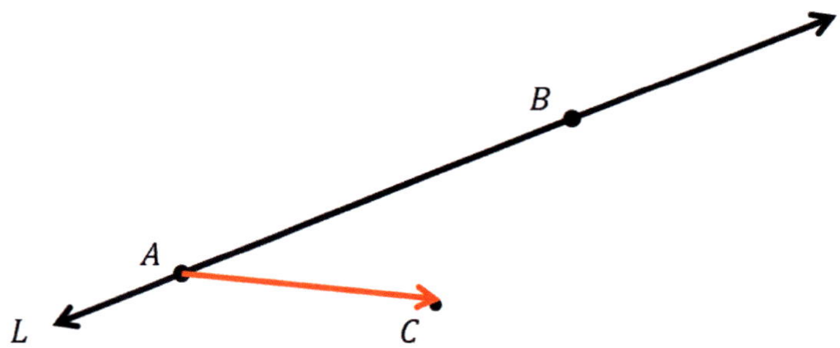

Exit Ticket Sample Solutions

1. Translate point Z along vector $\overrightarrow{AB}$. What do you know about the line containing vector $\overrightarrow{AB}$ and the line formed when you connect Z to its image Z'?

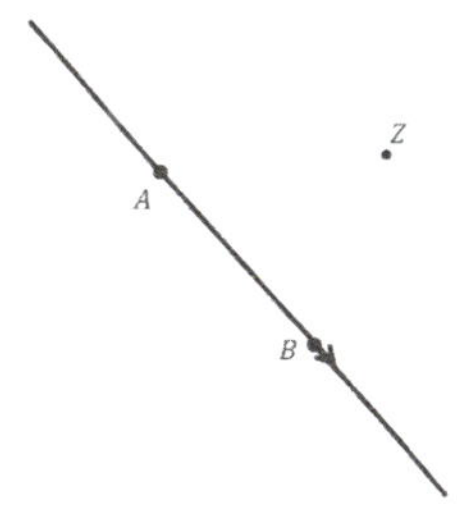

The line containing vector $\overrightarrow{AB}$ and ZZ' is parallel.

2. Using the above diagram, what do you know about the lengths of segment ZZ' and segment AB?

The lengths are equal: $|ZZ'| = |AB|$.

3. Let points A and B be on line L and the vector $\overrightarrow{AC}$ be given, as shown below. Translate line L along vector $\overrightarrow{AC}$. What do you know about line L and its image, L'? How many other lines can you draw through point C that have the same relationship as L and L'? How do you know?

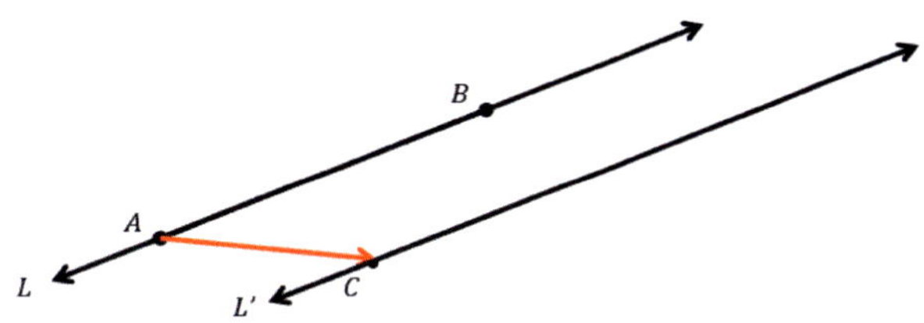

L and L' are parallel. There is only one line parallel to line L that goes through point C. The fact that there is only one line through a point parallel to a given line guarantees it.

Problem Set Sample Solutions

1. Translate $\angle XYZ$, point A, point B, and rectangle $HIJK$ along vector $\overrightarrow{EF}$. Sketch the images, and label all points using prime notation.

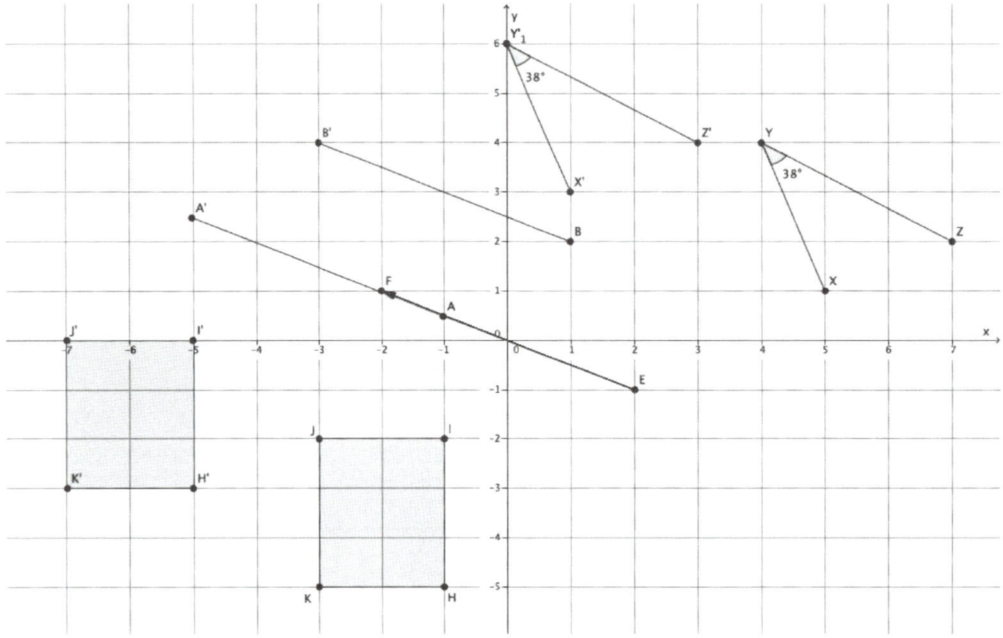

2. What is the measure of the translated image of $\angle XYZ$? How do you know?

   The measure is $38°$. Translations preserve angle measure.

3. Connect B to B'. What do you know about the line that contains the segment formed by BB' and the line containing the vector $\overrightarrow{EF}$?

   $\overleftrightarrow{BB'} \parallel \overleftrightarrow{EF}$.

4. Connect A to A'. What do you know about the line that contains the segment formed by AA' and the line containing the vector $\overrightarrow{EF}$?

   $\overleftrightarrow{AA'}$ and $\overleftrightarrow{EF}$ coincide.

5. Given that figure $HIJK$ is a rectangle, what do you know about lines that contain segments HI and JK and their translated images? Explain.

   Since $HIJK$ is a rectangle, I know that $\overleftrightarrow{HI} \parallel \overleftrightarrow{JK}$. Since translations map parallel lines to parallel lines, then $\overleftrightarrow{H'I'} \parallel \overleftrightarrow{J'K'}$.

Lesson 4: Definition of Reflection and Basic Properties

Student Outcomes

- Students know the definition of reflection and perform reflections across a line using a transparency.
- Students show that reflections share some of the same fundamental properties with translations (e.g., lines map to lines, angle- and distance-preserving motion). Students know that reflections map parallel lines to parallel lines.
- Students know that for the reflection across a line L and for every point P not on L, L is the bisector of the segment joining P to its reflected image P'.

Classwork

Example 1 (5 minutes)

MP.6

The reflection across a line L is defined by using the following example.

- Let L be a vertical line, and let P and A be two points not on L, as shown below. Also, let Q be a point on L. (The black rectangle indicates the border of the paper.)

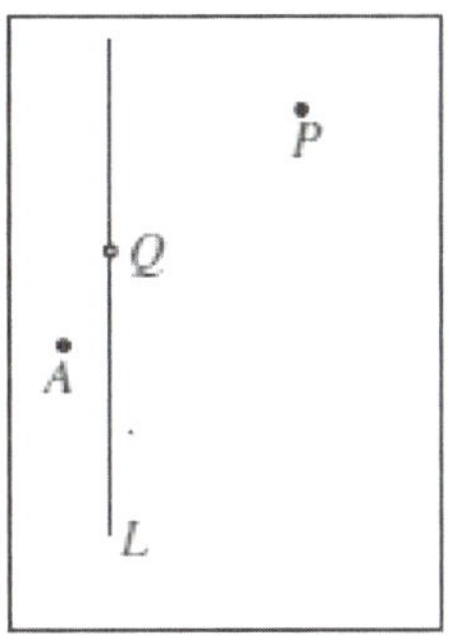

- The following is a description of how the reflection moves the points P, Q, and A by making use of the transparency.
- Trace the line L and three points onto the transparency exactly, using red. (Be sure to use a transparency that is the same size as the paper.)
- Keeping the paper fixed, flip the transparency across the vertical line (interchanging left and right) while keeping the vertical line and point Q on top of their black images.
- The position of the red figures on the transparency now represents the reflection of the original figure. $Reflection(P)$ is the point represented by the red dot to the left of L, $Reflection(A)$ is the red dot to the right of L, and point $Reflection(Q)$ is point Q itself.
- Note that point Q is unchanged by the reflection.

> *Scaffolding:*
> There are manipulatives, such as MIRA and Georeflector, which facilitate the learning of reflections by producing a reflected image.

- The red rectangle in the picture below represents the border of the transparency.

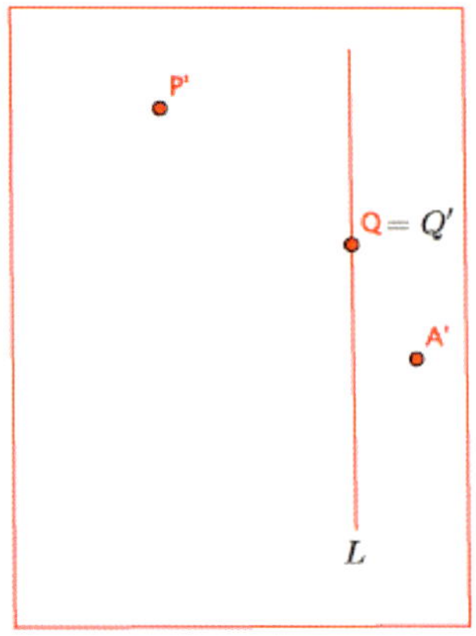

- In the picture above, the reflected image of the points is noted similar to how we represented translated images in Lesson 2. That is, the reflected point P is P'. More importantly, note that line L and point Q have reflected images in exactly the same location as the original; hence, $Reflection(L) = L$ and $Reflection(Q) = Q$, respectively.
- The figure and its reflected image are shown together, below.

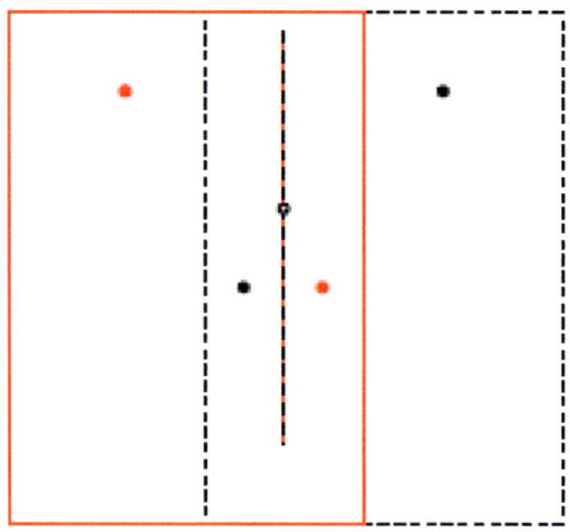

- Pictorially, reflection moves all of the points in the plane by *reflecting* them across L as if L were a mirror. The line L is called the *line of reflection.* A reflection across line L may also be noted as $Reflection_L$.

Video Presentation (2 minutes)

The following animation[1] of a reflection is helpful to beginners.

http://www.harpercollege.edu/~skoswatt/RigidMotions/reflection.html

[1]Animation developed by Sunil Koswatta.

Exercises 1–2 (3 minutes)

Students complete Exercises 1 and 2 independently.

Exercises

1. **Reflect $\triangle ABC$ and Figure D across line L. Label the reflected images.**

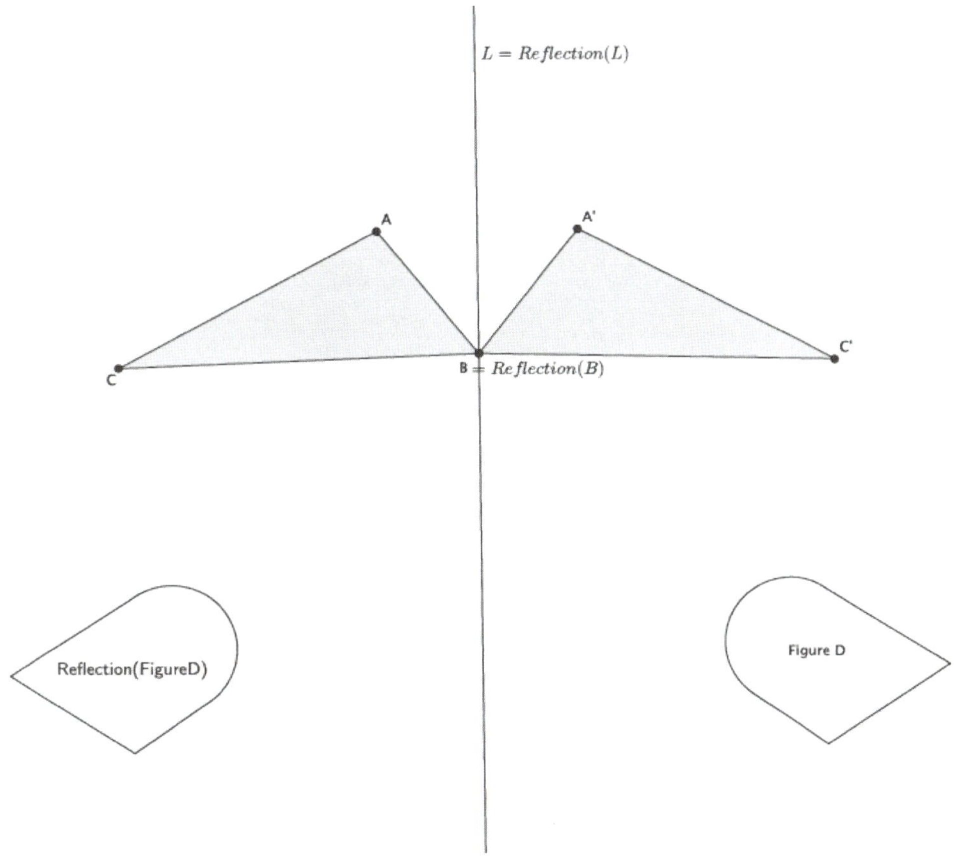

2. **Which figure(s) were not moved to a new location on the plane under this transformation?**

   Point B and line L were not moved to a new location on the plane under this reflection.

Example 2 (3 minutes)

Now the lesson looks at some features of reflected geometric figures in the plane.

- If we reflect across a vertical line l, then the reflected image of right-pointing figures, such as T below, are left-pointing. Similarly, the reflected image of a right-leaning figure, such as S below, becomes left-leaning.

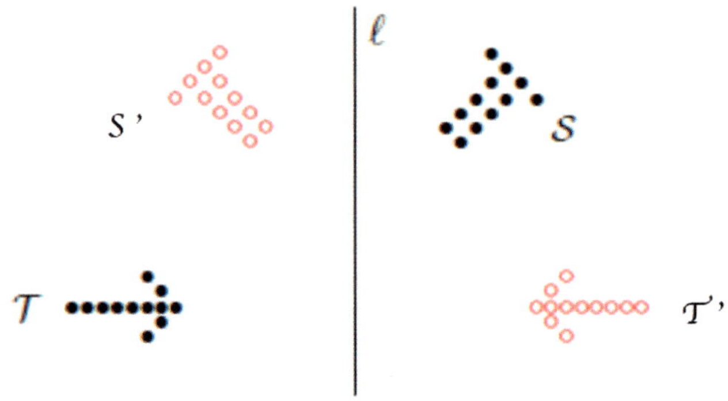

- Observe that *up* and *down* do not change in the reflection across a vertical line. Also observe that the horizontal figure T remains horizontal. This is similar to what a real mirror does.

Example 3 (2 minutes)

A line of reflection can be any line in the plane. This example looks at a horizontal line of reflection.

- Let l be the horizontal line of reflection, P be a point off of line l, and T be the figure above the line of reflection.
- Just as before, if we trace everything in red on the transparency and reflect across the horizontal line of reflection, we see the reflected images in red, as shown below.

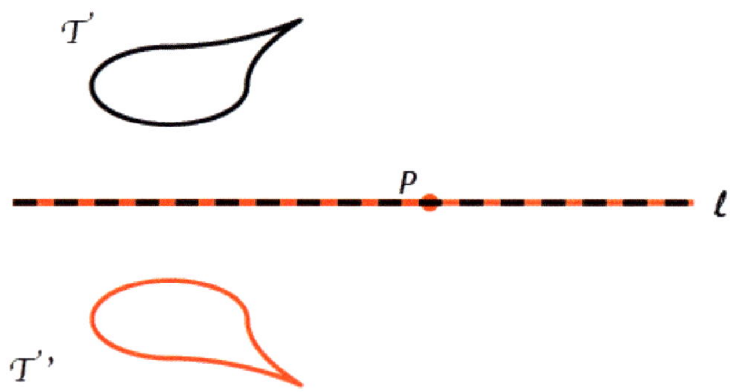

Exercises 3–5 (5 minutes)

Students complete Exercises 3–5 independently.

3. **Reflect the images across line L. Label the reflected images.**

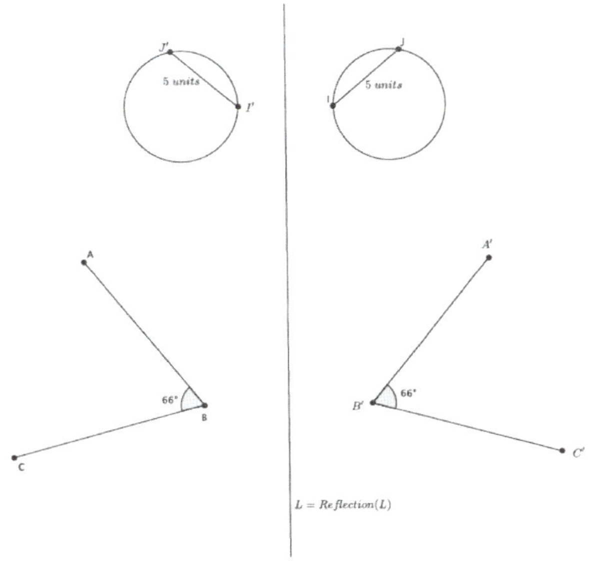

4. **Answer the questions about the image above.**

   a. **Use a protractor to measure the reflected $\angle ABC$. What do you notice?**

   The measure of the reflected image of $\angle ABC$ is $66°$.

   b. **Use a ruler to measure the length of IJ and the length of the image of IJ after the reflection. What do you notice?**

   The length of the reflected segment is the same as the original segment, 5 units.

   Note: This is not something students are expected to know, but it is a preview for what is to come later in this lesson.

5. **Reflect Figure R and $\triangle EFG$ across line L. Label the reflected images.**

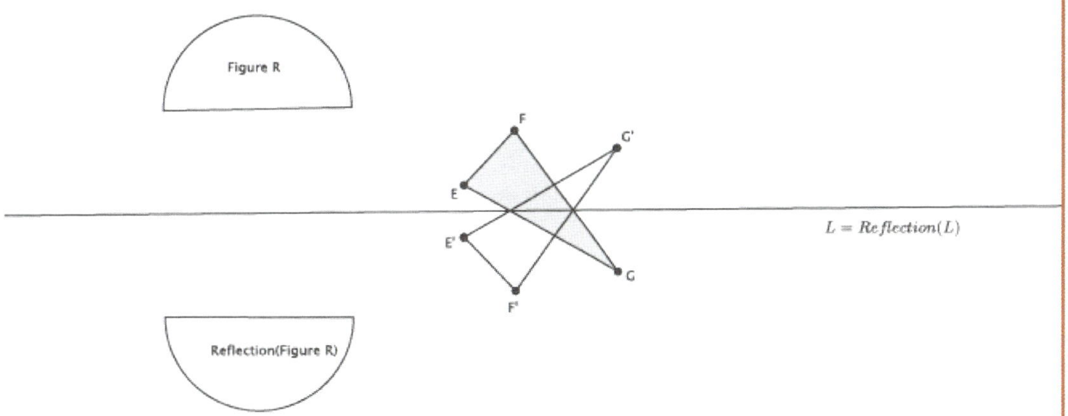

Discussion (3 minutes)

As with translation, a reflection has the same properties as (Translation 1)–(Translation 3) of Lesson 2. Precisely, lines, segments, and angles are moved by a reflection by moving their *exact* replicas (on the transparency) to another part of the plane. Therefore, distances and degrees are preserved.

(Reflection 1) A reflection maps a line to a line, a ray to a ray, a segment to a segment, and an angle to an angle.

(Reflection 2) A reflection preserves lengths of segments.

(Reflection 3) A reflection preserves degrees of angles.

These basic properties of reflections are taken for granted in all subsequent discussions of geometry.

Basic Properties of Reflections:

(Reflection 1) A reflection maps a line to a line, a ray to a ray, a segment to a segment, and an angle to an angle.

(Reflection 2) A reflection preserves lengths of segments.

(Reflection 3) A reflection preserves measures of angles.

If the reflection is across a line L and P is a point not on L, then L bisects and is perpendicular to the segment PP', joining P to its reflected image P'. That is, the lengths of OP and OP' are equal.

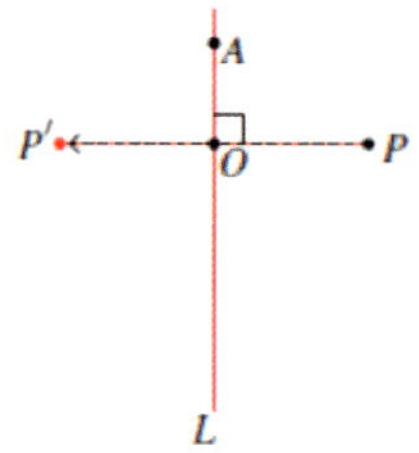

Example 4 (7 minutes)

A simple consequence of (Reflection 2) is that it gives a more precise description of the position of the reflected image of a point.

- Let there be a reflection across line L, let P be a point not on line L, and let P' represent $Reflection(P)$. Let the line PP' intersect L at O, and let A be a point on L distinct from O, as shown.

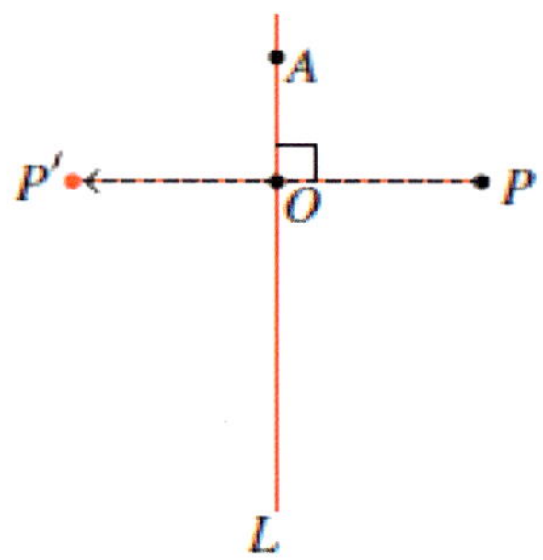

- What can we say about segments PO and OP'?
  - *Because $Reflection(PO) = P'O$, (Reflection 2) guarantees that segments PO and $P'O$ have the same length.*
- In other words, O is the *midpoint* (i.e., the point equidistant from both endpoints) of PP'.
- In general, the line passing through the midpoint of a segment is said to *bisect* the segment.
- What happens to point A under the reflection?
  - *Because the line of reflection maps to itself, point A remains unmoved, that is, $A = A'$.*
- As with translations, reflections map parallel lines to parallel lines (i.e., if $L_1 \parallel L_2$, and there is a reflection across a line, then $Reflection(L_1) \parallel Reflection(L_2)$).
- Let there be a reflection across line m. Given $L_1 \parallel L_2$, then $Reflection(L_1) \parallel Reflection(L_2)$. The reason is that any point A on line L_1 will be reflected across m to a point A' on $Reflection(L_1)$. Similarly, any point B on line L_2 will be reflected across m to a point B' on $Reflection(L_2)$. Since $L_1 \parallel L_2$, no point A on line L_1 will ever be on L_2, and no point B on L_2 will ever be on L_1. The same can be said for the reflections of those points. Then, since $Reflection(L_1)$ shares no points with $Reflection(L_2)$, $Reflection(L_1) \parallel Reflection(L_2)$.

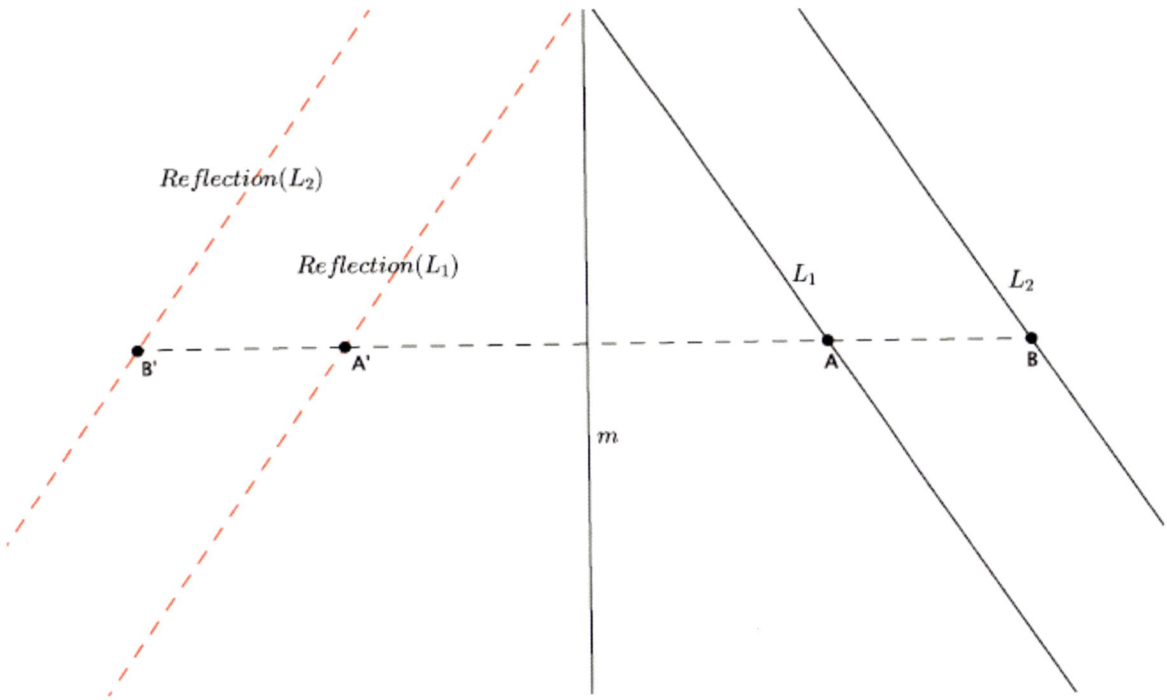

Exercises 6–9 (7 minutes)

Students complete Exercises 6–9 independently.

Use the picture below for Exercises 6–9.

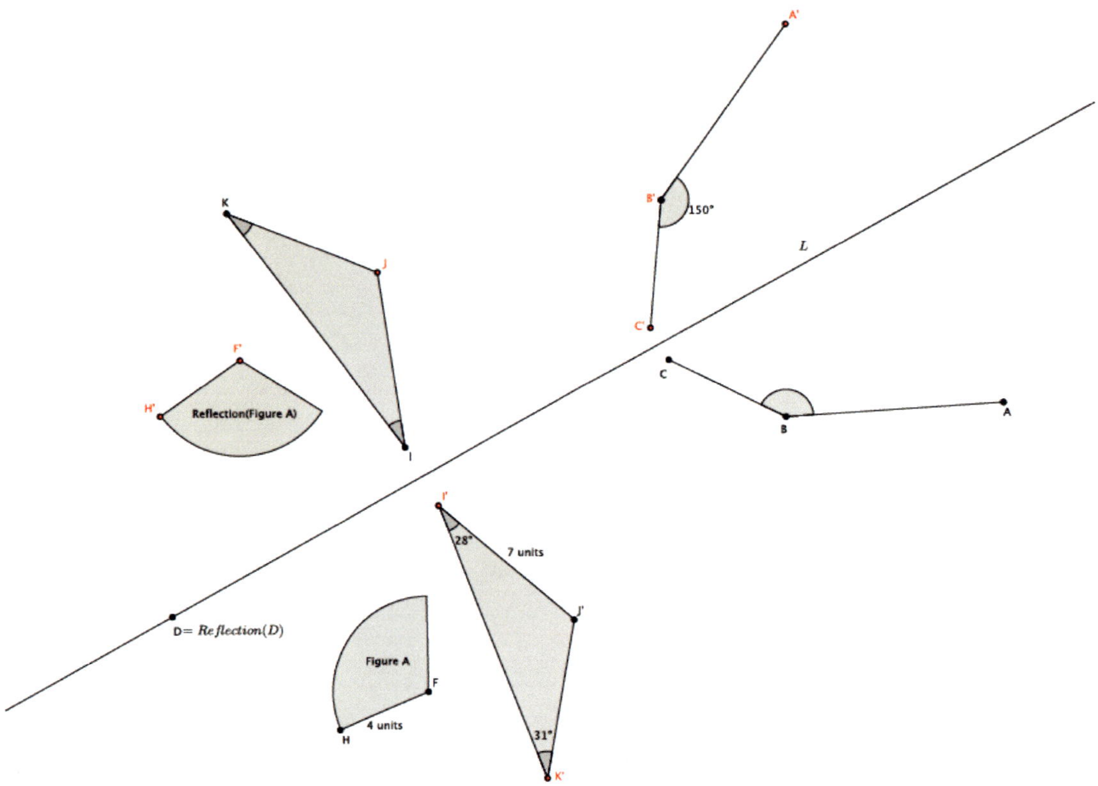

6. **Use the picture to label the unnamed points.**

   Points are labeled in red above.

7. **What is the measure of $\angle JKI$? $\angle KIJ$? $\angle ABC$? How do you know?**

   $m\angle JKI = 31°$, $m\angle KIJ = 28°$, *and* $m\angle ABC = 150°$. ***Reflections preserve angle measures.***

8. **What is the length of segment $Reflection(FH)$? IJ? How do you know?**

   $|Reflection(FH)| = 4$ *units, and* $IJ = 7$ *units.* ***Reflections preserve lengths of segments.***

9. **What is the location of $Reflection(D)$? Explain.**

   Point D and its image are in the same location on the plane. Point D was not moved to another part of the plane because it is on the line of reflection. The image of any point on the line of reflection will remain in the same location as the original point.

Closing (4 minutes)

Summarize, or have students summarize, the lesson.

- We know that a reflection across a line is a basic rigid motion.
- Reflections have the same basic properties as translations; reflections map lines to lines, rays to rays, segments to segments and angles to angles.
- Reflections have the same basic properties as translations because they, too, are distance- and angle-preserving.
- The line of reflection L is the bisector of the segment that joins a point not on L to its image.

Lesson Summary

- **A reflection is another type of basic rigid motion.**
- **A reflection across a line maps one half-plane to the other half-plane; that is, it maps points from one side of the line to the other side of the line. The reflection maps each point on the line to itself. The line being reflected across is called the line of reflection.**
- **When a point, P is joined with its reflection, P' to form segment PP', the line of reflection bisects and is perpendicular to the segment PP'.**

Terminology

REFLECTION (description): Given a line L in the plane, a *reflection across* L is the transformation of the plane that maps each point on the line L to itself, and maps each remaining point P of the plane to its image P' such that L is the perpendicular bisector of the segment PP'.

Exit Ticket (4 minutes)

Name ______________________________ Date ______________

Lesson 4: Definition of Reflection and Basic Properties

Exit Ticket

1. Let there be a reflection across line L_{AB}. Reflect $\triangle CDE$ across line L_{AB}. Label the reflected image.

Picture not drawn to scale.

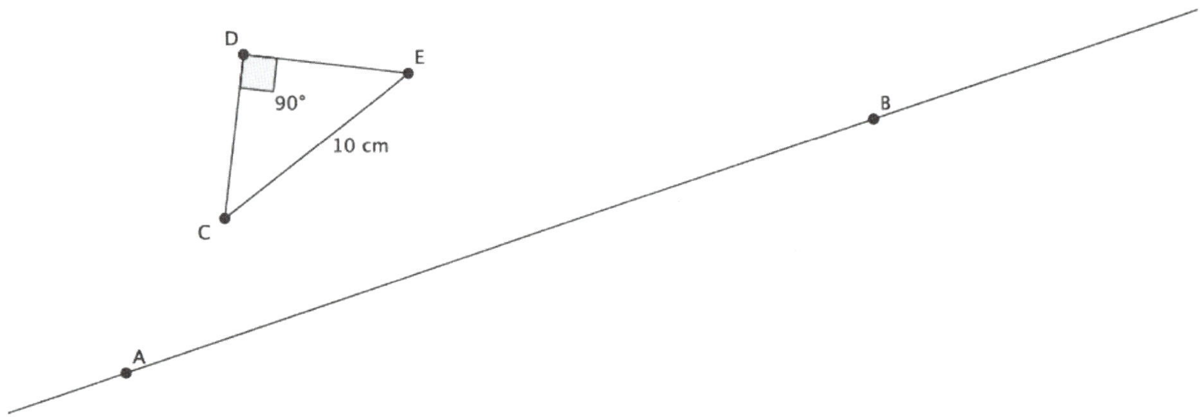

2. Use the diagram above to state the measure of $Reflection(\angle CDE)$. Explain.

3. Use the diagram above to state the length of segment $Reflection(CE)$. Explain.

4. Connect point C to its image in the diagram above. What is the relationship between line L_{AB} and the segment that connects point C to its image?

Exit Ticket Sample Solutions

1. Let there be a reflection across line L_{AB}. Reflect $\triangle CDE$ across line L_{AB}. Label the reflected image.

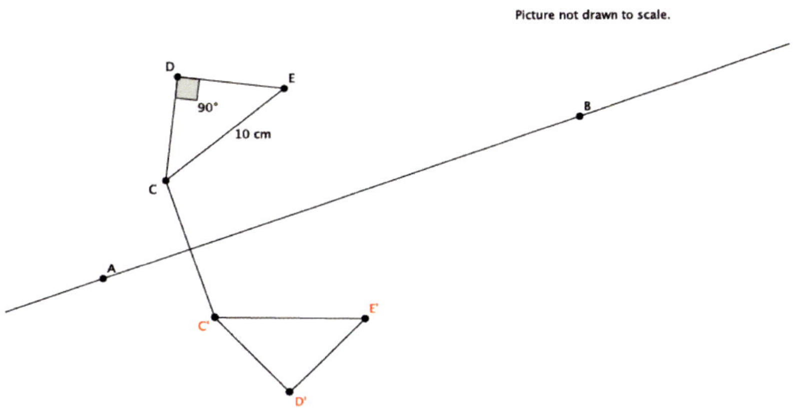

2. Use the diagram above to state the measure of $Reflection(\angle CDE)$. Explain.

   The measure of $Reflection(\angle CDE)$ is $90°$ because reflections preserve degrees of measures of angles.

3. Use the diagram above to state the length of segment $Reflection(CE)$. Explain.

   The length of $Reflection(CE)$ is 10 cm because reflections preserve segment lengths.

4. Connect point C to its image in the diagram above. What is the relationship between line L_{AB} and the segment that connects point C to its image?

   The line of reflection bisects the segment that connects C to its image.

Problem Set Sample Solutions

1. In the picture below, $\angle DEF = 56°$, $\angle ACB = 114°$, $AB = 12.6$ units, $JK = 5.32$ units, point E is on line L, and point I is off of line L. Let there be a reflection across line L. Reflect and label each of the figures, and answer the questions that follow.

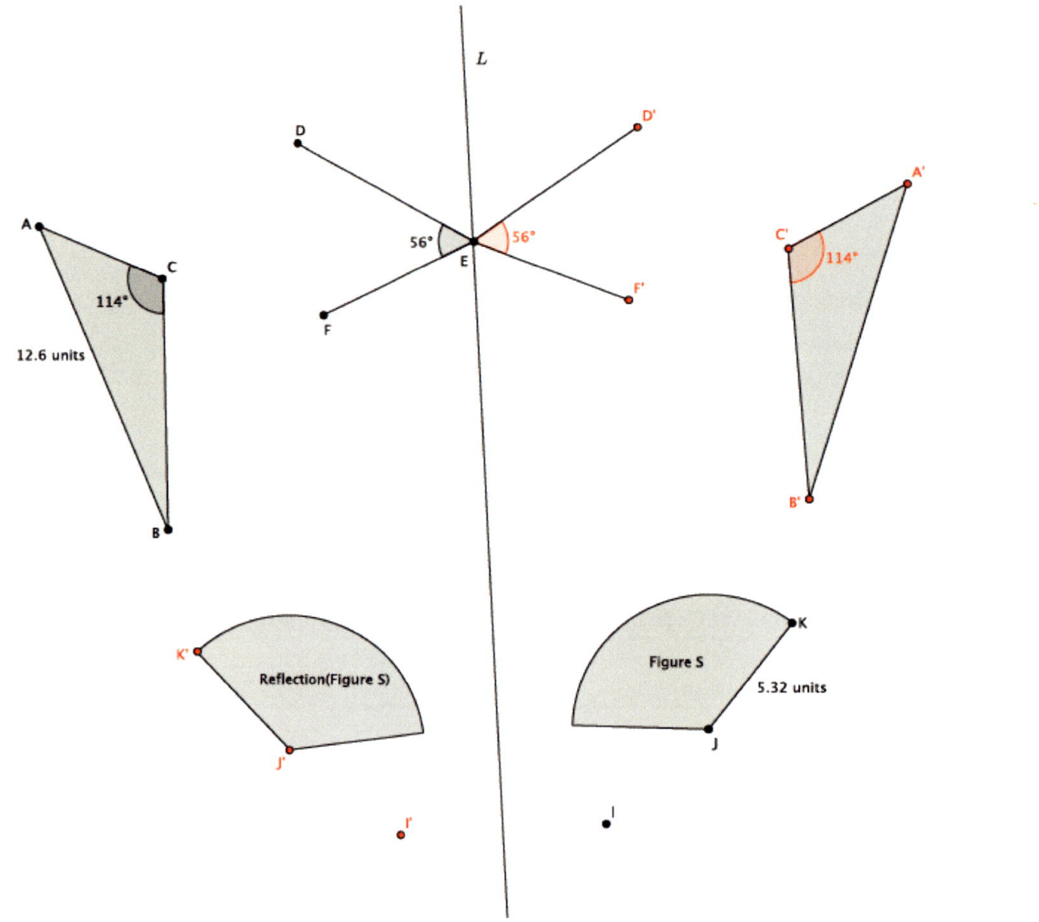

2. What is the measure of $Reflection(\angle DEF)$? Explain.

   The measure of $Reflection(\angle DEF)$ is $56°$. Reflections preserve degrees of angles.

3. What is the length of $Reflection(JK)$? Explain.

   The length of $Reflection(JK)$ is 5.32 units. Reflections preserve lengths of segments.

4. What is the measure of $Reflection(\angle ACB)$?

   The measure of $Reflection(\angle ACB)$ is $114°$.

5. What is the length of $Reflection(AB)$?

   The length of $Reflection(AB)$ is 12.6 units.

EUREKA MATH™

6. Two figures in the picture were not moved under the reflection. Name the two figures, and explain why they were not moved.

   Point E and line L were not moved. All of the points that make up the line of reflection remain in the same location when reflected. Since point E is on the line of reflection, it is not moved.

7. Connect points I and I'. Name the point of intersection of the segment with the line of reflection point Q. What do you know about the lengths of segments IQ and QI'?

   Segments IQ and QI' are equal in length. The segment II' connects point I to its image, I'. The line of reflection will go through the midpoint, or bisect, the segment created when you connect a point to its image.

Lesson 5: Definition of Rotation and Basic Properties

Student Outcomes

- Students know how to rotate a figure a given degree around a given center.
- Students know that rotations move lines to lines, rays to rays, segments to segments, and angles to angles. Students know that rotations preserve lengths of segments and degrees of measures of angles. Students know that rotations move parallel lines to parallel lines.

Lesson Notes

In general, students are not required to rotate a certain degree nor identify the degree of rotation. The only exceptions are when the rotations are multiples of $90°$. For this reason, it is recommended in the discussion following the video presentation that the teacher shows students how to use the transparency to rotate in multiples of $90°$, that is, turn the transparency one-quarter turn for each $90°$ rotation.

Classwork

Discussion (8 minutes)

- What is the simplest transformation that would map one of the following figures to the other?

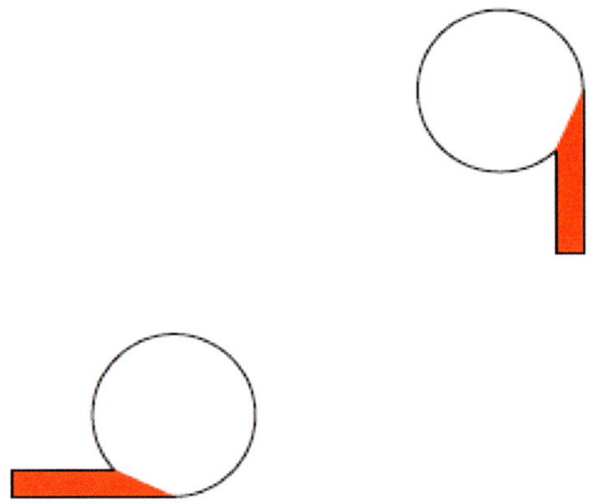

- Would a translation work? Would a reflection work?
  - *Because there seems to be no known simple transformation that would do the job, we will learn about a new transformation called rotation. Rotation is the transformation needed to map one of the figures onto the other.*

Let O be a point in the plane, and let d be a number between -360 and 360, or in the usual notation, $-360 < d < 360$.

- Why do you think the numbers -360 and 360 are used in reference to rotation?
  - *Rotating means that we are moving in a circular pattern, and circles have* $360°$.

The rotation of d degrees with center O is defined by using transparencies. On a piece of paper, fix a point O as the center of rotation, let P be a point in the plane, and let the ray $\overrightarrow{OP}$ be drawn. Let d be a number between -360 and 360.

Instructions for performing a rotation. If there is a rotation of d degrees with center O, the image $Rotation(P)$ is the point described as follows. On a piece of transparency, trace O, P, and $\overrightarrow{OP}$ in red. Now, use a pointed object (e.g., the leg-with-spike of a compass) to pin the transparency at the point O. First, suppose $d \geq 0$. Then, holding the paper in place, rotate the transparency counterclockwise so that if we denote the final position of the rotated red point (that was P) by P', then the $\angle P'OP$ is d degrees. For example, if $d = 30$, we have the following picture:

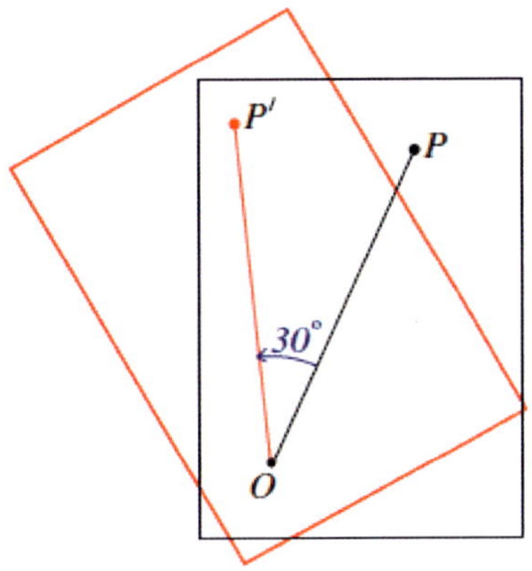

As before, the red rectangle represents the border of the rotated transparency. Then, by definition, $Rotation(P)$ is the point P'.

MP.6

If, however, $d < 0$, then holding the paper in place, we would now rotate the transparency clockwise so that if we denote the position of the red point (that was P) by P', then the angle $\angle POP'$ is d degrees. For example, if $d = -30$, we have the following picture:

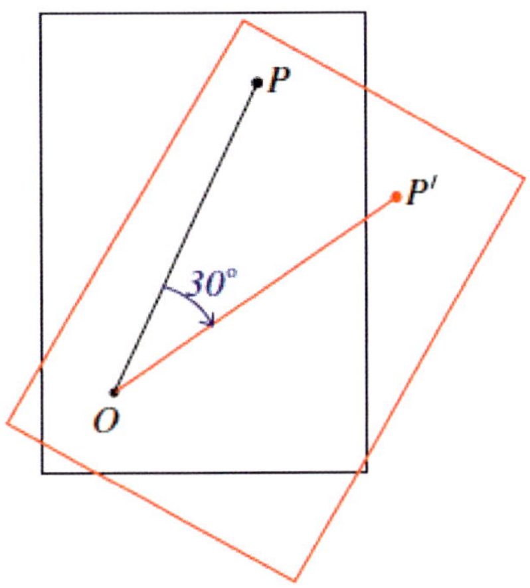

Again, we define $Rotation(P)$ to be P' in this case. Notice that the rotation moves the center of rotation O to itself, that is, $Rotation(O) = O$.

Exercises 1–4 (4 minutes)

Students complete Exercises 1–4 independently.

Exercises

1. **Let there be a rotation of d degrees around center O. Let P be a point other than O. Select d so that $d \geq 0$. Find P' (i.e., the rotation of point P) using a transparency.**

   Verify that students have rotated around center O in the counterclockwise direction.

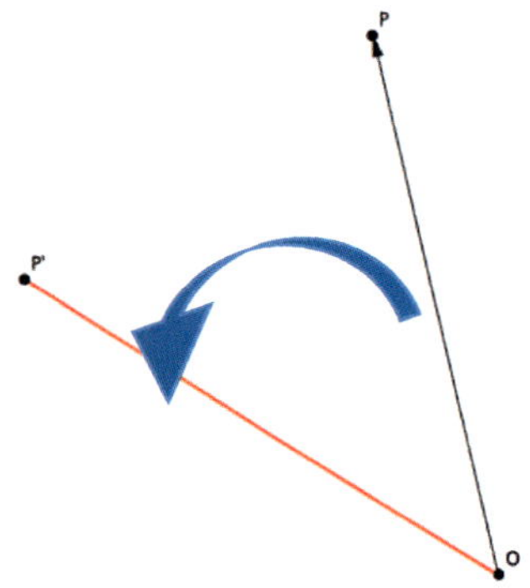

2. **Let there be a rotation of d degrees around center O. Let P be a point other than O. Select d so that $d < 0$. Find P' (i.e., the rotation of point P) using a transparency.**

   Verify that students have rotated around center O in the clockwise direction.

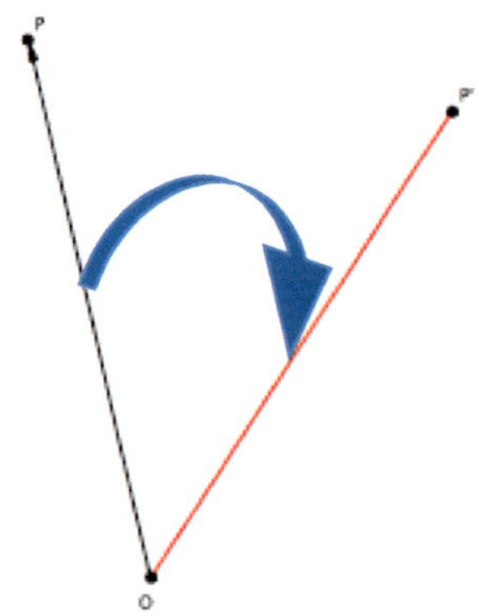

3. **Which direction did the point P rotate when $d \geq 0$?**

   It rotated counterclockwise, or to the left of the original point.

4. **Which direction did the point P rotate when $d < 0$?**

   It rotated clockwise, or to the right of the original point.

Discussion (5 minutes)

Observe that, with O as the center of rotation, the points P and $Rotation(P)$ lie on a circle whose center is O and whose radius is OP.

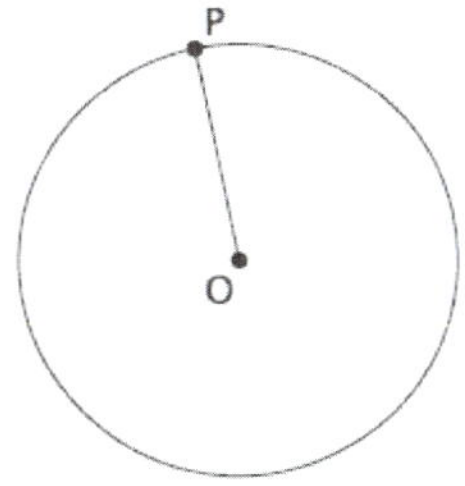

- Assume we rotate the plane d degrees around center O. Let P be a point other than O. Where do you think P' will be located?
  - *The points P and P' will be equidistant from O; that is, P' is on the circumference of the circle with center O and radius OP. The point P' would be clockwise from P if the degree of rotation is negative. The point P' would be counterclockwise from P if the degree of rotation is positive.*
- If we rotated P d degrees around center O several times, where would all of the images of P be located?
  - *All images of P will be on the circumference of the circle with radius OP.*

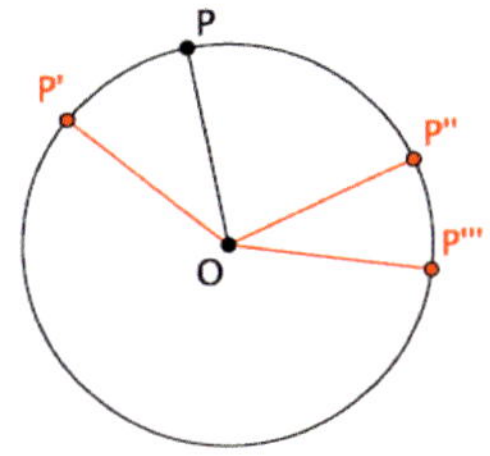

- Why do you think this happens?
  - *Because, like translations and reflections, rotations preserve lengths of segments. The segments of importance here are the segments that join the center O to the images of P. Each segment is the radius of the circle. We discuss this more later in the lesson.*
- Consider a rotation of point P, around center O, 180 degrees and -180 degrees. Where do you think the images of P will be located?
  - *Both rotations, although they are in opposite directions, will move the point P to the same location, P'. Further, the points P, O, and P' will always be collinear (i.e., they will lie on one line, for any point P). This concept is discussed in more detail in Lesson 6.*

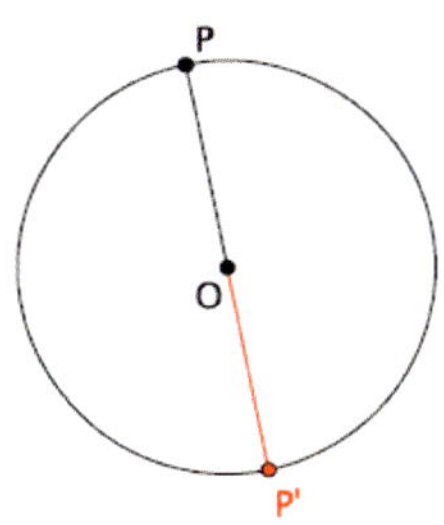

Concept Development (3 minutes)

- Now that we know how a point gets moved under a rotation, let us look at how a geometric figure gets moved under a rotation. Let S be the figure consisting of a vertical segment (not a line) and two points. Let the center of rotation be O, the lower endpoint of the segment, as shown.

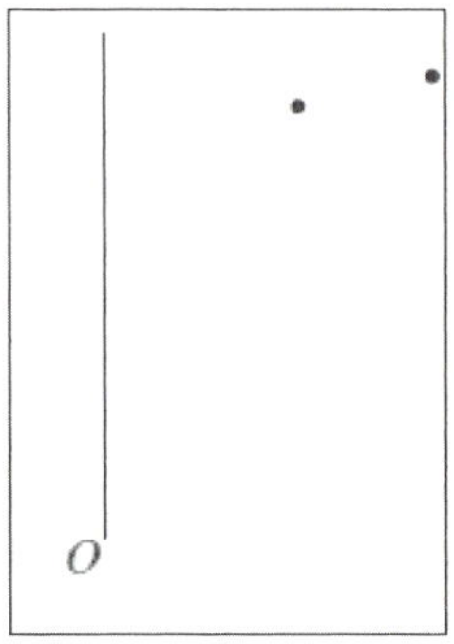

- Then, the rotation of 30 degrees with center O moves the point represented by the left black dot to the lower red dot, the point represented by the right black dot to the upper red dot, and the vertical black segment to the red segment to the left at an angle of 30 degrees, as shown.

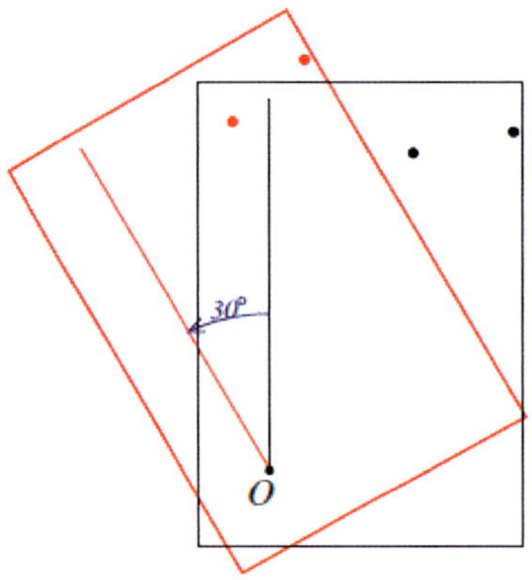

Video Presentation (2 minutes)

The following two videos[1] show how a rotation of 35 degrees and -35 degrees with center B, respectively, rotates a geometric figure consisting of three points and two line segments.

http://www.harpercollege.edu/~skoswatt/RigidMotions/rotateccw.html

http://www.harpercollege.edu/~skoswatt/RigidMotions/rotatecw.html

[1]The videos were developed by Sunil Koswatta.

Discussion (2 minutes)

Revisit the question posed at the beginning of the lesson and ask students:

- What is the simplest transformation that would map one of the following figures to the other?

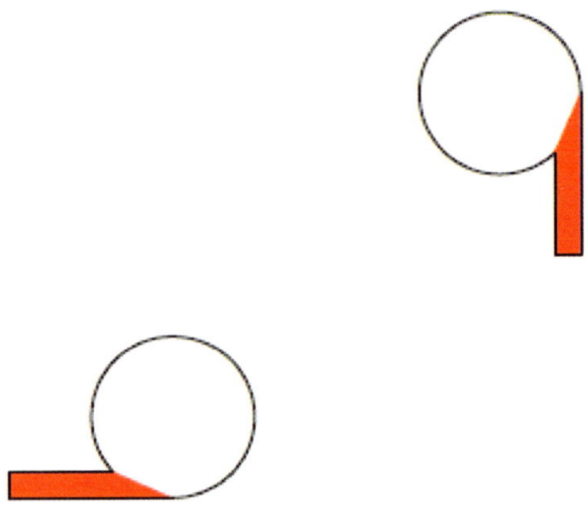

  - *We now know that the answer is a rotation.*

Show students how a rotation of approximately 90 degrees around a point O, chosen on the perpendicular bisector ($\perp$ bisector) of the segment joining the centers of the two circles in the figures, would map the figure on the left to the figure on the right. Similarly, a rotation of -90 degrees would map the figure on the right to the figure on the left.

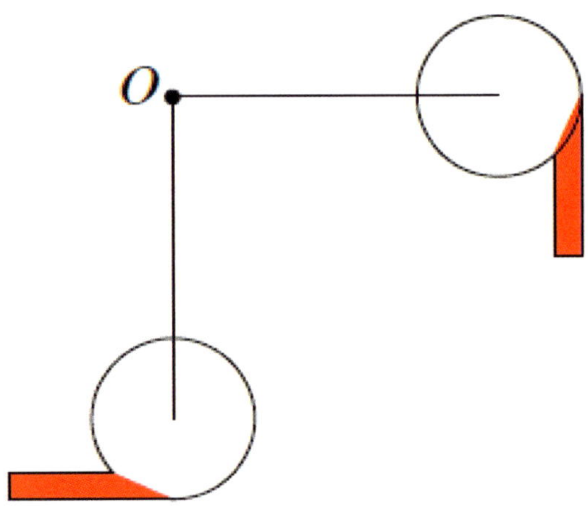

> *Note to Teacher:*
> Continue to remind students that a positive degree of rotation moves the figure counterclockwise, and a negative degree of rotation moves the figure clockwise.

Exercises 5–6 (4 minutes)

Students complete Exercises 5 and 6 independently.

5. Let L be a line, $\overrightarrow{AB}$ be a ray, $\overline{CD}$ be a segment, and $\angle EFG$ be an angle, as shown. Let there be a rotation of d degrees around point O. Find the images of all figures when $d \geq 0$.

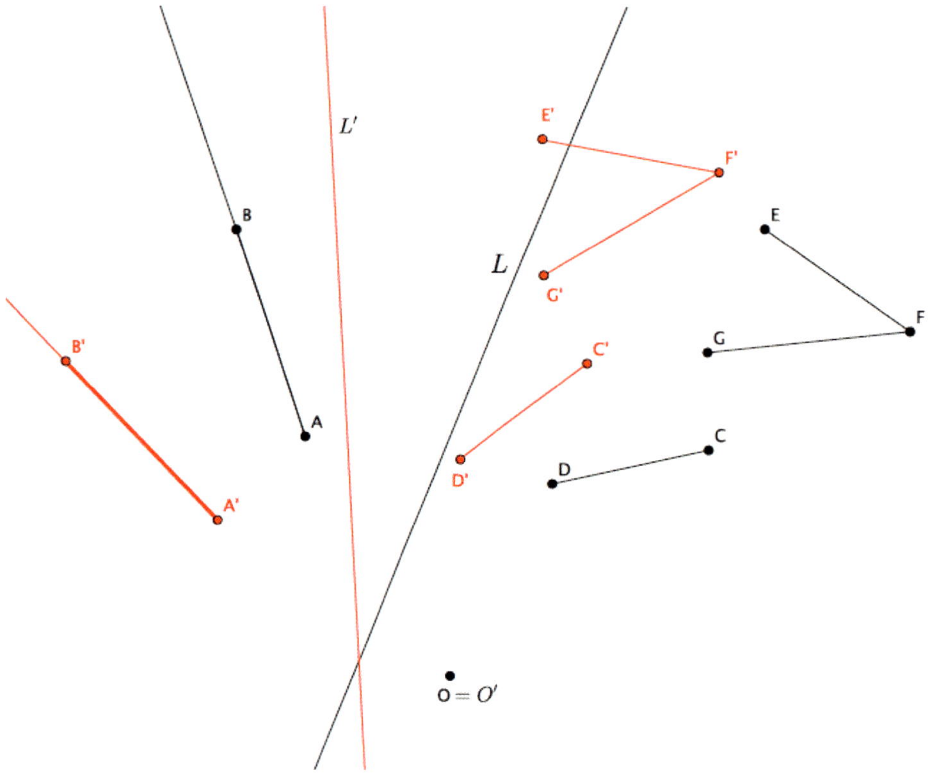

Verify that students have rotated around center O in the counterclockwise direction.

6. Let $\overline{AB}$ be a segment of length 4 units and $\angle CDE$ be an angle of size $45°$. Let there be a rotation by d degrees, where $d < 0$, about O. Find the images of the given figures. Answer the questions that follow.

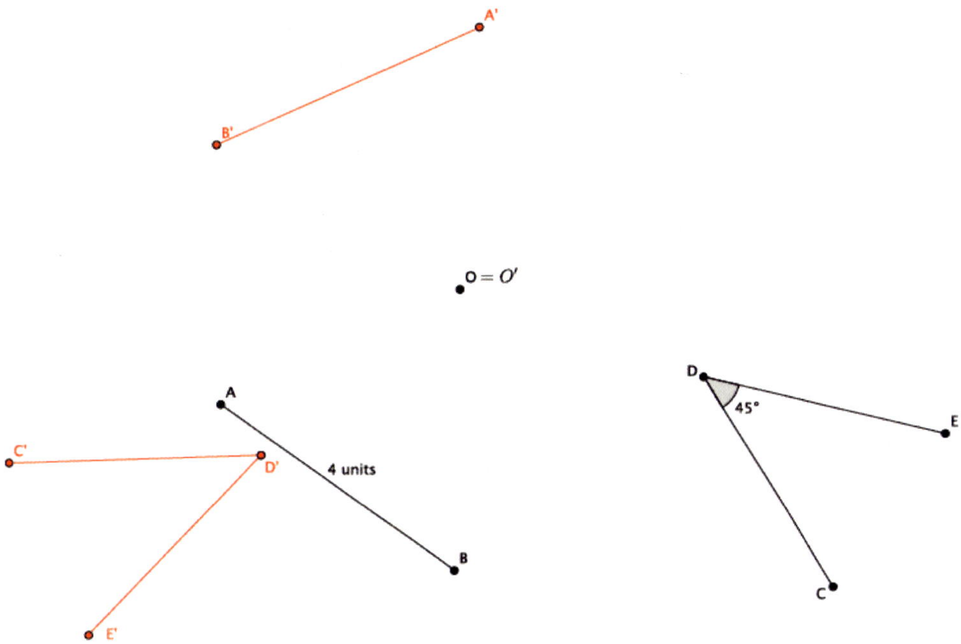

Verify that students have rotated around center O in the clockwise direction.

a. What is the length of the rotated segment $Rotation(AB)$?

The length of the rotated segment is 4 units.

b. What is the degree of the rotated angle $Rotation\ (\angle CDE)$?

The degree of the rotated angle is $45°$.

Concept Development (4 minutes)

Based on the work completed during the lesson, and especially in Exercises 5 and 6, we can now state that rotations have properties similar to translations with respect to (Translation 1)–(Translation 3) of Lesson 2 and reflections with respect to (Reflection 1)–(Reflection 3) of Lesson 4:

(Rotation 1) A rotation maps a line to a line, a ray to a ray, a segment to a segment, and an angle to an angle.

(Rotation 2) A rotation preserves lengths of segments.

(Rotation 3) A rotation preserves measures of angles.

Also, as with translations and reflections, if L_1 and L_2 are parallel lines and if there is a rotation, then the lines $Rotation(L_1)$ and $Rotation(L_2)$ are also parallel. However, if there is a rotation of degree $d \neq 180$, and L is a line, L and $Rotation(L)$ are not parallel. (Note to teacher: Exercises 7 and 8 illustrate these two points.)

Exercises 7–8 (5 minutes)

Students complete Exercises 7 and 8 independently.

7. Let L_1 and L_2 be parallel lines. Let there be a rotation by d degrees, where $-360 < d < 360$, about O. Is $(L_1)' \parallel (L_2)'$?

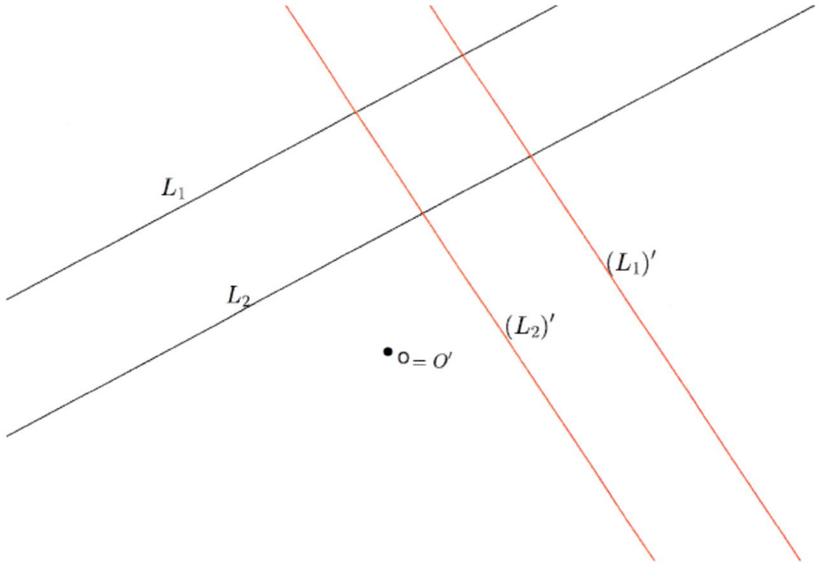

Verify that students have rotated around center O in either direction. Students should respond that $(L_1)' \parallel (L_2)'$.

8. Let L be a line and O be the center of rotation. Let there be a rotation by d degrees, where $d \neq 180$ about O. Are the lines L and L' parallel?

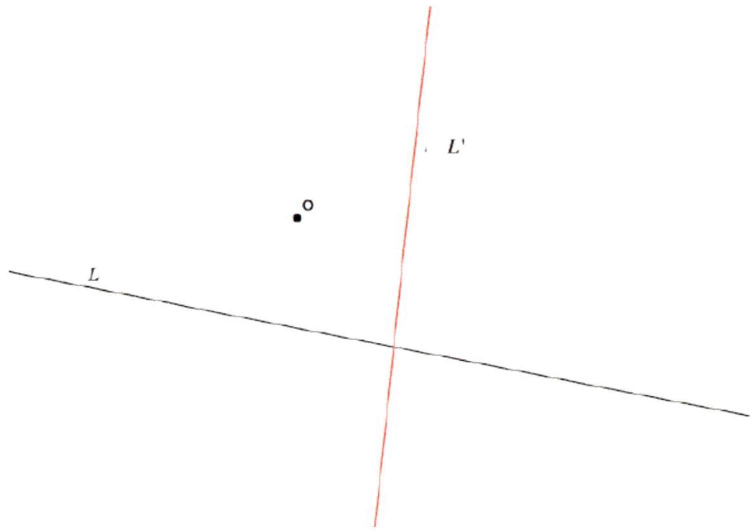

Verify that students have rotated around center O in either direction any degree other than 180. Students should respond that L and L' are not parallel.

Closing (3 minutes)

Summarize, or have students summarize, what we know of rigid motions to this point:

- We now have definitions for all three rigid motions: translations, reflections, and rotations.
- Rotations move lines to lines, rays to rays, segments to segments, angles to angles, and parallel lines to parallel lines, similar to translations and reflections.
- Rotations preserve lengths of segments and degrees of measures of angles similar to translations and reflections.
- Rotations require information about the center and degree of rotation, whereas translations require only a vector, and reflections require only a line of reflection.

Lesson Summary

Rotations require information about the center of rotation and the degree in which to rotate. Positive degrees of rotation move the figure in a counterclockwise direction. Negative degrees of rotation move the figure in a clockwise direction.

Basic Properties of Rotations:

- **(Rotation 1) A rotation maps a line to a line, a ray to a ray, a segment to a segment, and an angle to an angle.**
- **(Rotation 2) A rotation preserves lengths of segments.**
- **(Rotation 3) A rotation preserves measures of angles.**

When parallel lines are rotated, their images are also parallel. A line is only parallel to itself when rotated exactly $180°$.

Terminology

ROTATION (description): For a number d between 0 and 180, the *rotation of d degrees around center O* is the transformation of the plane that maps the point O to itself, and maps each remaining point P of the plane to its image P' in the counterclockwise half-plane of ray $\overrightarrow{OP}$ so that P and P' are the same distance away from O and the measurement of $\angle P'OP$ is d degrees.
The *counterclockwise half-plane* is the half-plane that lies to the left of $\overrightarrow{OP}$ while moving along $\overrightarrow{OP}$ in the direction from O to P.

Exit Ticket (5 minutes)

Name ______________________________ Date ______________

Lesson 5: Definition of Rotation and Basic Properties

Exit Ticket

1. Given the figure H, let there be a rotation by d degrees, where $d \geq 0$, about O. Let $Rotation(H)$ be H'.

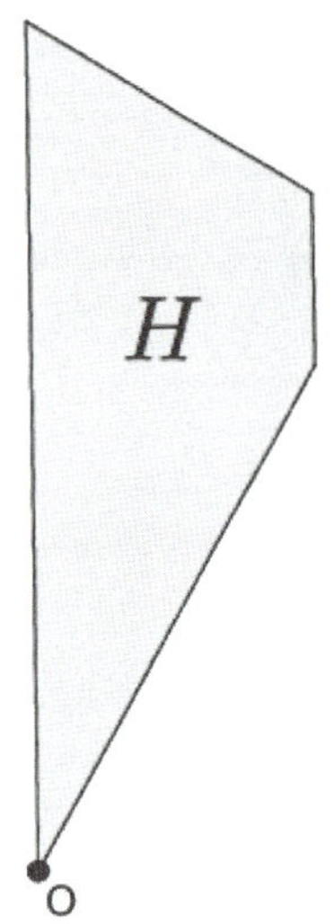

2. Using the drawing above, let $Rotation_1$ be the rotation d degrees with $d < 0$, about O. Let $Rotation_1(H)$ be H''.

Exit Ticket Sample Solutions

1. Given the figure H, let there be a rotation by d degrees, where $d \geq 0$, about O. Let $Rotation(H)$ be H'.

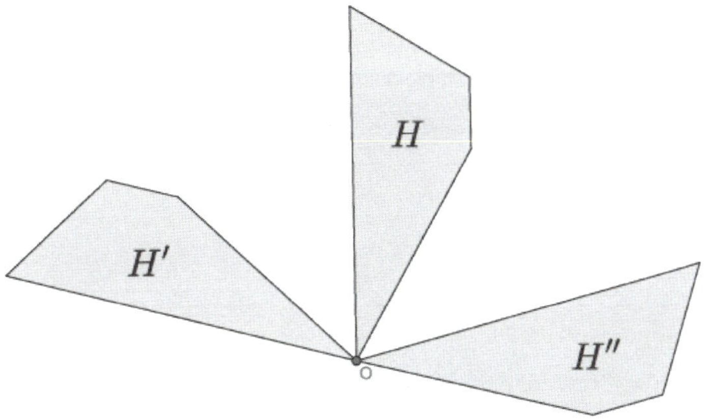

Sample rotation shown above. Verify that the figure H' has been rotated counterclockwise with center O.

2. Using the drawing above, let $Rotation_1$ be the rotation d degrees with $d < 0$, about O. Let $Rotation_1(H)$ be H''.

Sample rotation shown above. Verify that the figure H'' has been rotated clockwise with center O.

Problem Set Sample Solutions

1. Let there be a rotation by $-90°$ around the center O.

Rotated figures are shown in red.

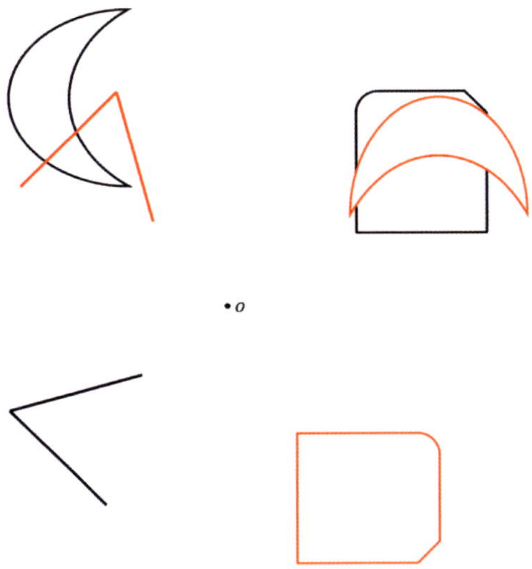

2. Explain why a rotation of 90 degrees around any point O never maps a line to a line parallel to itself.

A 90-degree rotation around point O will move a given line L to L'. Parallel lines never intersect, so it is obvious that a 90-degree rotation in either direction does not make lines L and L' parallel. Additionally, we know that there exists just one line parallel to the given line L that goes through a point not on L. If we let P be a point not on L, the line L' must go through it in order to be parallel to L. L' does not go through point P; therefore, L and L' are not parallel lines. Assume we rotate line L first and then place a point P on line L' to get the desired effect (a line through P). This contradicts our definition of parallel (i.e., parallel lines never intersect); so, again, we know that line L is not parallel to L'.

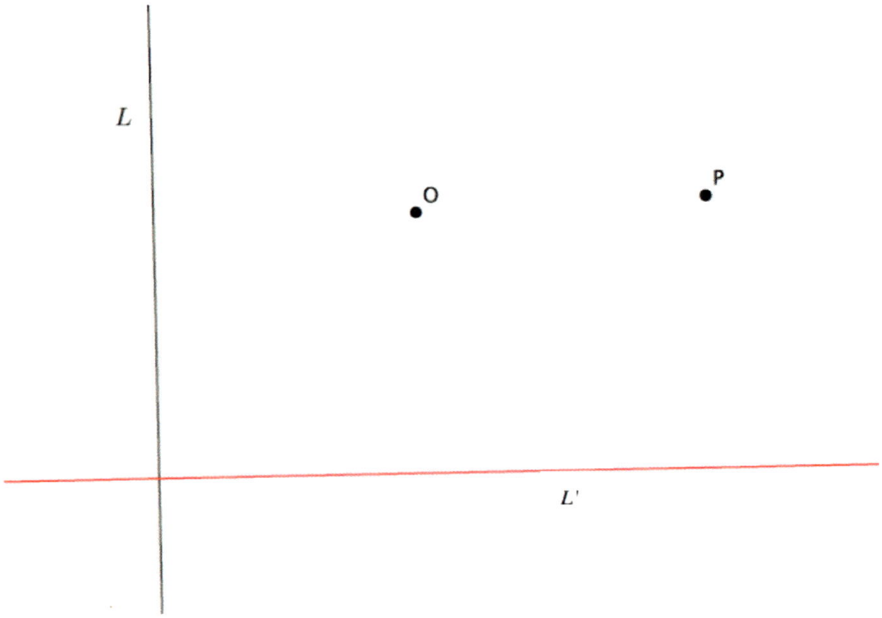

3. A segment of length 94 cm has been rotated d degrees around a center O. What is the length of the rotated segment? How do you know?

The rotated segment will be 94 cm in length. (Rotation 2) states that rotations preserve lengths of segments, so the length of the rotated segment will remain the same as the original.

4. An angle of size $124°$ has been rotated d degrees around a center O. What is the size of the rotated angle? How do you know?

The rotated angle will be $124°$. (Rotation 3) states that rotations preserve the degrees of angles, so the rotated angle will be the same size as the original.

Lesson 6: Rotations of 180 Degrees

Student Outcomes

- Students learn that a rotation of 180 degrees moves a point on the coordinate plane (a, b) to $(-a, -b)$.
- Students learn that a rotation of 180 degrees around a point, not on the line, produces a line parallel to the given line.

Classwork

Example 1 (5 minutes)

- Rotations of 180 degrees are special. Recall, a rotation of 180 degrees around O is a rigid motion so that if P is any point in the plane, P, O, and $Rotation(P)$ are collinear (i.e., they lie on the same line).
- Rotations of 180 degrees occur in many situations. For example, the frequently cited fact that vertical angles (vert. $\angle$'s) at the intersection of two lines are equal, follows immediately from the fact that 180-degree rotations are angle-preserving. More precisely, let two lines L_1 and L_2 intersect at O, as shown:

Example 1

The picture below shows what happens when there is a rotation of $180°$ around center O.

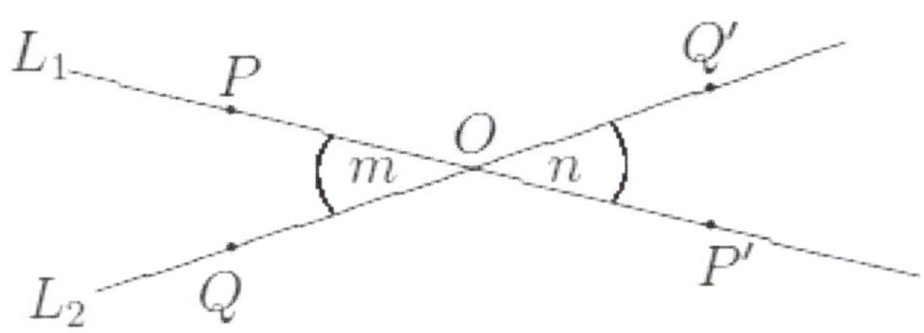

- We want to show that the vertical angles (vert. $\angle$'s), $\angle m$ and $\angle n$, are equal in measure (i.e., $\angle m = \angle n$). If we let $Rotation_0$ be the 180-degree rotation around O, then $Rotation_0$ maps $\angle m$ to $\angle n$. More precisely, if P and Q are points on L_1 and L_2, respectively (as shown above), let $Rotation_0(P) = P'$ and $Rotation_0(Q) = Q'$. Then, $Rotation_0$ maps $\angle POQ$ ($\angle m$) to $\angle Q'OP'$ ($\angle n$), and since $Rotation_0$ is angle-preserving, we have $\angle m = \angle n$.

Example 2 (5 minutes)

- Let's look at a 180-degree rotation, $Rotation_0$ around the origin O of a coordinate system. If a point P has coordinates (a, b), it is generally said that $Rotation_0(P)$ is the point with coordinates $(-a, -b)$.
- Suppose the point P has coordinates $(-4, 3)$; we show that the coordinates of $Rotation_0(P)$ are $(4, -3)$.

Example 2

The picture below shows what happens when there is a rotation of $180°$ around center O, the origin of the coordinate plane.

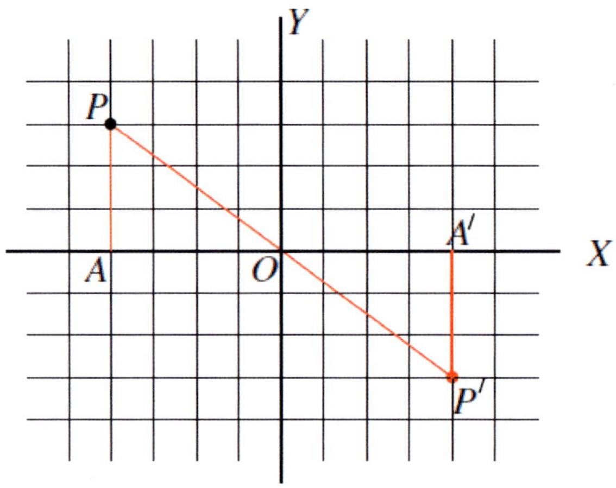

- Let $P' = Rotation_0(P)$. Let the *vertical line* (i.e., the line parallel to the y-axis) through P meet the x-axis at a point A. Because the coordinates of P are $(-4, 3)$, the point A has coordinates $(-4, 0)$ by the way coordinates are defined. In particular, A is of distance 4 from O, and since $Rotation_0$ is length-preserving, the point $A' = Rotation_0(A)$ is also of distance 4 from O. However, $Rotation_0$ is a 180-degree rotation around O, so A' also lies on the x-axis but on the opposite side of the x-axis from A. Therefore, the coordinates of A' are $(4,0)$. Now, $\angle PAO$ is a right angle and—since $Rotation_0$ maps it to $\angle P'A'O$ and also preserves degrees—we see that $\angle P'A'O$ is also a right angle. This means that A' is the point of intersection of the vertical line through P' and the x-axis. Since we already know that A' has coordinates of $(4, 0)$, then the x-coordinate of P' is 4, by definition.
- Similarly, the y-coordinate of P being 3 implies that the y-coordinate of P' is -3. Altogether, we have proved that the 180-degree rotation of a point of coordinates $(-4, 3)$ is a point with coordinates $(4, -3)$.

The reasoning is perfectly general: The same logic shows that the 180-degree rotation around the origin of a point of coordinates (a, b) is the point with coordinates $(-a, -b)$, as desired.

Exercises 1–9 (20 minutes)

Students complete Exercises 1–2 independently. Check solutions. Then, let students work in pairs on Exercises 3–4. Students complete Exercises 5–9 independently in preparation for the example that follows.

Exercises 1–9

1. Using your transparency, rotate the plane 180 degrees, about the origin. Let this rotation be $Rotation_0$. What are the coordinates of $Rotation_0(2, -4)$?

   $Rotation_0(2, -4) = (-2, 4)$

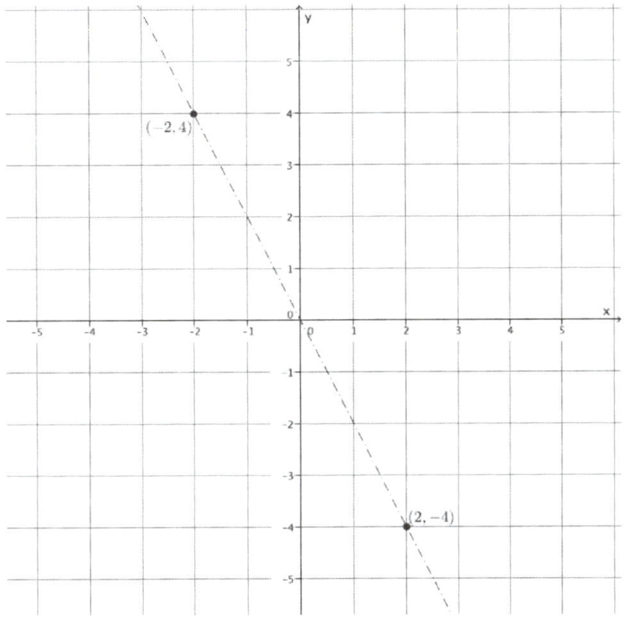

2. Let $Rotation_0$ be the rotation of the plane by 180 degrees, about the origin. <u>Without</u> using your transparency, find $Rotation_0(-3, 5)$.

   $Rotation_0(-3, 5) = (3, -5)$

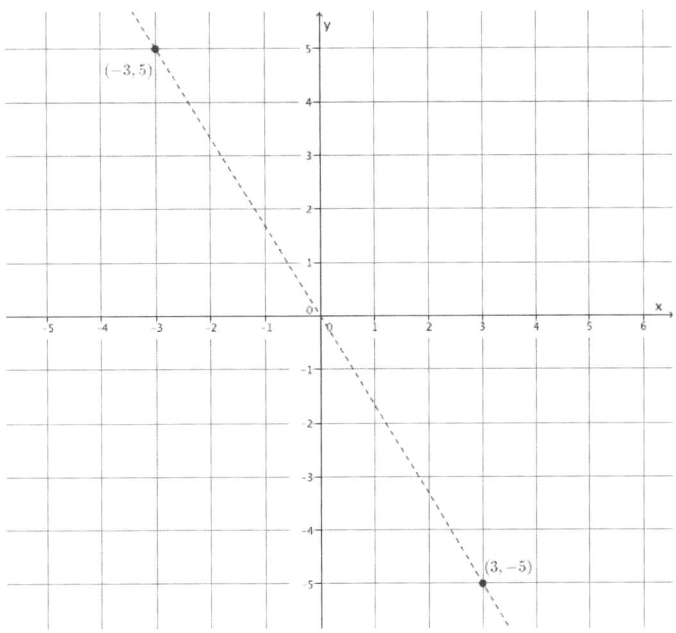

3. Let $Rotation_0$ be the rotation of 180 degrees around the origin. Let L be the line passing through $(-6, 6)$ parallel to the x-axis. Find $Rotation_0(L)$. Use your transparency if needed.

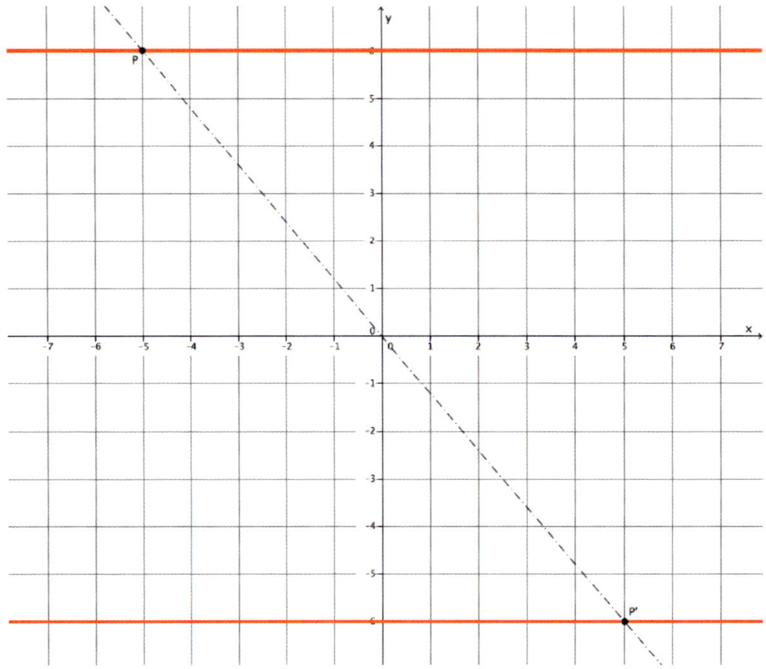

4. Let $Rotation_0$ be the rotation of 180 degrees around the origin. Let L be the line passing through $(7, 0)$ parallel to the y-axis. Find $Rotation_0(L)$. Use your transparency if needed.

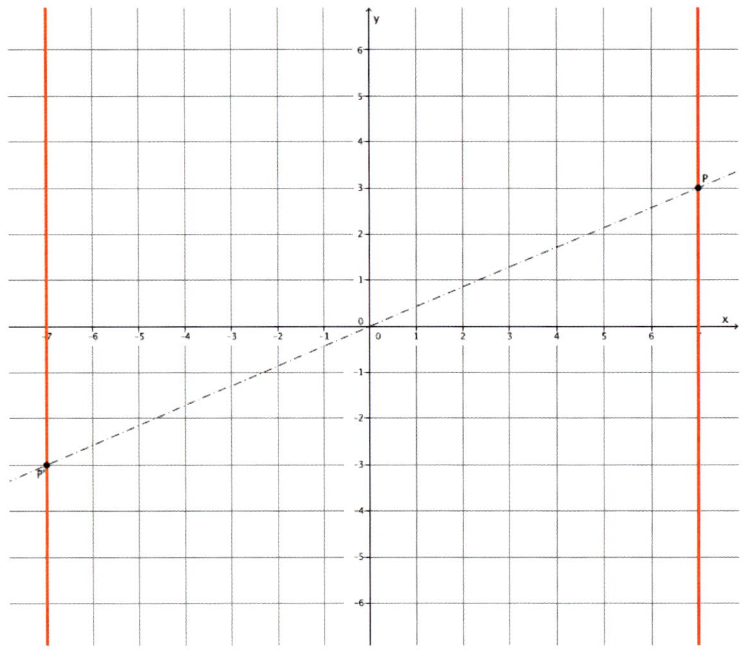

5. Let $Rotation_0$ be the rotation of 180 degrees around the origin. Let L be the line passing through $(0, 2)$ parallel to the x-axis. Is L parallel to $Rotation_0(L)$?

Yes, $L \parallel Rotation_0(L)$.

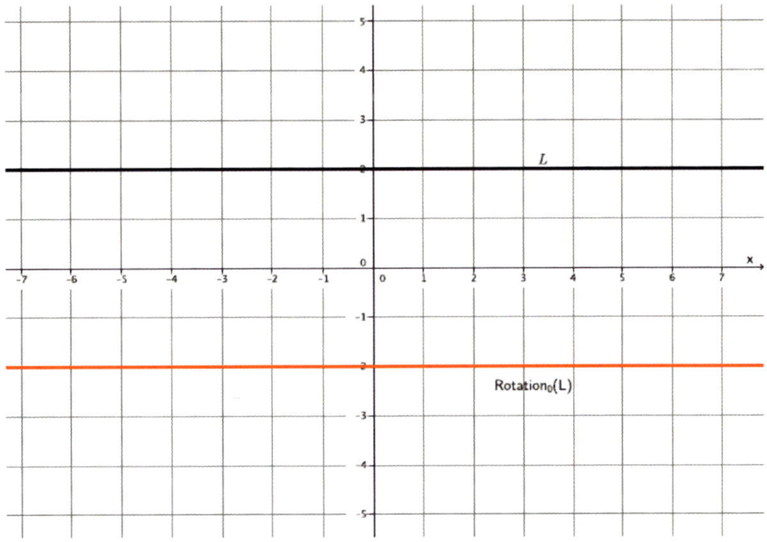

6. Let $Rotation_0$ be the rotation of 180 degrees around the origin. Let L be the line passing through $(4, 0)$ parallel to the y-axis. Is L parallel to $Rotation_0(L)$?

Yes, $L \parallel Rotation_0(L)$.

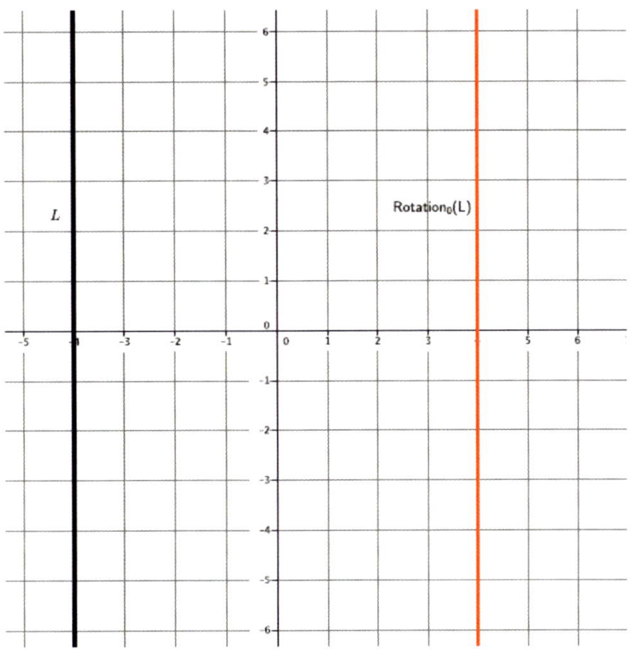

7. Let $Rotation_0$ be the rotation of 180 degrees around the origin. Let L be the line passing through $(0, -1)$ parallel to the x-axis. Is L parallel to $Rotation_0(L)$?

Yes, $L \parallel Rotation_0(L)$.

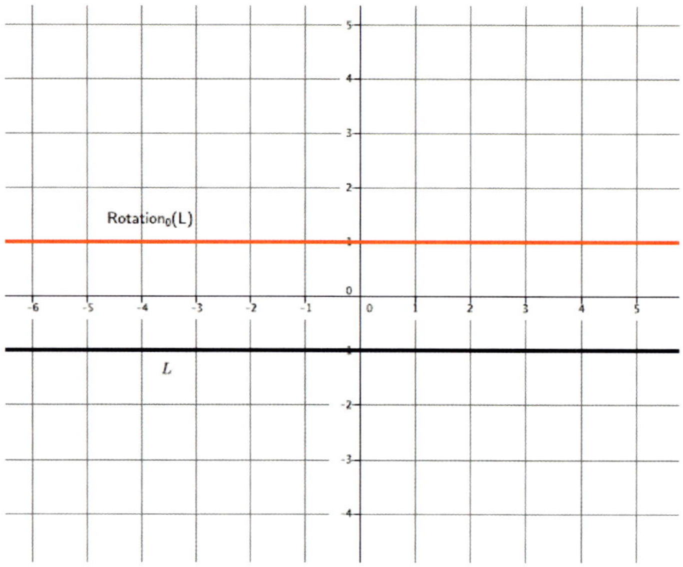

8. Let $Rotation_0$ be the rotation of 180 degrees around the origin. Is L parallel to $Rotation_0(L)$? Use your transparency if needed.

Yes, $L \parallel Rotation_0(L)$.

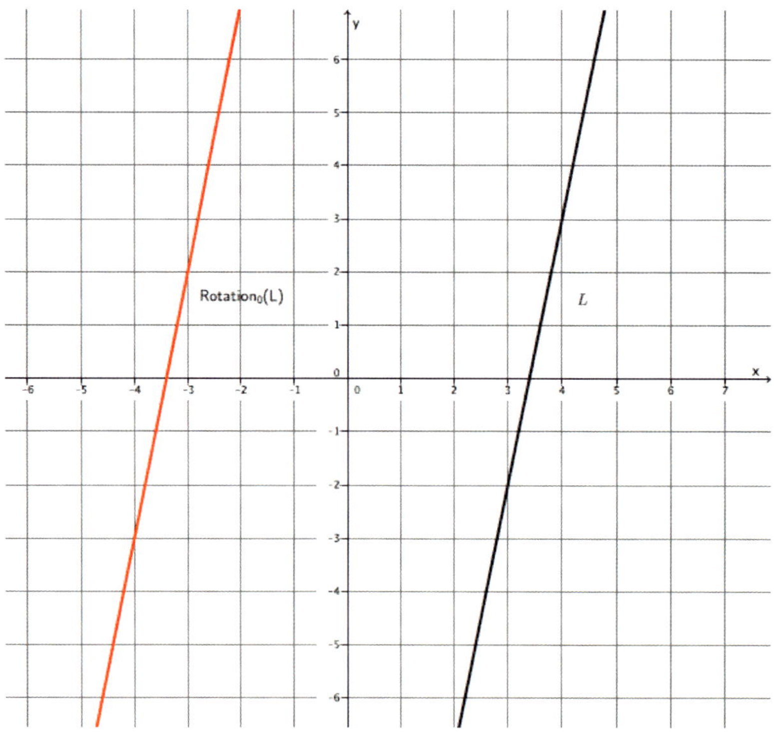

9. **Let $Rotation_0$ be the rotation of 180 degrees around the center O. Is L parallel to $Rotation_0(L)$? Use your transparency if needed.**

Yes, $L \parallel Rotation_0(L)$

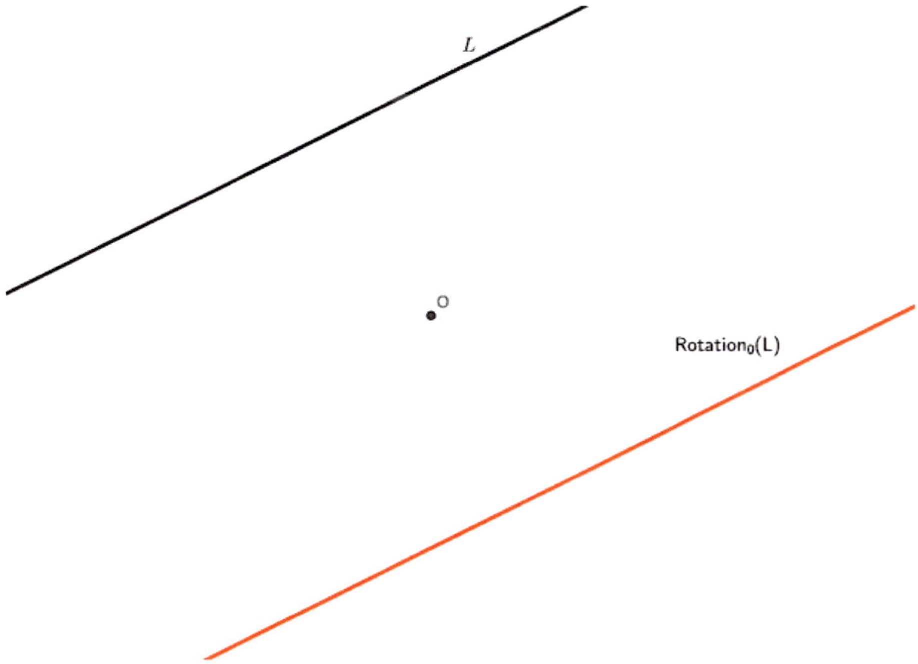

Example 3 (5 minutes)

MP.2 **THEOREM:** Let O be a point not lying on a given line L. Then, the 180-degree rotation around O maps L to a line parallel to L.

PROOF: Let $Rotation_0$ be the 180-degree rotation around O, and let P be a point on L. As usual, denote $Rotation_0(P)$ by P'. Since $Rotation_0$ is a 180-degree rotation, P, O, and P' lie on the same line (denoted by ℓ).

Scaffolding:

After completing Exercises 5–9, students should be convinced that the theorem is true. Make it clear that their observations can be proven (by contradiction) if we assume something different will happen (e.g., the lines will intersect).

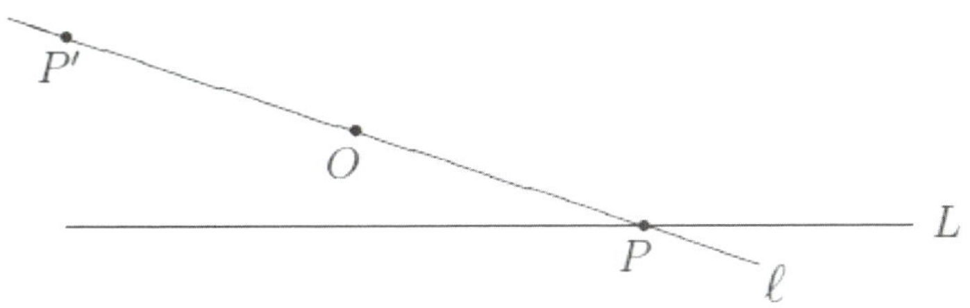

We want to investigate whether P' lies on L or not. Keep in mind that we want to show that the 180-degree rotation maps L to a line parallel to L. If the point P' lies on L, then at some point, the line L and $Rotation_0(L)$ intersect, meaning they are not parallel. If we can eliminate the possibility that P' lies on L, then we have to conclude that P' does not lie on L (rotations of 180 degrees make points that are collinear). If P' lies on L, then ℓ is a line that joins two points, P' and P, on L. However, L is already a line that joins P' and P, so ℓ and L must be the same line (i.e., $\ell = L$). This is trouble because we know O lies on ℓ, so $\ell = L$ implies that O lies on L. Look at the hypothesis of the theorem: *Let O be a point not lying on a given line L.* We have a contradiction. So, the possibility that P' lies on L is nonexistent. As we said, this means that P' does not lie on L.

What we have proved is that no matter which point P we take from L, we know $Rotation_0(P)$ does not lie on L. But $Rotation_0(L)$ consists of *all* the points of the form $Rotation_0(P)$ where P lies on L, so what we have proved is that no point of $Rotation_0(L)$ lies on L. In other words, L and $Rotation_0(L)$ have no point in common (i.e., $L \parallel Rotation_0(L)$). The theorem is proved.

Closing (5 minutes)

Summarize, or have students summarize, the lesson.

- Rotations of 180 degrees are special:
  - A point, P, that is rotated 180 degrees around a center O, produces a point P' so that P, O, and P' are collinear.
  - When we rotate around the origin of a coordinate system, we see that the point with coordinates (a, b) is moved to the point $(-a, -b)$.
- We now know that when a line is rotated 180 degrees around a point not on the line, it maps to a line parallel to the given line.

Lesson Summary

- **A rotation of 180 degrees around O is the rigid motion so that if P is any point in the plane, P, O, and $Rotation(P)$ are *collinear* (i.e., lie on the same line).**
- **Given a 180-degree rotation around the origin O of a coordinate system, R_0, and a point P with coordinates (a, b), it is generally said that $R_0(P)$ is the point with coordinates $(-a, -b)$.**

THEOREM: Let O be a point not lying on a given line L. Then, the 180-degree rotation around O maps L to a line parallel to L.

Exit Ticket (5 minutes)

Name ______________________________ Date ____________

Lesson 6: Rotations of 180 Degrees

Exit Ticket

Let there be a rotation of 180 degrees about the origin. Point A has coordinates $(-2, -4)$, and point B has coordinates $(-3, 1)$, as shown below.

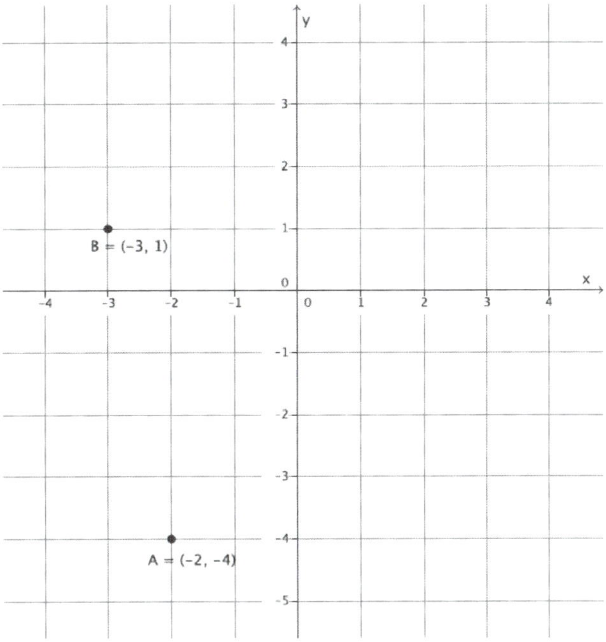

1. What are the coordinates of $Rotation(A)$? Mark that point on the graph so that $Rotation(A) = A'$. What are the coordinates of $Rotation(B)$? Mark that point on the graph so that $Rotation(B) = B'$.

2. What can you say about the points A, A', and O? What can you say about the points B, B', and O?

3. Connect point A to point B to make the line L_{AB}. Connect point A' to point B' to make the line $L_{A'B'}$. What is the relationship between L_{AB} and $L_{A'B'}$?

Exit Ticket Sample Solutions

Let there be a rotation of 180 degrees about the origin. Point A has coordinates $(-2,-4)$, and point B has coordinates $(-3,1)$, as shown below.

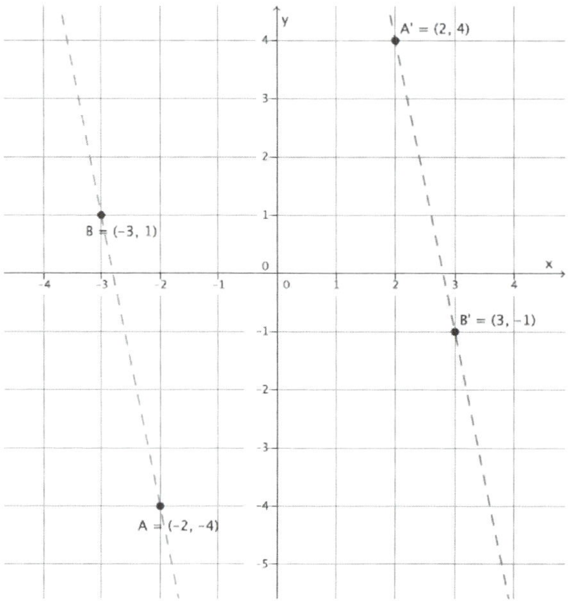

1. What are the coordinates of $Rotation(A)$? Mark that point on the graph so that $Rotation(A) = A'$. What are the coordinates of $Rotation(B)$? Mark that point on the graph so that $Rotation(B) = B'$.

   $A' = (2,4)$, $B' = (3,-1)$

2. What can you say about the points A, A', and O? What can you say about the points B, B', and O?

   The points A, A', and O are collinear. The points B, B', and O are collinear.

3. Connect point A to point B to make the line L_{AB}. Connect point A' to point B' to make the line $L_{A'B'}$. What is the relationship between L_{AB} and $L_{A'B'}$?

   $L_{AB} \parallel L_{A'B'}$

Problem Set Sample Solutions

Use the following diagram for Problems 1–5. Use your transparency as needed.

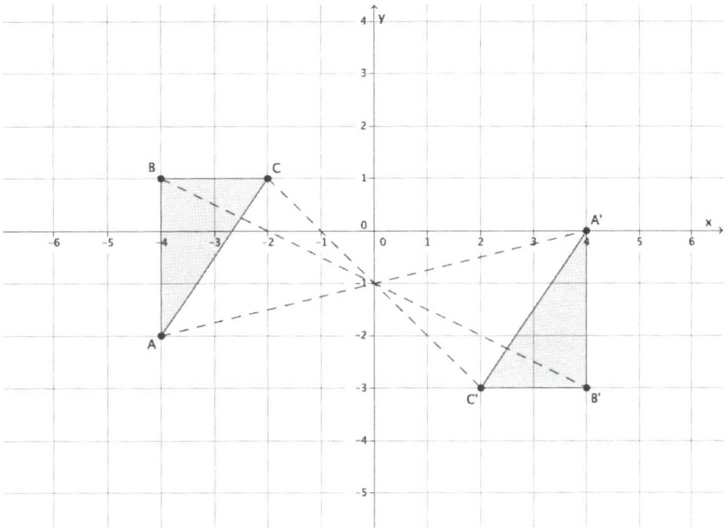

1. **Looking only at segment BC, is it possible that a $180°$ rotation would map segment BC onto segment $B'C'$? Why or why not?**

   It is possible because the segments are parallel.

2. **Looking only at segment AB, is it possible that a $180°$ rotation would map segment AB onto segment $A'B'$? Why or why not?**

   It is possible because the segments are parallel.

3. **Looking only at segment AC, is it possible that a $180°$ rotation would map segment AC onto segment $A'C'$? Why or why not?**

   It is possible because the segments are parallel.

4. **Connect point B to point B', point C to point C', and point A to point A'. What do you notice? What do you think that point is?**

   All of the lines intersect at one point. The point is the center of rotation. I checked by using my transparency.

5. **Would a rotation map triangle ABC onto triangle $A'B'C'$? If so, define the rotation (i.e., degree and center). If not, explain why not.**

   Let there be a rotation $180°$ around point $(0, -1)$. Then, $Rotation(\triangle ABC) = \triangle A'B'C'$.

6. The picture below shows right triangles ABC and $A'B'C'$, where the right angles are at B and B'. Given that $AB = A'B' = 1$, and $BC = B'C' = 2$, and that $\overline{AB}$ is not parallel to $\overline{A'B'}$, is there a $180°$ rotation that would map $\triangle ABC$ onto $\triangle A'B'C'$? Explain.

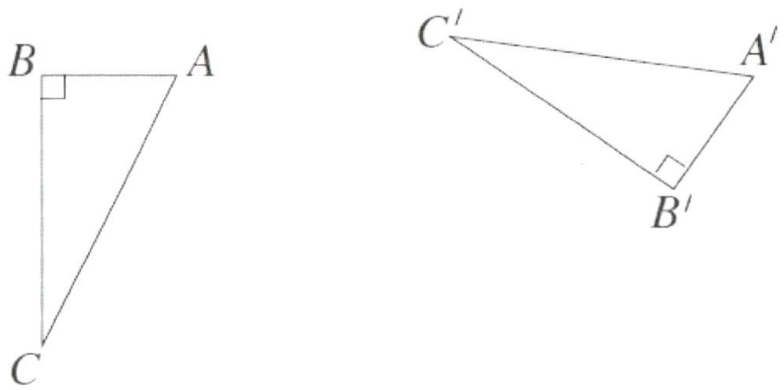

No, because a $180°$ rotation of a segment maps to a segment that is parallel to the given one. It is given that $\overline{AB}$ is not parallel to $\overline{A'B'}$; therefore, a rotation of $180°$ does not map $\triangle ABC$ onto $\triangle A'B'C'$.

Topic B

Sequencing the Basic Rigid Motions

8.G.A.2

Focus Standard:	8.G.A.2	Understand that a two-dimensional figure is congruent to another if the second can be obtained from the first by a sequence of rotations, reflections, and translations; given two congruent figures, describe a sequence that exhibits the congruence between them.
Instructional Days:	4	
Lesson 7:	Sequencing Translations (E)[1]	
Lesson 8:	Sequencing Reflections and Translations (S)	
Lesson 9:	Sequencing Rotations (E)	
Lesson 10:	Sequences of Rigid Motions (P)	

Topic B focuses on the first part of **8.G.A.2** in the respect that students learn how to sequence rigid motions. Lesson 7 begins with the concept of composing translations and introduces the idea that translations can be undone. In Lesson 8, students explore images of figures under a sequence of reflections and translations. In Lesson 9, students explore with sequences of rotations around the same center and rotations around different centers. In each of Lessons 7–9, students verify that the basic properties of individual rigid motions remain intact and describe the sequences using precise language. In Lesson 10, students perform sequences of translations, rotations, and reflections as a prelude to learning about congruence.

[1]Lesson Structure Key: **P**-Problem Set Lesson, **M**-Modeling Cycle Lesson, **E**-Exploration Lesson, **S**-Socratic Lesson

Lesson 7: Sequencing Translations

Student Outcomes

- Students learn about the sequence of transformations (one move on the plane followed by another) and that a sequence of translations enjoys the same properties as a single translation with respect to lengths of segments and degrees of angles.
- Students learn that a translation along a vector followed by another translation along a vector of the same length in the opposite direction can move all points of a plane back to their original positions.

Classwork

Discussion (5 minutes)

- Is it possible to translate a figure more than one time? That is, translating a figure along one vector, and then taking that figure and translating it along another vector?

MP.6

  - *The simple answer is yes. It is called a* sequence of transformations *or, more specifically, a sequence of translations.*
- Suppose we have two transformations of the plane, F and G. A point P, under transformation F, will be assigned to a new location, *Transformation* $F(P)$ denoted by P'. Transformation G will assign P' to a new location, *Transformation* $G(P')$ denoted by P''. This is true for every point P in the plane.
- In the picture below, the point P and ellipse E in black have undergone a sequence of transformations, first along the red vector where images are shown in red, and then along the blue vector where images are shown in blue.

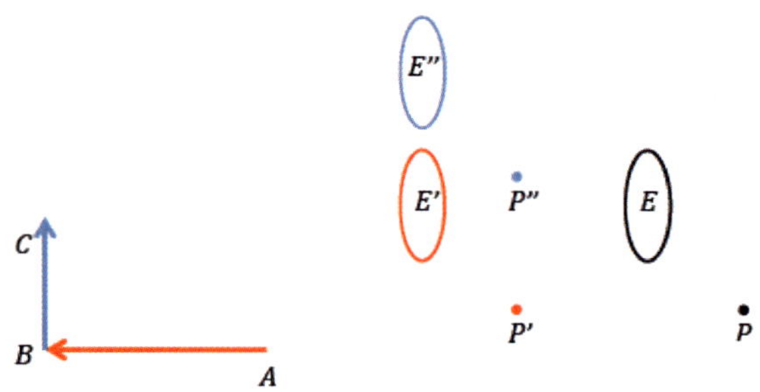

Exploratory Challenge/Exercises 1–4 (22 minutes)

Students complete Exercises 1–4 individually or in pairs.

Exploratory Challenge/Exercises 1–4

1.

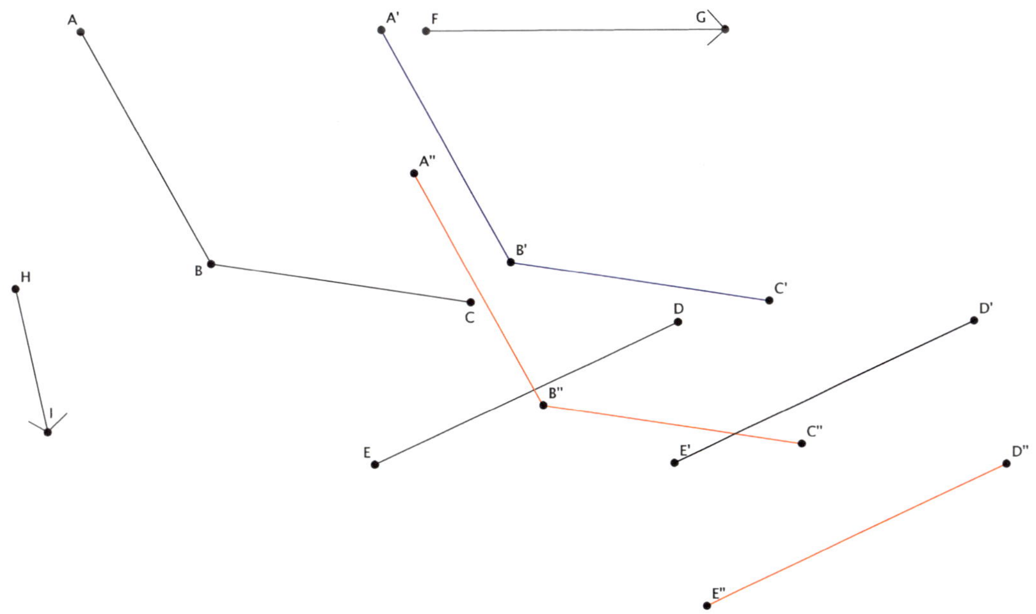

a. **Translate $\angle ABC$ and segment ED along vector $\overrightarrow{FG}$. Label the translated images appropriately, that is, $\angle A'B'C'$ and segment $E'D'$.**

Images are shown above in blue.

b. **Translate $\angle A'B'C'$ and segment $E'D'$ along vector $\overrightarrow{HI}$. Label the translated images appropriately, that is, $\angle A''B''C''$ and segment $E''D''$.**

Images are shown above in red.

c. **How does the size of $\angle ABC$ compare to the size of $\angle A''B''C''$?**

The measure of $\angle ABC$ is equal to the size of $\angle A''B''C''$.

d. **How does the length of segment ED compare to the length of the segment $E''D''$?**

The length of $\overline{ED}$ is equal to the length of $\overline{E''D''}$.

e. **Why do you think what you observed in parts (d) and (e) were true?**

One translation of the plane moved the angle and the segment to a new location. We know that translations preserve lengths of segments and degrees of measures of angles. The second translation moved the images to another new location, also preserving the length of the segment and the measure of the angle. Therefore, performing two translations, or a sequence of translations, keeps lengths of segments and size of angles rigid.

2. Translate $\triangle ABC$ along vector $\overrightarrow{FG}$, and then translate its image along vector $\overrightarrow{JK}$. Be sure to label the images appropriately.

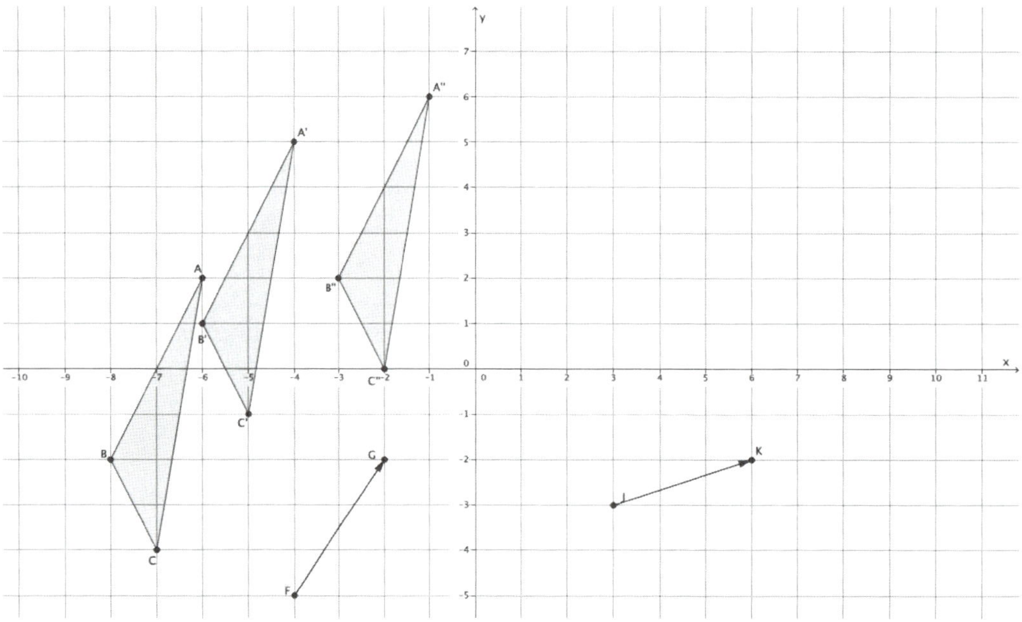

3. Translate figure $ABCDEF$ along vector $\overrightarrow{GH}$. Then translate its image along vector $\overrightarrow{JI}$. Label each image appropriately.

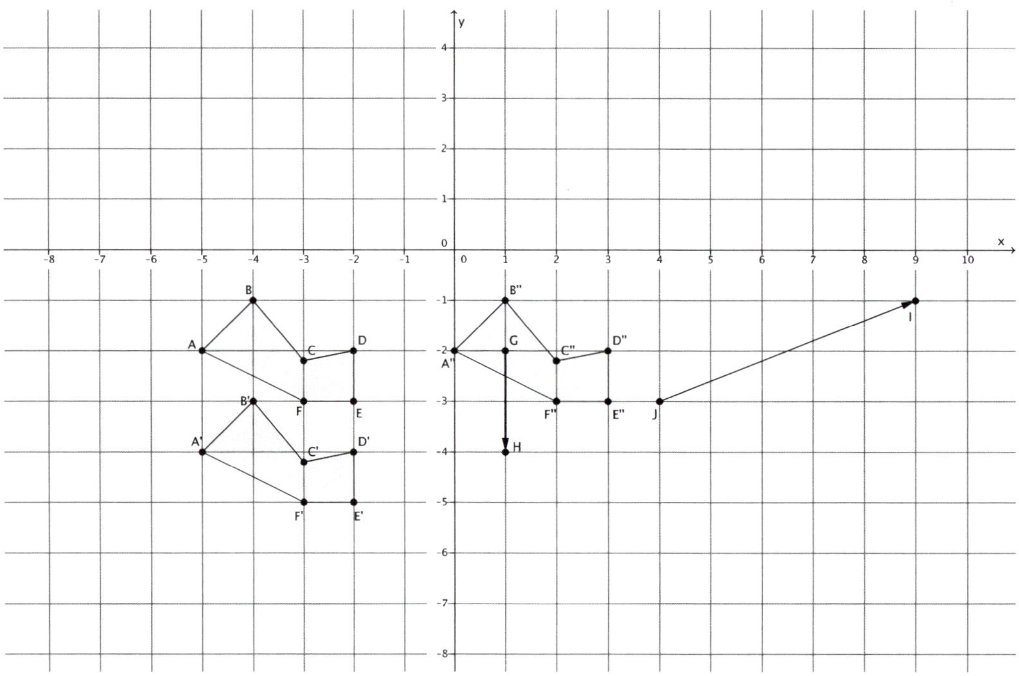

4.

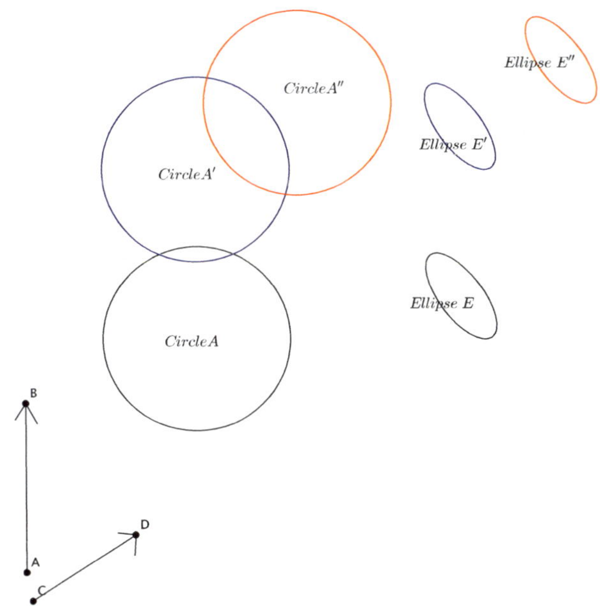

a. **Translate Circle A and Ellipse E along vector $\overrightarrow{AB}$. Label the images appropriately.**

Images are shown above in blue.

b. **Translate Circle A' and Ellipse E' along vector $\overrightarrow{CD}$. Label each image appropriately.**

Images are shown above in red.

c. **Did the size or shape of either figure change after performing the sequence of translations? Explain.**

The circle and the ellipse remained the same in size and shape after the sequence of translations. Since a translation is a basic rigid motion, a sequence of translations will maintain the shape and size of the figures.

Discussion (5 minutes)

- What need is there for sequencing transformations?
- Imagine life without an undo button on your computer or smartphone. If we move something in the plane, it would be nice to know we can move it back to its original position.
- Specifically, if a figure undergoes two transformations F and G and ends up in the same place as it was originally, then the figure has been mapped onto itself.

- Suppose we translate figure D along vector $\overrightarrow{AB}$.

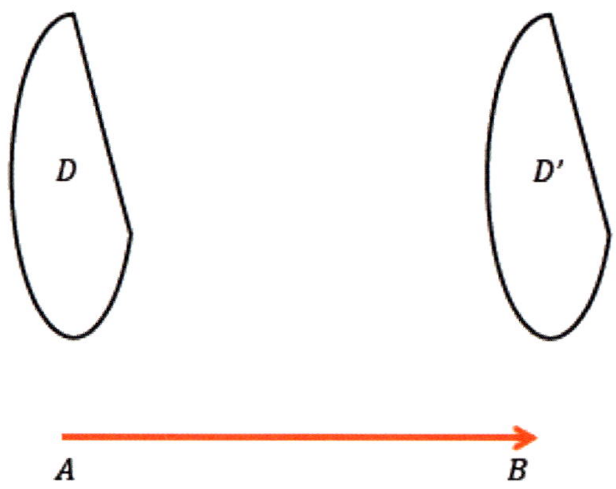

- How do we undo this move? That is, what translation of figure D along vector $\overrightarrow{AB}$ would bring D' back to its original position?
  - *We translate D' (the image of D) along the vector $\overrightarrow{BA}$.*

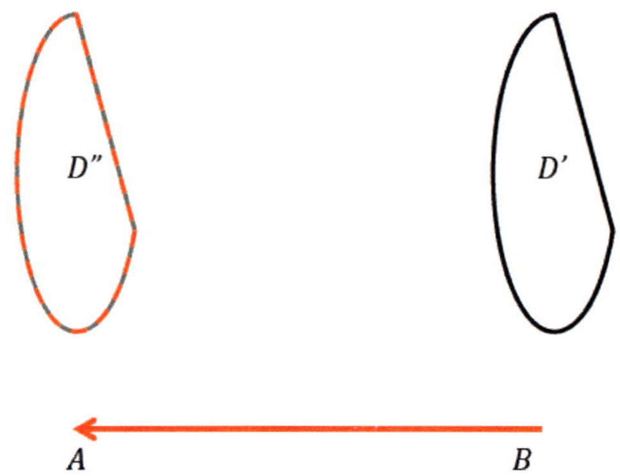

- All of the points in D were translated to D' and then translated again to D''. Because all of the points in D (shown in grey under the dashed red lines of D'') are also in D'' (shown as the figure with the dashed red lines), we can be sure that we have performed a sequence of translations that map the figure back onto itself.
- The ability to undo something or put it back in its original place is obviously very desirable. We see in the next few lessons that every basic rigid motion can be undone. That is one of the reasons we want to learn about basic rigid motions and their properties.

Exercises 5–6 (3 minutes)

Students continue with the Exploratory Challenge to complete Exercises 5 and 6 independently.

5. **The picture below shows the translation of Circle A along vector $\overrightarrow{CD}$. Name the vector that maps the image of Circle A back to its original position.**

   $\overrightarrow{DC}$

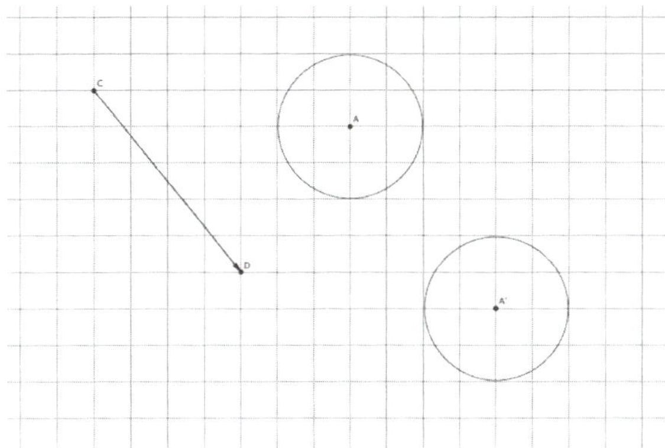

6. **If a figure is translated along vector $\overrightarrow{QR}$, what translation takes the figure back to its original location?**

   A translation along vector $\overrightarrow{RQ}$ would take the figure back to its original location.

Closing (5 minutes)

Summarize, or have students summarize, the lesson.

- We know that we can sequence translations, and the figure remains rigid, that is, lengths of segments and degrees of measures of angles are preserved.
- Any translation of the plane can be undone, and figures can be mapped onto themselves.

Lesson Summary

- **Translating a figure along one vector and then translating its image along another vector is an example of a sequence of transformations.**
- **A sequence of translations enjoys the same properties as a single translation. Specifically, the figures' lengths and degrees of angles are preserved.**
- **If a figure undergoes two transformations, F and G, and is in the same place it was originally, then the figure has been mapped onto itself.**

Exit Ticket (5 minutes)

Name ______________________________ Date ______________

Lesson 7: Sequencing Translations

Exit Ticket

Use the picture below to answer Problems 1 and 2.

1. Describe a sequence of translations that would map Figure H onto Figure K.

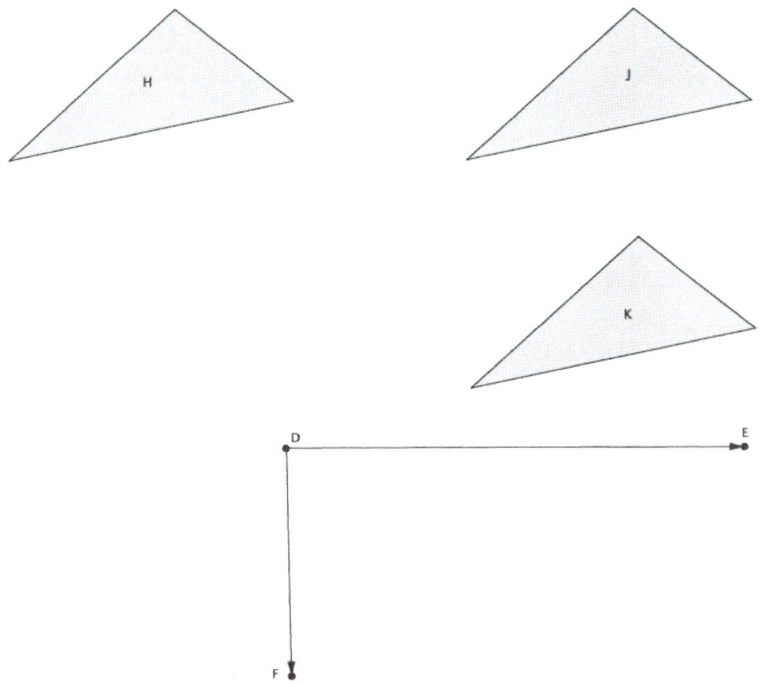

2. Describe a sequence of translations that would map Figure J onto itself.

Exit Ticket Sample Solutions

Use the picture below to answer Problems 1 and 2.

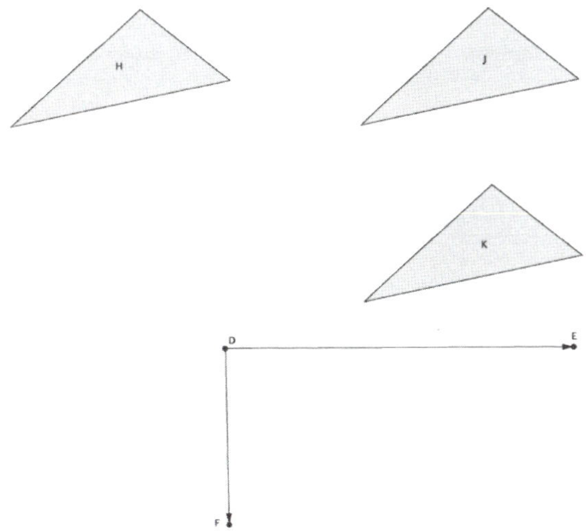

1. **Describe a sequence of translations that would map Figure H onto Figure K.**

   Translate Figure H along vector $\overrightarrow{DE}$, and then translate the image along vector $\overrightarrow{DF}$.

   Translate Figure H along vector $\overrightarrow{DF}$, and then translate the image along vector $\overrightarrow{DE}$.

2. **Describe a sequence of translations that would map Figure J onto itself.**

   Translate Figure J along vector $\overrightarrow{DE}$, and then translate the image along vector $\overrightarrow{ED}$.

   Translate Figure J along vector $\overrightarrow{DF}$, and then translate the image along vector $\overrightarrow{FD}$.

Problem Set Sample Solutions

1. **Sequence translations of parallelogram $ABCD$ (a quadrilateral in which both pairs of opposite sides are parallel) along vectors $\overrightarrow{HG}$ and $\overrightarrow{FE}$. Label the translated images.**

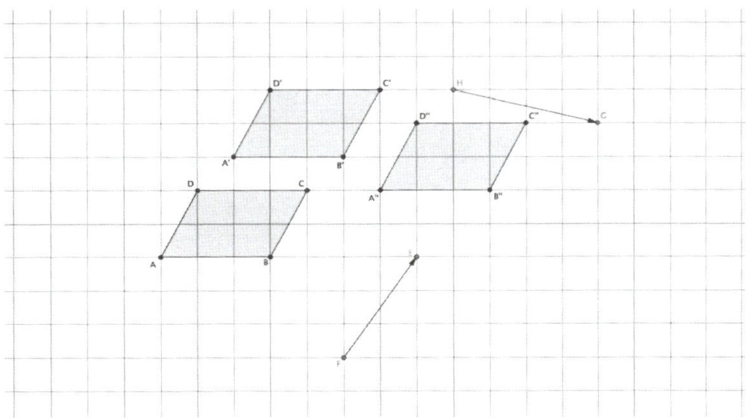

2. **What do you know about $\overline{AD}$ and $\overline{BC}$ compared with $\overline{A'D'}$ and $\overline{B'C'}$? Explain.**

   By the definition of a parallelogram, $\overline{AD} \parallel \overline{BC}$. Since translations map parallel lines to parallel lines, I know that $\overline{A'D'} \parallel \overline{B'C'}$.

3. **Are the segments $A'B'$ and $A''B''$ equal in length? How do you know?**

   Yes, $|A'B'| = |A''B''|$. Translations preserve lengths of segments.

4. Translate the curved shape ABC along the given vector. Label the image.

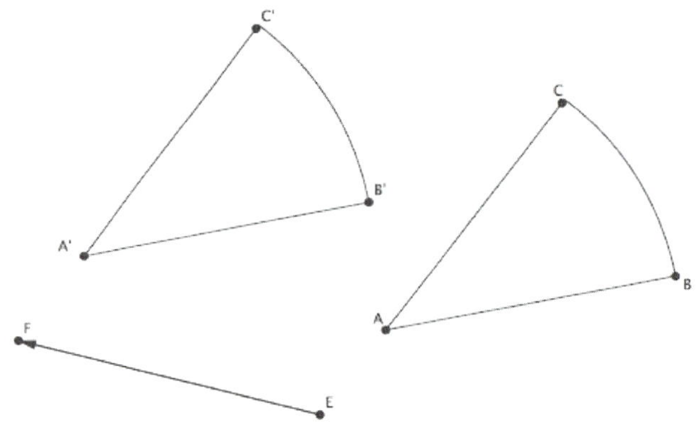

5. What vector would map the shape $A'B'C'$ back onto shape ABC?

Translating the image along vector $\overrightarrow{FE}$ would map the image back onto its original position.

Lesson 8: Sequencing Reflections and Translations

Student Outcomes

- Students learn that the reflection is its own inverse transformation.
- Students understand that a sequence of a reflection followed by a translation is not necessarily equal to a translation followed by a reflection.

Classwork

Discussion (10 minutes)

- Lesson 7 was an introduction to sequences of translations. It was clear that when a figure was translated along a vector, we could undo the move by translating along the same vector, but in the opposite direction, creating an inverse transformation.
- Note that not all transformations can be undone. For that reason, we will investigate sequences of reflections.
- Let there be a reflection across line L. What would undo this action? What is the inverse of this transformation?
  - *A reflection is always its own inverse.*
- Consider the picture below of a reflection across a vertical line L.

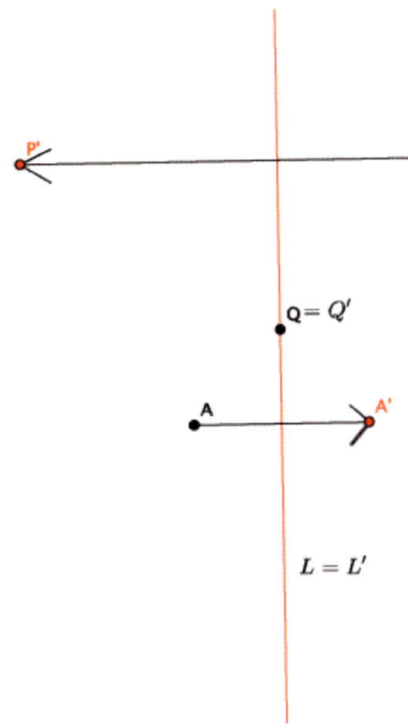

- Trace this picture of the line L and the points P, A, and Q as shown. Create a reflection across line L followed by another reflection across line L. Is the transformation corresponding to flipping the transparency once across L and then flipping it once more across L? Obviously, the red figure on the transparency would be right back on top of the original black figure. *Everything stays the same.*

Let us take this opportunity to reason through the preceding fact without a transparency.

- For a point P not on line L, what would the reflection of the reflection of point P be?
- The picture shows $Reflection(P) = P'$ is a point to the left of L, and if we reflect the point P' across L, clearly we get back to P itself. Thus, the reflection of the reflection of point P is P itself. The same holds true for A: the reflection of the reflection of point A is A.
- For point Q on the line L, what would the reflection of the reflection of point Q be?
- The lesson on reflection showed us that a point on the line of reflection is equal to itself, that is, $Reflection(Q) = Q$. Then, the reflection of the reflection of point Q is Q. No matter how many times a point on the line of reflection is reflected, it will be equal to itself.
- Based on the last two statements, we can say that the reflection of the reflection of P is P for any point P in the plane. Further,

  $$\text{the reflection of } P \text{ followed by the reflection of } P = I, \quad (4)$$

  where I denotes the identity transformation (Lesson 1). In terms of transparencies, equation (4) says that if we flip the transparency (on which we have traced a given figure in red) across the line of reflection L, then flipping it once more across L brings the red figure to coincide completely with the original figure. In this light, equation (4) is hardly surprising.

Exercises 1–3 (3 minutes)

Students complete Exercises 1–3 independently.

Exercises 1–3

Use the figure below to answer Exercises 1–3.

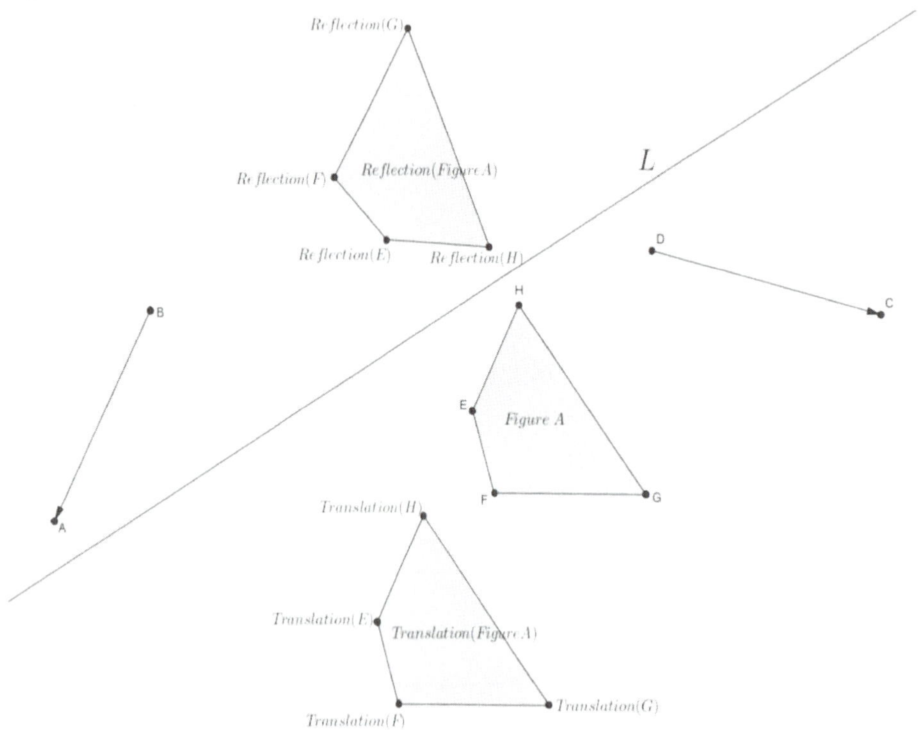

1. Figure A was translated along vector $\overrightarrow{BA}$, resulting in $Translation(Figure\ A)$. Describe a sequence of translations that would map Figure A back onto its original position.

   Translate Figure A along vector $\overrightarrow{BA}$; then, translate the image of Figure A along vector $\overrightarrow{AB}$.

2. Figure A was reflected across line L, resulting in $Reflection(Figure\ A)$. Describe a sequence of reflections that would map Figure A back onto its original position.

   Reflect Figure A across line L; then, reflect Figure A across line L again.

3. Can $Translation_{\overrightarrow{BA}}$ of Figure A undo the transformation of $Translation_{\overrightarrow{DC}}$ of Figure A? Why or why not?

   No. To undo the transformation, you would need to move the image of Figure A after the translations back to Figure A. The listed translations do not do that.

Discussion (10 minutes)

- Does the order in which we sequence rigid motions really matter?
- Consider a reflection followed by a translation. Would a figure be in the same final location if the translation was done first and then followed by the reflection?
- Let there be a reflection across line L, and let T be the translation along vector $\overrightarrow{AB}$. Let E represent the ellipse. The following picture shows the reflection of E followed by the translation of E.
- Before showing the picture, ask students which transformation happens first: the reflection or the translation?
  - *Reflection*

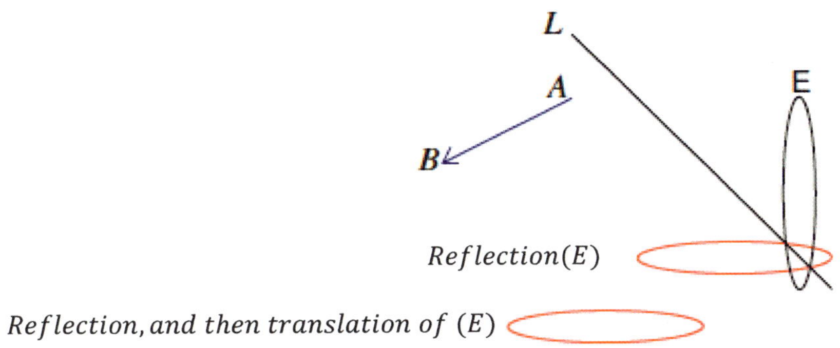

- Ask students again if they think the image of the ellipse will be in the same place if we translate first and then reflect. The following picture shows a translation of E followed by the reflection of E.

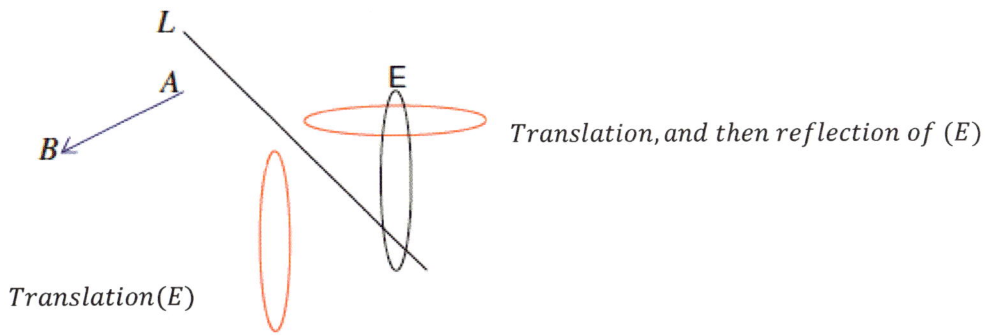

- It must be clear now that the order in which the rigid motions are performed matters. In the above example, we saw that the reflection followed by the translation of E is not the same as the translation followed by the reflection of E; therefore, a translation followed by a reflection and a reflection followed by a translation are not equal.

Video Presentation (2 minutes)

Show the video[1] on the sequence of basic rigid motions located at http://youtu.be/O2XPy3ZLU7Y. Note that this video makes use of rotation, which is not defined until Lesson 9. The video, however, does clearly convey the general idea of sequencing.

Exercises 4–7 (10 minutes)

Students complete Exercises 4, 5, and 7 independently. Students complete Exercise 6 in pairs.

Exercises 4–7

Let S be the black figure.

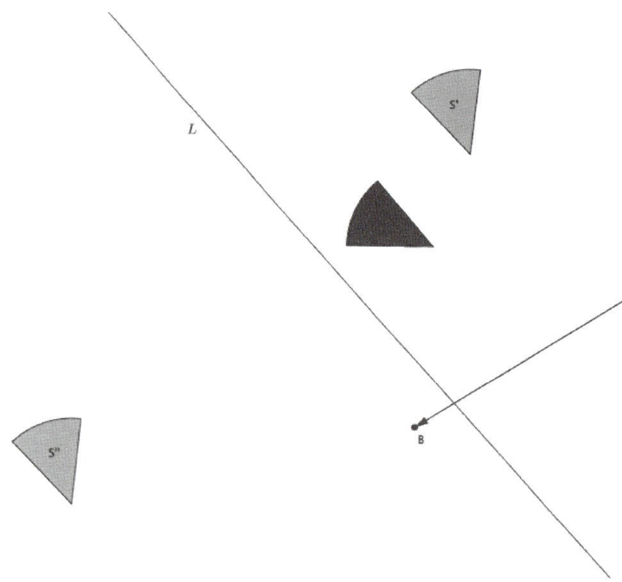

4. **Let there be the translation along vector $\overrightarrow{AB}$ and a reflection across line L.**

   Use a transparency to perform the following sequence: Translate figure S; then, reflect figure S. Label the image S'.

   The solution is on the diagram above.

5. **Let there be the translation along vector $\overrightarrow{AB}$ and a reflection across line L.**

   Use a transparency to perform the following sequence: Reflect figure S; then, translate figure S. Label the image S''.

   The solution is on the diagram above.

[1]The video was developed by Larry Francis.

6. Using your transparency, show that under a sequence of any two translations, $Translation$ and $Translation_0$ (along different vectors), that the sequence of the $Translation$ followed by the $Translation_0$ is equal to the sequence of the $Translation_0$ followed by the $Translation$. That is, draw a figure, A, and two vectors. Show that the translation along the first vector, followed by a translation along the second vector, places the figure in the same location as when you perform the translations in the reverse order. (This fact is proven in high school Geometry.) Label the transformed image A'. Now, draw two new vectors and translate along them just as before. This time, label the transformed image A''. Compare your work with a partner. Was the statement "the sequence of the $Translation$ followed by the $Translation_0$ is equal to the sequence of the $Translation_0$ followed by the $Translation$" true in all cases? Do you think it will always be true?

Sample Student Work:

First, let T be the translation along vector $\overrightarrow{AB}$, and let T_0 be the translation along vector $\overrightarrow{CD}$.

Then, let T be the translation along vector $\overrightarrow{EF}$, and let T_0 be the translation along vector $\overrightarrow{GH}$.

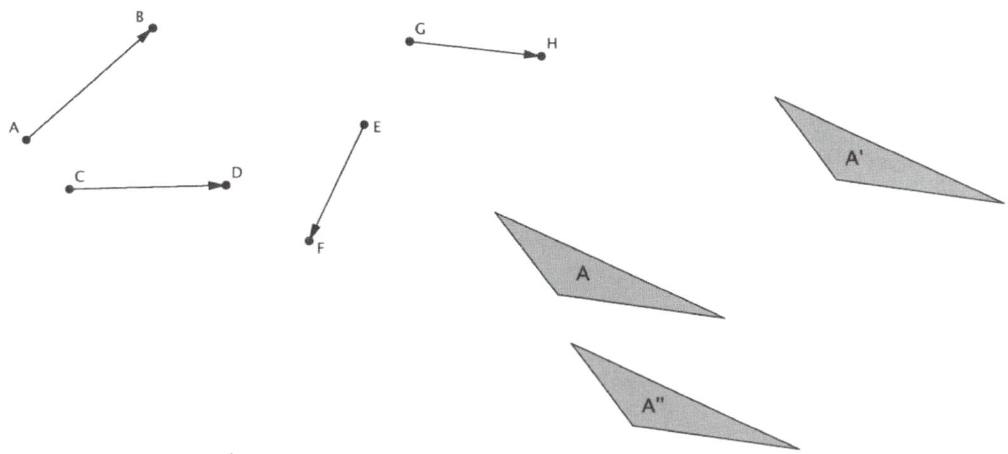

7. Does the same relationship you noticed in Exercise 6 hold true when you replace one of the translations with a reflection. That is, is the following statement true: A translation followed by a reflection is equal to a reflection followed by a translation?

No. The translation followed by a reflection would put a figure in a different location in the plane when compared to the same reflection followed by the same translation.

Closing (5 minutes)

Summarize, or have students summarize, the lesson.

- We can sequence rigid motions.
- We have notation related to sequences of rigid motions.
- The sequence of a reflection followed by the same reflection is the identity transformation, and the order in which we sequence rigid motions matters.

Lesson Summary

- **A reflection across a line followed by a reflection across the same line places all figures in the plane back onto their original position.**
- **A reflection followed by a translation does not necessarily place a figure in the same location in the plane as a translation followed by a reflection. The order in which we perform a sequence of rigid motions matters.**

Exit Ticket (5 minutes)

Name ____ Date ____________________

Lesson 8: Sequencing Reflections and Translations

Exit Ticket

Draw a figure, A, a line of reflection, L, and a vector $\overrightarrow{FG}$ in the space below. Show that under a sequence of a translation and a reflection, that the sequence of the reflection followed by the translation is not equal to the translation followed by the reflection. Label the figure as A' after finding the location according to the sequence reflection followed by the translation, and label the figure A'' after finding the location according to the composition translation followed by the reflection. If A' is not equal to A'', then we have shown that the sequence of the reflection followed by a translation is not equal to the sequence of the translation followed by the reflection. (This is proven in high school Geometry.)

Exit Ticket Sample Solutions

Draw a figure, A, a line of reflection, L, and a vector $\overrightarrow{FG}$ in the space below. Show that under a sequence of a translation and a reflection, that the sequence of the reflection followed by the translation is not equal to the translation followed by the reflection. Label the figure as A' after finding the location according to the sequence reflection followed by the translation, and label the figure A''after finding the location according to the composition translation followed by the reflection. If A' is not equal to A'', then we have shown that the sequence of the reflection followed by a translation is not equal to the sequence of the translation followed by the reflection. (This is proven in high school Geometry.)

Sample student drawing:

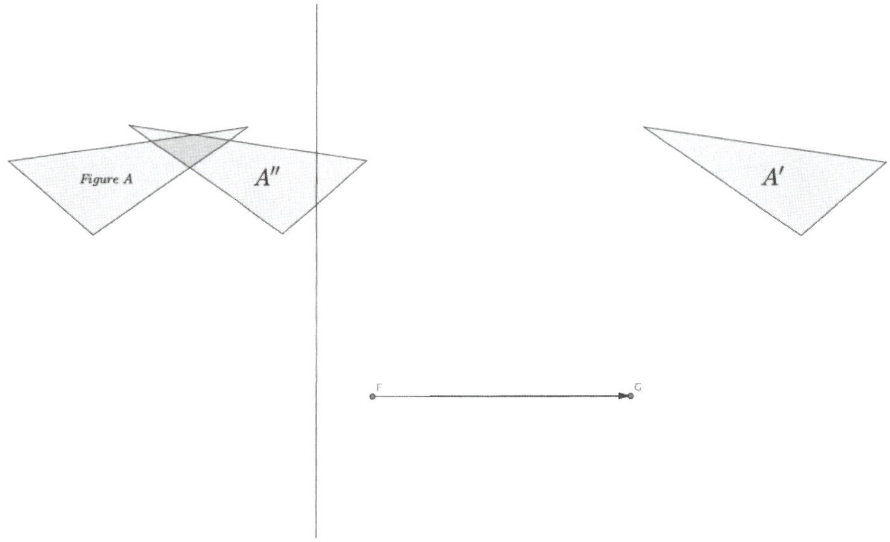

Note: If the figure lies completely on the line of reflection—a point, for example—then the image after the reflection followed by the translation would be in the same as the image after the translation followed by the reflection.

Problem Set Sample Solutions

1. Let there be a reflection across line L, and let there be a translation along vector $\overrightarrow{AB}$, as shown. If S denotes the black figure, compare the translation of S followed by the reflection of S with the reflection of S followed by the translation of S.

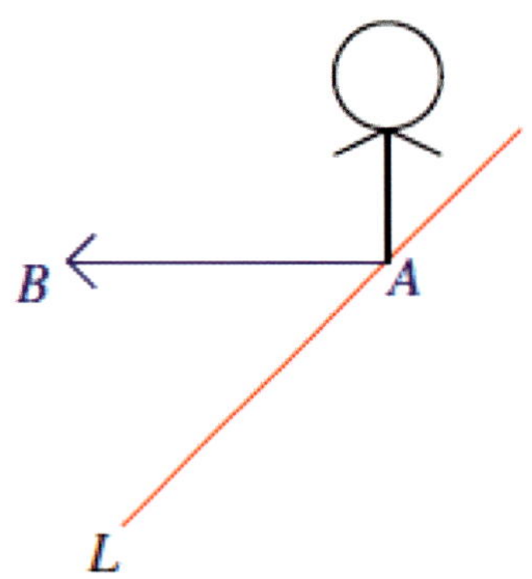

Students should notice that the two sequences place figure S in different locations in the plane.

2. Let L_1 and L_2 be parallel lines, and let $Reflection_1$ and $Reflection_2$ be the reflections across L_1 and L_2, respectively (in that order). Show that a $Reflection_2$ followed by $Reflection_1$ is not equal to a $Reflection_1$ followed by $Reflection_2$. (Hint: Take a point on L_1 and see what each of the sequences does to it.)

Let D be a point on L_1, as shown, and let D' be the transformation given by $Reflection_2$ followed by $Reflection_1$. Notice where D' is.

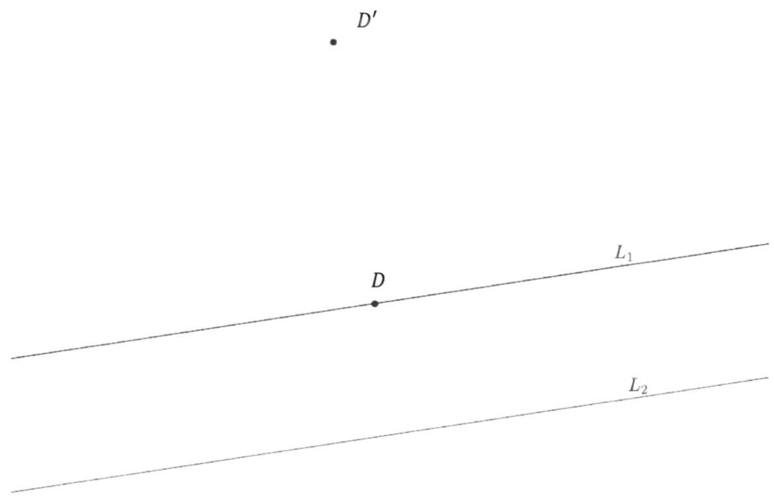

Let D'' be the transformation given by $Reflection_1$ followed by $Reflection_2$. Notice where the D'' is.

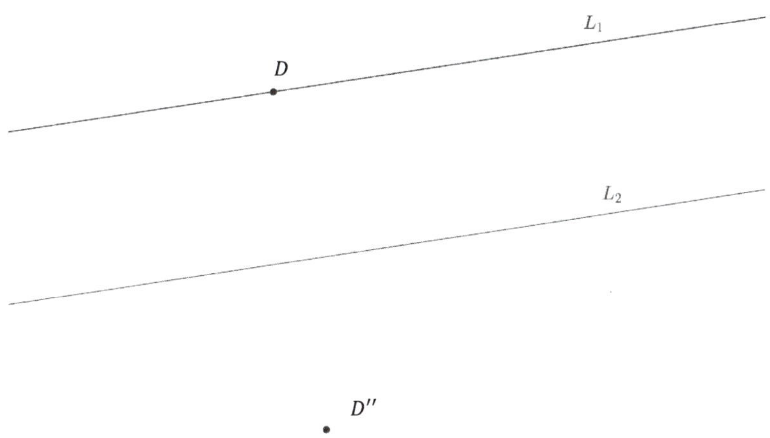

Since $D' \neq D''$, the sequences are not equal.

3. Let L_1 and L_2 be parallel lines, and let $Reflection_1$ and $Reflection_2$ be the reflections across L_1 and L_2, respectively (in that order). Can you guess what $Reflection_1$ followed by $Reflection_2$ is? Give as persuasive an argument as you can. (Hint: Examine the work you just finished for the last problem.)

The sequence $Reflection_1$ followed by $Reflection_2$ is just like the translation along a vector $\overrightarrow{AB}$, as shown below, where $AB \perp L_1$. The length of AB is equal to twice the distance between L_1 and L_2.

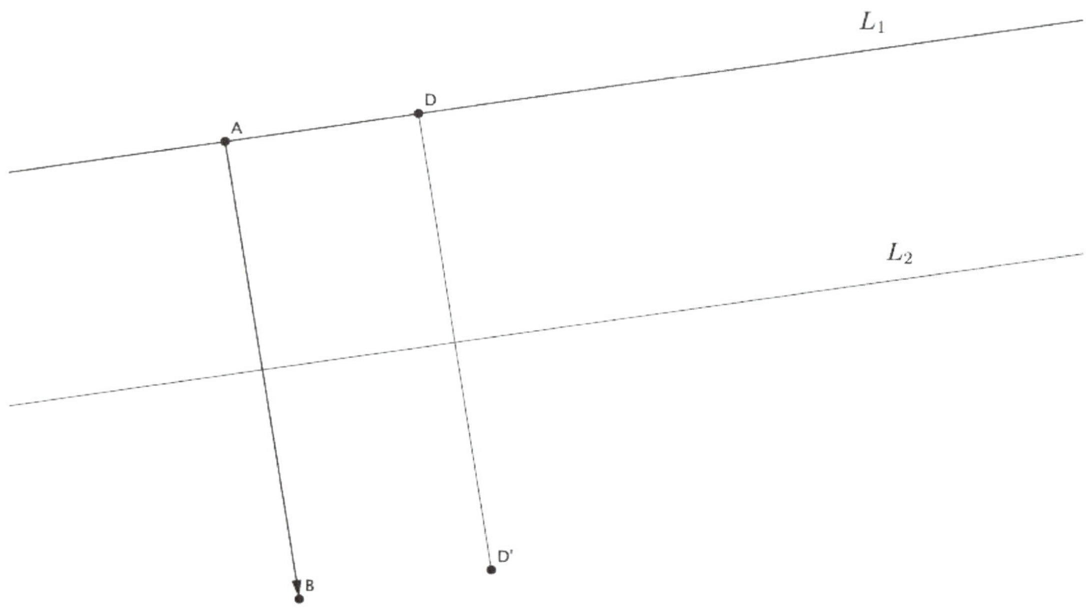

Lesson 9: Sequencing Rotations

Student Outcomes

- Students learn that sequences of rotations preserve lengths of segments as well as degrees of measures of angles.
- Students describe a sequence of rigid motions that would map a triangle back to its original position after being rotated around two different centers.

Classwork

Exploratory Challenge (35 minutes)

> *Scaffolding:*
> It may be beneficial to check student work periodically throughout the Exploratory Challenge. As necessary, provide teacher modeling.

Exploratory Challenge

1.

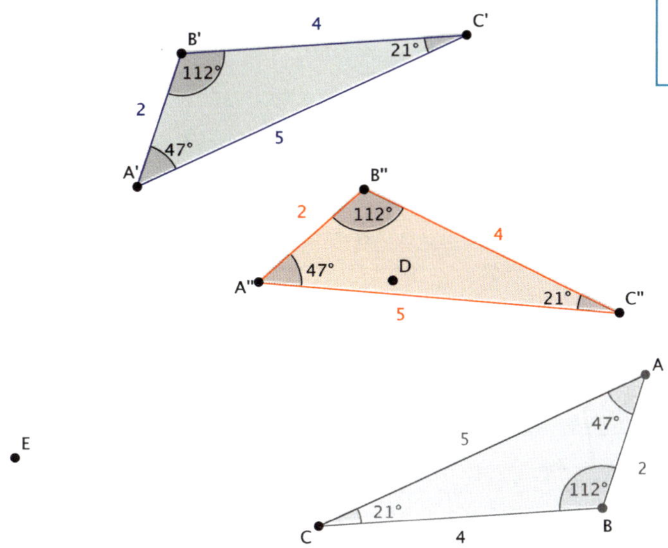

a. Rotate $\triangle ABC$ d degrees around center D. Label the rotated image as $\triangle A'B'C'$.

Sample student work is shown in blue.

b. Rotate $\triangle A'B'C'$ d degrees around center E. Label the rotated image as $\triangle A''B''C''$.

Sample student work is shown in red.

c. Measure and label the angles and side lengths of $\triangle ABC$. How do they compare with the images $\triangle A'B'C'$ and $\triangle A''B''C''$?

Measures of corresponding sides and measures of corresponding angles of three triangles are equal.

d. How can you explain what you observed in part (c)? What statement can you make about properties of sequences of rotations as they relate to a single rotation?

We already knew that a single rotation would preserve the lengths of segments and degrees of angles. Performing one rotation after the other does not change the lengths of segments or degrees of angles. That means that sequences of rotations enjoy the same properties as a single rotation.

2.

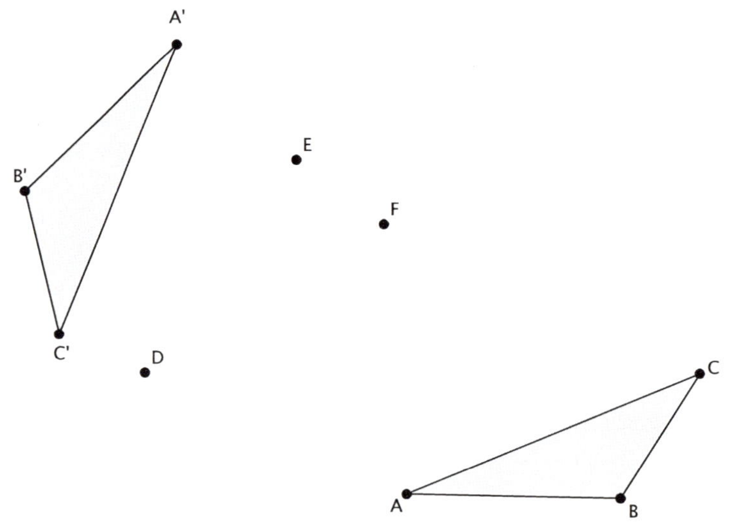

a. Rotate $\triangle ABC$ d degrees around center D, and then rotate again d degrees around center E. Label the image as $\triangle A'B'C'$ after you have completed both rotations.

Possible student solution is shown in diagram as $\triangle A'B'C'$.

b. Can a single rotation around center D map $\triangle A'B'C'$ onto $\triangle ABC$?

No, a single rotation around center D will not map $\triangle A'B'C'$ onto $\triangle ABC$.

c. Can a single rotation around center E map $\triangle A'B'C'$ onto $\triangle ABC$?

No, a single rotation around center E will not map $\triangle A'B'C'$ onto $\triangle ABC$.

d. Can you find a center that would map $\triangle A'B'C'$ onto $\triangle ABC$ in one rotation? If so, label the center F.

Yes, a d-degree rotation around center F will map $\triangle A'B'C'$ onto $\triangle ABC$.

Note: Students can only find the center F through trial and error at this point. Finding the center of rotation for two congruent figures is a skill that will be formalized in high school Geometry.

3.

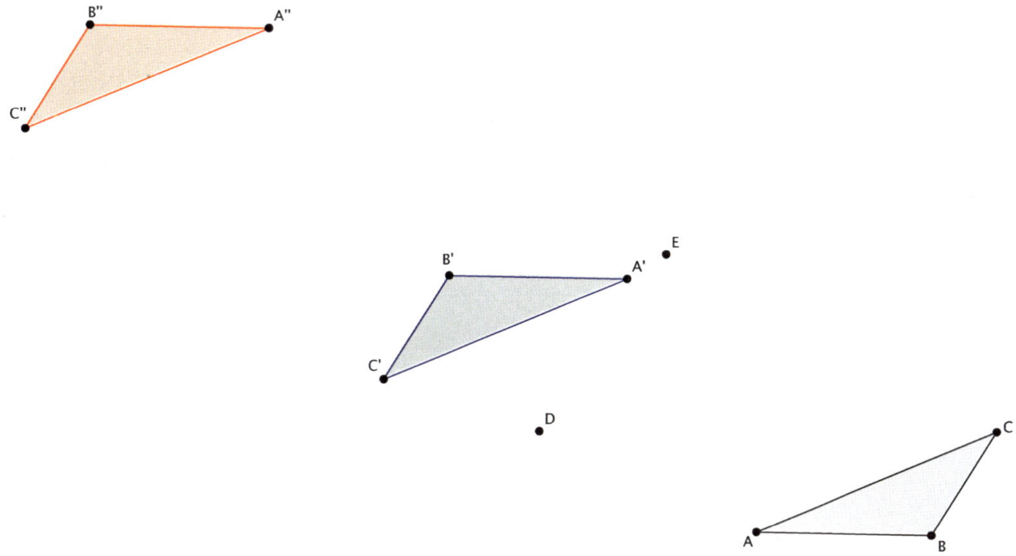

a. Rotate $\triangle ABC$ $90°$ (counterclockwise) around center D, and then rotate the image another $90°$ (counterclockwise) around center E. Label the image $\triangle A'B'C'$.

Sample student work is shown in blue.

b. Rotate $\triangle ABC$ $90°$ (counterclockwise) around center E, and then rotate the image another $90°$ (counterclockwise) around center D. Label the image $\triangle A''B''C''$.

Sample student work is shown in red.

c. What do you notice about the locations of $\triangle A'B'C'$ and $\triangle A''B''C''$? Does the order in which you rotate a figure around different centers have an impact on the final location of the figure's image?

The triangles are in two different locations. Yes, the order in which we rotate a figure around two different centers must matter because the triangles are not in the same location after rotating around center D and then center E compared to rotating around center E and then center D.

4.

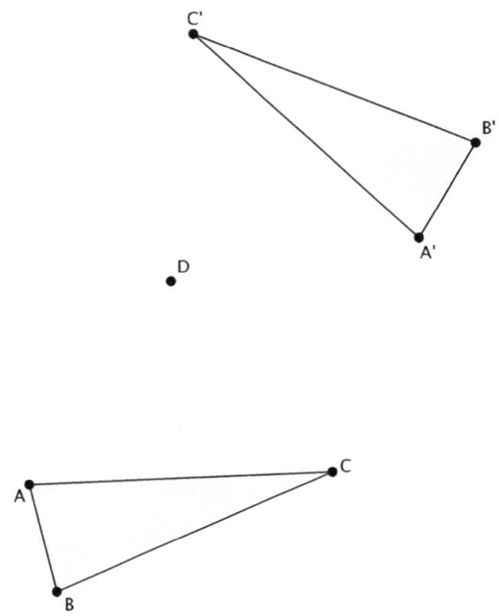

a. Rotate $\triangle ABC$ $90°$ (counterclockwise) around center D, and then rotate the image another $45°$ (counterclockwise) around center D. Label the image $\triangle A'B'C'$.

Rotated triangle is shown above.

b. Rotate $\triangle ABC$ $45°$ (counterclockwise) around center D, and then rotate the image another $90°$ (counterclockwise) around center D. Label the image $\triangle A''B''C''$.

Rotated triangle is shown above.

c. What do you notice about the locations of $\triangle A'B'C'$ and $\triangle A''B''C''$? Does the order in which you rotate a figure around the same center have an impact on the final location of the figure's image?

The triangles are in the same location. This indicates that when a figure is rotated twice around the same center, it does not matter in which order you perform the rotations.

5. $\triangle ABC$ has been rotated around two different centers, and its image is $\triangle A'B'C'$. Describe a sequence of rigid motions that would map $\triangle ABC$ onto $\triangle A'B'C'$.

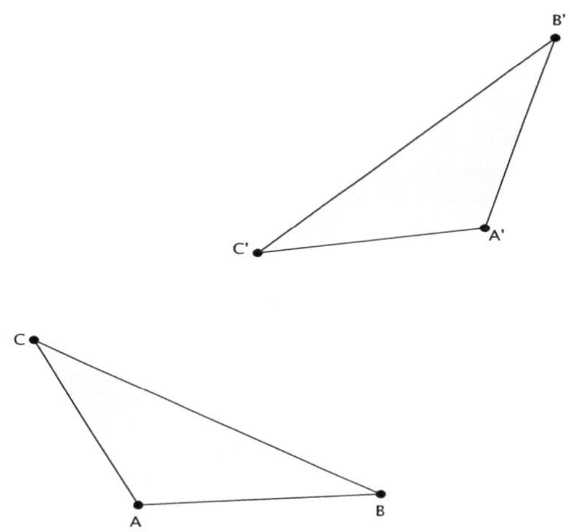

Translate $\triangle ABC$ along vector $\overrightarrow{CC'}$. Then, rotate $\triangle ABC$ around point C' until $\triangle ABC$ maps onto $\triangle A'B'C'$.

Closing (5 minutes)

Summarize, or have students summarize, the lesson.

- Sequences of rotations enjoy the same properties as single rotations. That is, a sequence of rotations preserves lengths of segments and degrees of measures of angles.
- The order in which a sequence of rotations around two different centers is performed matters. The order in which a sequence of rotations around the same center is performed does not matter.
- When a figure is rotated around two different centers, we can describe a sequence of rigid motions that would map the original figure onto the resulting image.

Lesson Summary

- **Sequences of rotations have the same properties as a single rotation:**
  - **A sequence of rotations preserves degrees of measures of angles.**
  - **A sequence of rotations preserves lengths of segments.**
- **The order in which a sequence of rotations around different centers is performed matters with respect to the final location of the image of the figure that is rotated.**
- **The order in which a sequence of rotations around the same center is performed does not matter. The image of the figure will be in the same location.**

Exit Ticket (5 minutes)

Name ____ Date ________________

Lesson 9: Sequencing Rotations

Exit Ticket

1. Let $Rotation_1$ be the rotation of a figure d degrees around center O. Let $Rotation_2$ be the rotation of the same figure d degrees around center P. Does the $Rotation_1$ of the figure followed by the $Rotation_2$ equal a $Rotation_2$ of the figure followed by the $Rotation_1$? Draw a picture if necessary.

2. Angle ABC underwent a sequence of rotations. The original size of $\angle ABC$ Is $37°$. What was the size of the angle after the sequence of rotations? Explain.

3. Triangle ABC underwent a sequence of rotations around two different centers. Its image is $\triangle A'B'C'$. Describe a sequence of rigid motions that would map $\triangle ABC$ onto $\triangle A'B'C'$.

Exit Ticket Sample Solutions

1. Let $Rotation_1$ be the rotation of a figure d degrees around center O. Let $Rotation_2$ be the rotation of the same figure d degrees around center P. Does the $Rotation_1$ of the figure followed by the $Rotation_2$ equal a $Rotation_2$ of the figure followed by the $Rotation_1$? Draw a picture if necessary.

   No. If the sequence of rotations were around the same center, then it would be true. However, when the sequence involves two different centers, the order in which they are performed matters because the images are not in the same location in the plane.

2. Angle ABC underwent a sequence of rotations. The original size of $\angle ABC$ is $37°$. What was the size of the angle after the sequence of rotations? Explain.

   Since sequences of rotations enjoy the same properties as a single rotation, then the measure of any image of $\angle ABC$ under any sequence of rotations remains $37°$. Rotations and sequences of rotations preserve the measure of degrees of angles.

3. Triangle ABC underwent a sequence of rotations around two different centers. Its image is $\triangle A'B'C'$. Describe a sequence of rigid motions that would map $\triangle ABC$ onto $\triangle A'B'C'$.

   Translate $\triangle ABC$ along vector $\overrightarrow{BB'}$. Then, rotate $\triangle ABC$ d degrees around point B' until $\triangle ABC$ maps onto $\triangle A'B'C'$.

Problem Set Sample Solutions

1. Refer to the figure below.

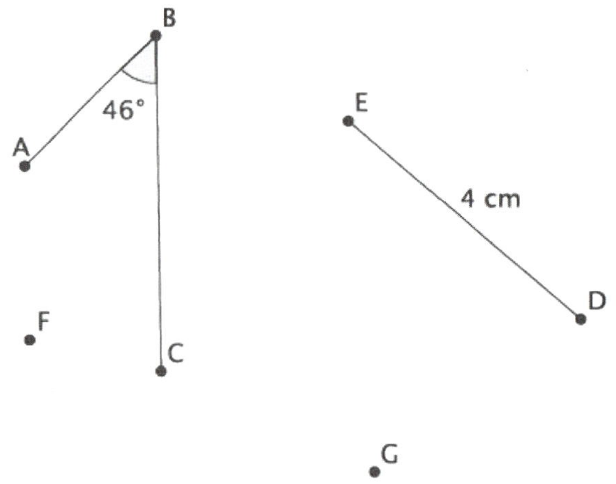

a. Rotate $\angle ABC$ and segment DE d degrees around center F and then d degrees around center G. Label the final location of the images as $\angle A'B'C'$ and segment $D'E'$.

b. What is the size of $\angle ABC$, and how does it compare to the size of $\angle A'B'C'$? Explain.

The measure of $\angle ABC$. Is $46°$. The measure of $\angle A'B'C'$ is $46°$. The angles are equal in measure because a sequence of rotations preserves the degrees of an angle.

c. What is the length of segment DE, and how does it compare to the length of segment $D'E'$? Explain.

The length of segment DE is 4 cm. The length of segment $D'E'$ is also 4 cm. The segments are equal in length because a sequence of rotations preserves the length of segments.

2. Refer to the figure given below.

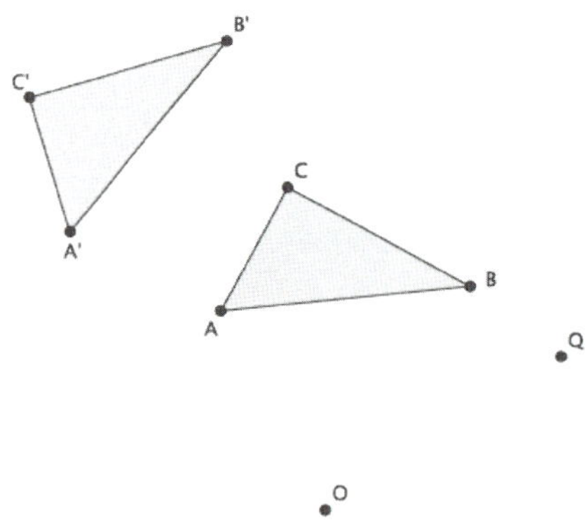

a. Let $Rotation_1$ be a counterclockwise rotation of $90°$ around the center O. Let $Rotation_2$ be a clockwise rotation of $(-45)°$ around the center Q. Determine the approximate location of $Rotation_1(\triangle ABC)$ followed by $Rotation_2$. Label the image of $\triangle ABC$ as $\triangle A'B'C'$.

b. **Describe the sequence of rigid motions that would map $\triangle ABC$ onto $\triangle A'B'C'$.**

The image of $\triangle ABC$ is shown above. Translate $\triangle ABC$ along vector $\overrightarrow{AA'}$. Rotate $\triangle ABC$ d degrees around center A'. Then, $\triangle ABC$ will map onto $\triangle A'B'C'$.

3. **Refer to the figure given below.**

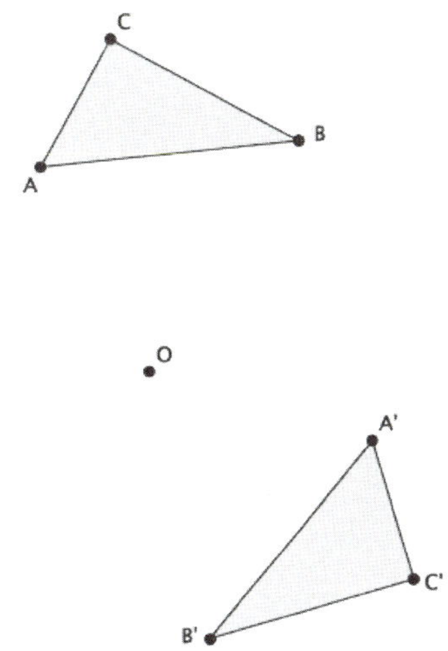

Let R be a rotation of $(-90)°$ around the center O. Let $Rotation_2$ be a rotation of $(-45)°$ around the same center O. Determine the approximate location of $Rotation_1(\triangle ABC)$ followed by $Rotation_2(\triangle ABC)$. Label the image of $\triangle ABC$ as $\triangle A'B'C'$.

The image of $\triangle ABC$ is shown above.

Lesson 10: Sequences of Rigid Motions

Student Outcomes

- Students describe a sequence of rigid motions that maps one figure onto another.

Classwork

Example 1 (8 minutes)

So far, we have seen how to sequence translations, sequence reflections, sequence translations and reflections, and sequence translations and rotations. Now that we know about rotation, we can move geometric figures around the plane by sequencing a combination of translations, reflections, and rotations.

Let's examine the following sequence:

- Let E denote the ellipse in the coordinate plane as shown.

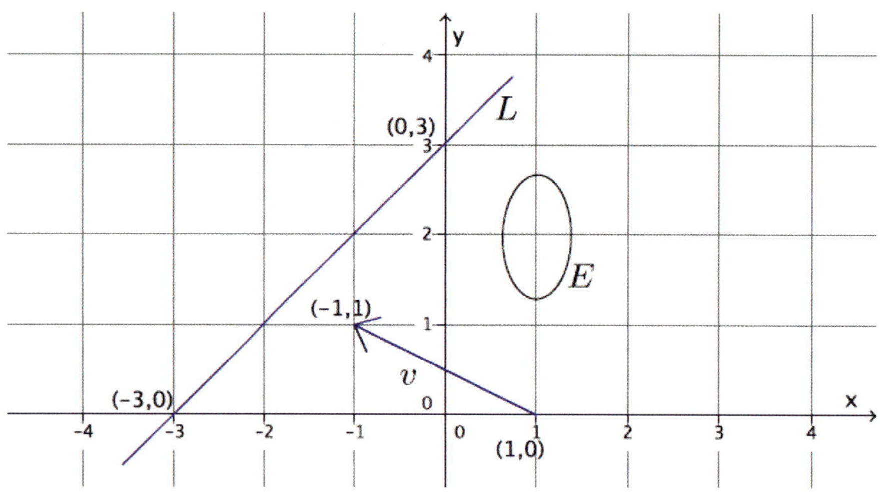

- Let $Translation_1$ be the translation along the vector $\vec{v}$ from $(1,0)$ to $(-1,1)$, let $Rotation_2$ be the 90-degree rotation around $(-1,1)$, and let $Reflection_3$ be the reflection across line L joining $(-3,0)$ and $(0,3)$. What is the $Translation_1(E)$ followed by the $Rotation_2(E)$ followed by the $Reflection_3(E)$?

- Which transformation do we perform first, the translation, the reflection, or the rotation? How do you know? Does it make a difference?
  - *The order in which transformations are performed makes a difference. Therefore, we perform the translation first. So now, we let* E_1 *be* $Translation_1(E)$.

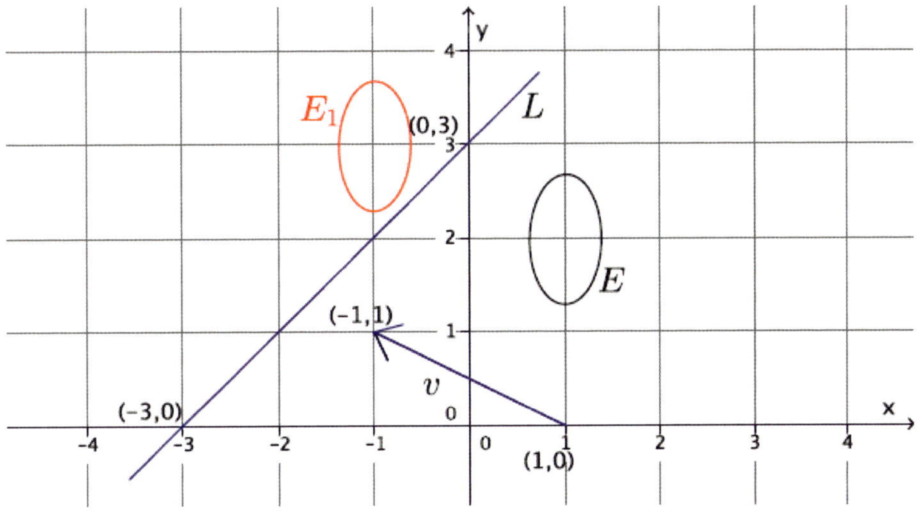

- Which transformation do we perform next?
  - *The rotation is next. So now, we let* E_2 *be the image of* E *after the* $Translation_1(E)$ *followed by the* $Rotation_2(E)$.

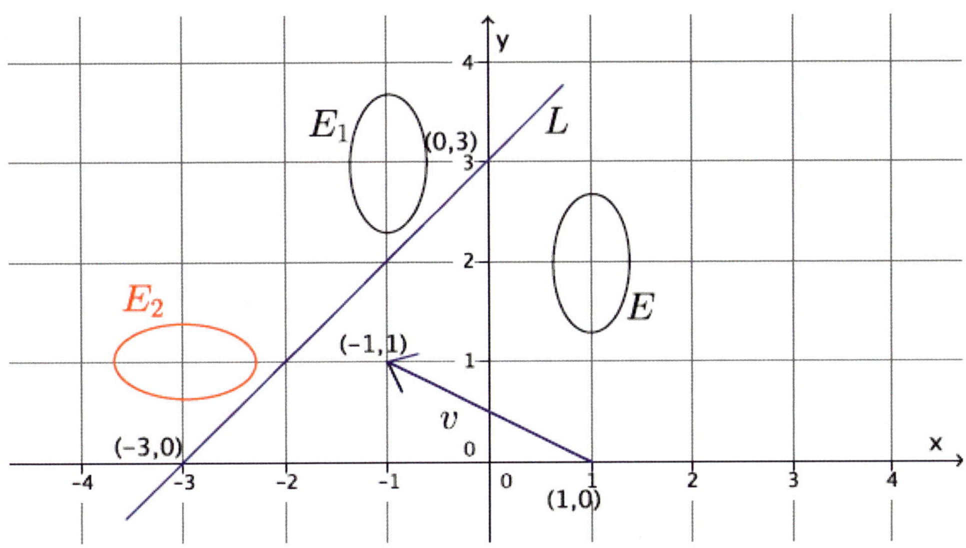

- Now, the only transformation left is $Reflection_3$. So now, we let E_3 be the image of E after the $Translation_1(E)$ followed by the $Rotation_2(E)$ followed by the $Reflection_3(E)$.

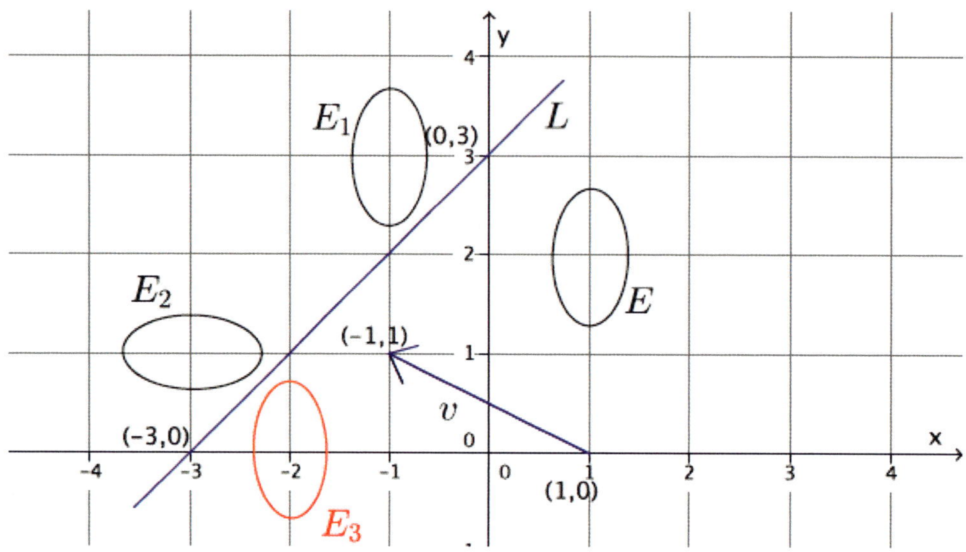

Video Presentation (2 minutes)

Students have seen this video[1] in an earlier lesson, but now that they know about rotation, it is worthwhile to watch it again.

http://youtu.be/O2XPy3ZLU7Y

Exercises 1–5 (25 minutes)

Give students one minute or less to work independently on Exercise 1. Then have the discussion with them that follows the likely student response. Repeat this process for Exercises 2 and 3. For Exercise 4, have students work in pairs. One student can work on Scenario 1 and the other on Scenario 2, or each student can do both scenarios and then compare with a partner.

[1]The video was developed by Larry Francis.

Exercises

1. **In the following picture, triangle ABC can be traced onto a transparency and mapped onto triangle $A'B'C'$. Which basic rigid motion, or sequence of, would map one triangle onto the other?**

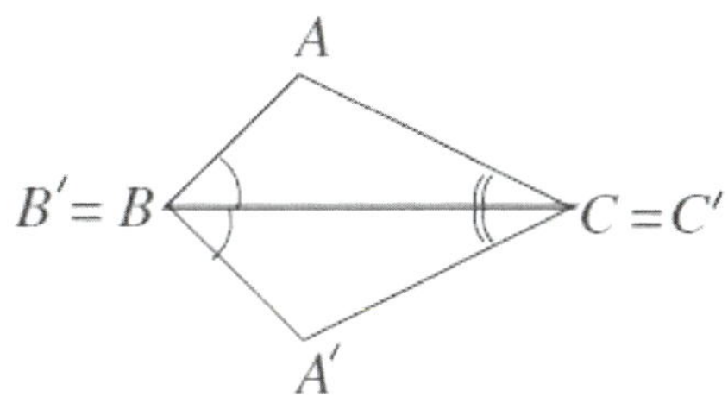

Solution provided below with likely student responses.

Reflection

Elicit more information from students by asking the following:

- Reflection requires some information about which line to reflect over; can you provide a clearer answer?
  - *Reflect over line L_{BC}*

Expand on their answer: Let there be a reflection across the line L_{BC}. We claim that the reflection maps $\triangle ABC$ onto $\triangle A'B'C'$. We can trace $\triangle A'B'C'$ on the transparency and see that when we reflect across line L_{BC}, $\triangle A'B'C'$ maps onto $\triangle ABC$. The reason is because $\angle B$ and $\angle B'$ are equal in measure, and the ray $\overrightarrow{B'A'}$ on the transparency falls on the ray $\overrightarrow{BA}$. Similarly, the ray $\overrightarrow{C'A'}$ falls on the ray $\overrightarrow{CA}$. By the picture, it is obvious that A' on the transparency falls exactly on A, so the reflection of $\triangle A'B'C'$ across L_{BC} is exactly $\triangle ABC$.

Note to Teacher: Here is the precise reasoning without appealing to a transparency. Since a reflection does not move any point on L_{BC}, we already know that $Reflection(B') = B$ and $Reflection(C') = C$. It remains to show that the reflection maps A' to A. The hypothesis says $\angle A'BC$ is equal to $\angle ABC$ in measure; therefore, the ray $\overrightarrow{BC}$ is the angle bisector ($\angle$ bisector) of $\angle ABA'$. The reflection maps the ray $\overrightarrow{BA'}$ to the ray $\overrightarrow{BA}$. Similarly, the reflection maps the ray $\overrightarrow{CA'}$ to the ray $\overrightarrow{CA}$. Therefore, the reflection maps the intersection of the rays $\overrightarrow{BA'}$ and $\overrightarrow{CA'}$, which is of course just A', to the intersection of rays $\overrightarrow{BA}$ and $\overrightarrow{CA}$, which is, of course, just A. So, $Reflection(A') = A$; therefore, $Reflection(\triangle A'B'C') = \triangle ABC$.

2. **In the following picture, triangle ABC can be traced onto a transparency and mapped onto triangle $A'B'C'$. Which basic rigid motion, or sequence of, would map one triangle onto the other?**

Rotation

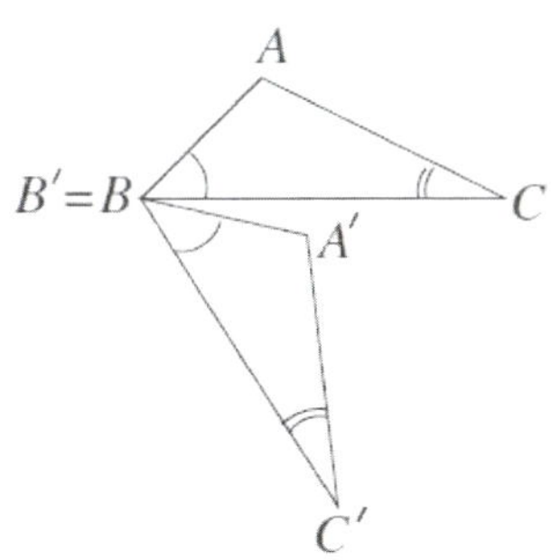

Elicit more information from students by asking:

- Rotation requires some information about what point to rotate around (the center) and how many degrees. If we say we need to rotate d degrees, can you provide a clearer answer?
  - *Rotate around point B as the center, d degrees*

Expand on their answer: Let there be the (counterclockwise) rotation of d degrees around B, where d is the (positive) degree of the $\angle CBC'$. We claim that the rotation maps $\triangle A'B'C'$ to $\triangle ABC$. We can trace $\triangle A'B'C'$ on the transparency and see that when we pin the transparency at B' (same point as B) and perform a counterclockwise rotation of d degrees, the segment $B'C'$ on the transparency maps onto segment BC (both are equal in length because we can trace one on the transparency and show it is the same length as the other). The point A' on the transparency and A are on the same side (half-plane) of line L_{BC}. Now, we are at the same point we were in the end of Exercise 1; therefore, $\triangle A'B'C'$ and $\triangle ABC$ completely coincide.

Note to Teacher: Here is the precise reasoning without appealing to a transparency. By definition of rotation, rotation maps the ray $\overrightarrow{BC'}$ to the ray $\overrightarrow{BC}$. However, by hypothesis, $BC = BC'$, so $Rotation(C') = C$. Now, the picture implies that after the rotation, A and $Rotation(A)$ lie on the same side of line L_{BC}. If we compare the triangles ABC and $Rotation(A'B'C')$, we are back to the situation at the end of Exercise 1; therefore, the reasoning given here shows that the two triangles coincide.

3. In the following picture, triangle ABC can be traced onto a transparency and mapped onto triangle $A'B'C'$. Which basic rigid motion, or sequence of, would map one triangle onto the other?

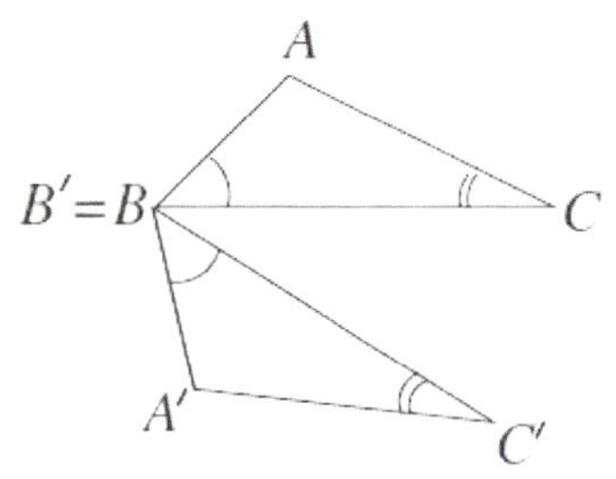

Rotation and reflection

Elicit more information from students. Prompt students to think back to what was needed in the last two examples.

- What additional information do we need to provide?
  - *Rotate around point B as the center d degrees; then, reflect across line L_{BC}.*

Expand on their answer: We need a sequence this time. Let there be the (counterclockwise) rotation of d degrees around B, where d is the (positive) degree of the $\angle CBC'$, and let there be the reflection across the line L_{BC}. We claim that the sequence rotation and then reflection maps $\triangle A'B'C'$ to $\triangle ABC$. We can trace $\triangle A'B'C'$ on the transparency and see that when we pin the transparency at B' (same point as B) and perform a counterclockwise rotation of d degrees, that the segment $B'C'$ on the transparency maps onto segment BC. Now, $\triangle ABC$ and $\triangle A'B'C'$ are in the exact position as they were in the beginning of Example 2 (Exercise 1); therefore, the reflection across L_{BC} would map $\triangle A'B'C'$ on the transparency to $\triangle ABC$.

Students may say that they want to reflect first and then rotate. The sequence can be completed in that order, but point out that we need to state which line to reflect across. In that case, we would have to find the appropriate line of reflection. For that reason, it makes more sense to bring a pair of sides together first, that is, segment BC and segment BC', by a rotation, and then reflect across the common side of the two triangles. When the rotation is performed first, we can use what we know about Exercise 1.

Note to Teacher: Without appealing to a transparency, the reasoning is as follows. By definition of rotation, rotation maps the ray $\overrightarrow{BC'}$ to the ray $\overrightarrow{BC}$. However, by hypothesis, $BC = BC'$, so $Rotation(C') = C$. Now, when comparing the triangles ABC and $Rotation(A'B'C')$, we see that we are back to the situation in Exercise 1; therefore, the reflection maps the $Rotation(\triangle A'B'C')$ to $\triangle ABC$. This means that rotation then reflection maps $\triangle A'B'C'$ to $\triangle ABC$.

4. In the following picture, we have two pairs of triangles. In each pair, triangle ABC can be traced onto a transparency and mapped onto triangle $A'B'C'$.

Which basic rigid motion, or sequence of, would map one triangle onto the other?

Scenario 1:

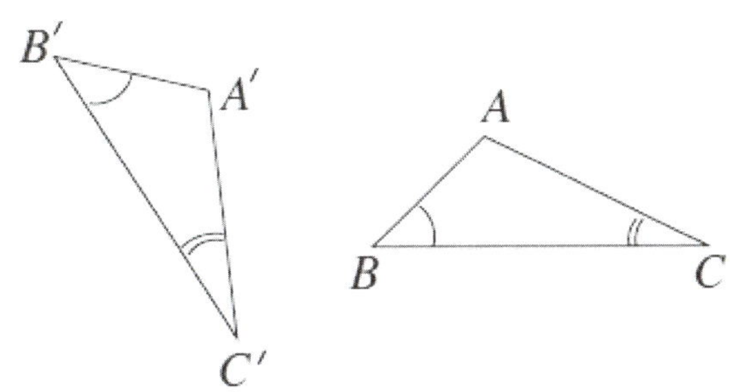

Scenario 2:

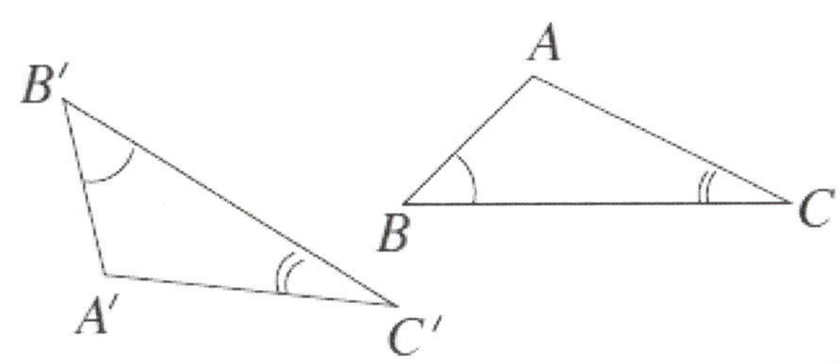

In Scenario 1, a translation and a rotation; in Scenario 2, a translation, a reflection, and then a rotation

Elicit more information from students by asking the following:

- What additional information is needed for a translation?
  - *We need to translate along a vector.*
- Since there is no obvious vector in our picture, which vector should we draw and then use to translate along?

When they do not respond, prompt them to select a vector that would map a point from $\triangle A'B'C'$ to a corresponding point in $\triangle ABC$. Students will likely respond,

- *Draw vector $\overrightarrow{B'B}$ (or $\overrightarrow{A'A}$ or $\overrightarrow{C'C}$).*

Make it clear to students that we can use any of the vectors they just stated, but using $\overrightarrow{B'B}$ makes the most sense because we can use the reasoning given in the previous exercises rather than constructing the reasoning from the beginning. (For example, in Exercises 1–3, $B = B'$.)

Expand on their answer: Let there be the translation along vector $\overrightarrow{B'B}$. In Scenario 1, the triangles ABC and $Translation(A'B'C')$ would be similar to the situation of Exercise 2. In Scenario 2, the triangles ABC and $Translation(A'B'C')$ would be similar to the situation of Exercise 3. Based on the work done in Exercises 2 and 3, we can conclude the following: In Scenario 1, the sequence of a translation along $\overrightarrow{B'B}$ followed by a rotation around B would map $\triangle A'B'C'$ to $\triangle ABC$, and in Scenario 2, the sequence of a translation along $\overrightarrow{B'B}$ followed by a rotation around B and finally followed by the reflection across line L_{BC} would map $\triangle A'B'C'$ to $\triangle ABC$.

Students complete Exercise 5 independently or in pairs.

5. **Let two figures ABC and $A'B'C'$ be given so that the length of curved segment AC equals the length of curved segment $A'C'$, $|\angle B| = |\angle B'| = 80°$, and $|AB| = |A'B'| = 5$. With clarity and precision, describe a sequence of rigid motions that would map figure ABC onto figure $A'B'C'$.**

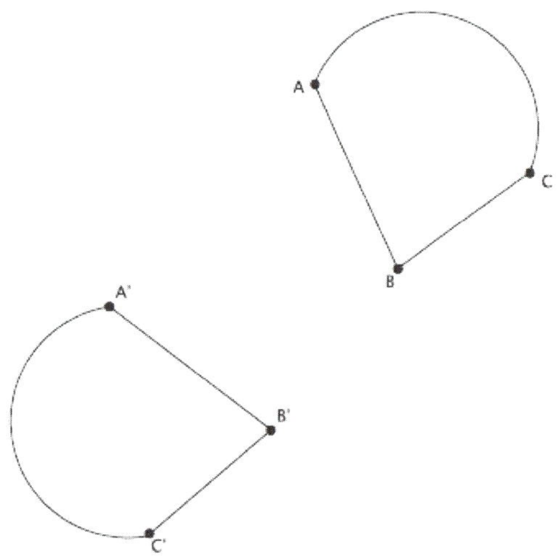

Let there be the translation along vector $\overrightarrow{AA'}$, let there be the rotation around point A d degrees, and let there be the reflection across line L_{AB}. Translate so that $Translation(A') = A$. Rotate so that $Rotation(B') = B$, and $Rotation(\overline{A'B'})$ coincides with $\overline{AB}$ (by hypothesis, they are the same length, so we know they will coincide). Reflect across L_{AB} so that $Reflection(C') = C$ and $Reflection(\overline{C'B'})$ coincides with $\overline{CB}$ (by hypothesis, $|\angle B| = |\angle B'| = 80°$, so we know that segment $C'B'$ will coincide with segment CB). By hypothesis, the length of the curved segment $A'C'$ is the same as the length of the curved segment AC, so they will coincide. Therefore, a sequence of translation, then rotation, and then reflection will map figure $A'B'C'$ onto figure ABC.

Closing (5 minutes)

Summarize, or have students summarize, the lesson and what they know of rigid motions to this point:

- We can now describe, using precise language, how to sequence rigid motions so that one figure maps onto another.

Exit Ticket (5 minutes)

Name ___________________________________ Date ________________

Lesson 10: Sequences of Rigid Motions

Exit Ticket

Triangle ABC has been moved according to the following sequence: a translation followed by a rotation followed by a reflection. With precision, describe each rigid motion that would map $\triangle ABC$ onto $\triangle A'B'C'$. Use your transparency and add to the diagram if needed.

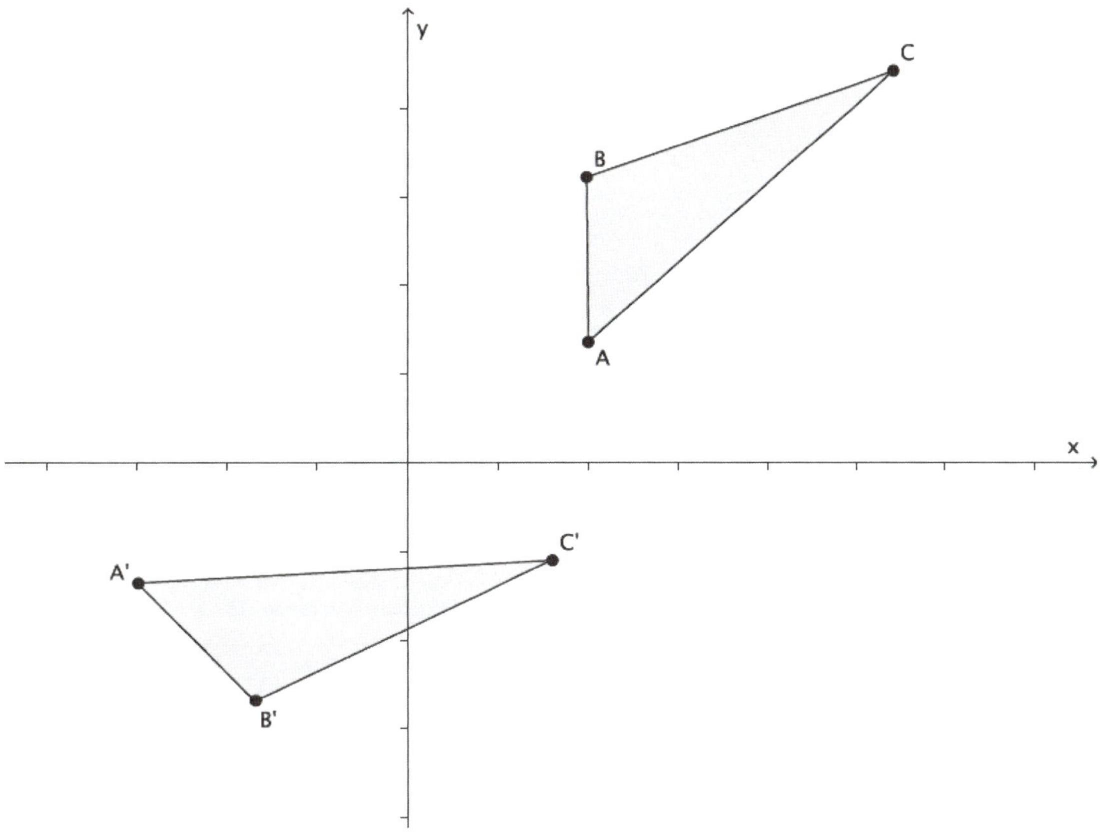

Exit Ticket Sample Solutions

Triangle ABC has been moved according to the following sequence: a translation followed by a rotation followed by a reflection. With precision, describe each rigid motion that would map $\triangle ABC$ onto $\triangle A'B'C'$. Use your transparency and add to the diagram if needed.

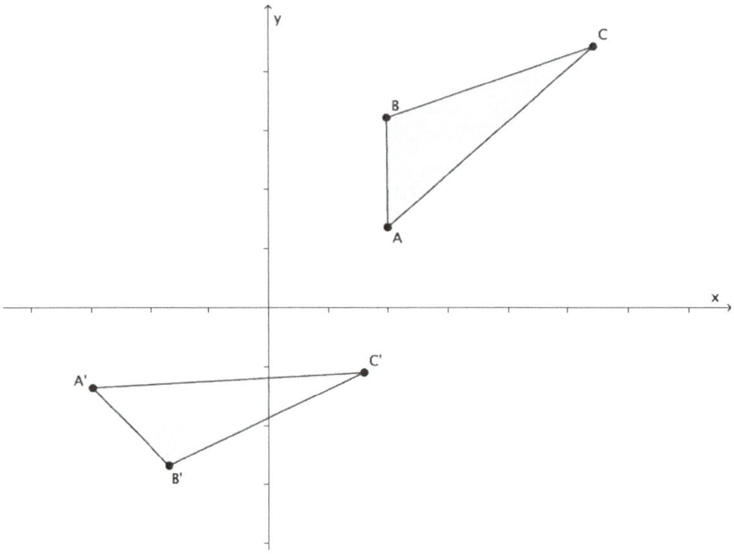

Let there be the translation along vector $\overrightarrow{AA'}$ so that A is mapped to A'. Let there be the clockwise rotation by d degrees around point A' so that C is mapped to C' and $AC = A'C'$. Let there be the reflection across $L_{A'C'}$ so that B is mapped to B'.

Problem Set Sample Solutions

1. Let there be the translation along vector $\vec{v}$, let there be the rotation around point A, -90 degrees (clockwise), and let there be the reflection across line L. Let S be the figure as shown below. Show the location of S after performing the following sequence: a translation followed by a rotation followed by a reflection.

   Solution is shown in red.

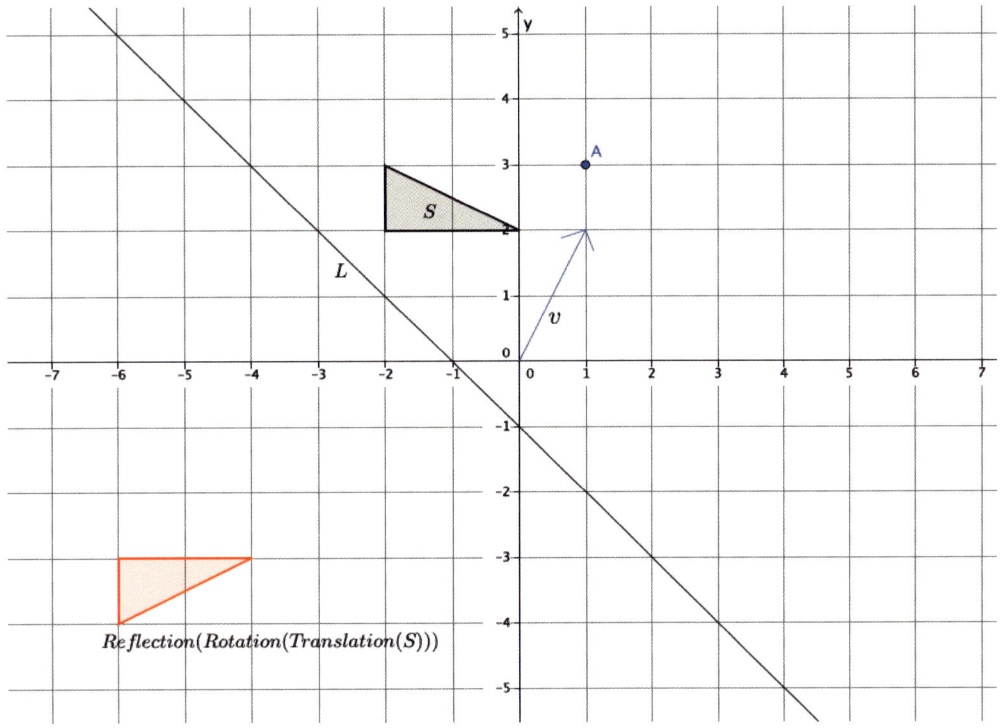

2. Would the location of the image of S in the previous problem be the same if the translation was performed last instead of first; that is, does the sequence, translation followed by a rotation followed by a reflection, equal a rotation followed by a reflection followed by a translation? Explain.

   No, the order of the transformation matters. If the translation was performed last, the location of the image of S, after the sequence, would be in a different location than if the translation was performed first.

3. Use the same coordinate grid to complete parts (a)–(c).

a. Reflect triangle ABC across the vertical line, parallel to the y-axis, going through point $(1, 0)$. Label the transformed points A, B, C as A', B', C', respectively.

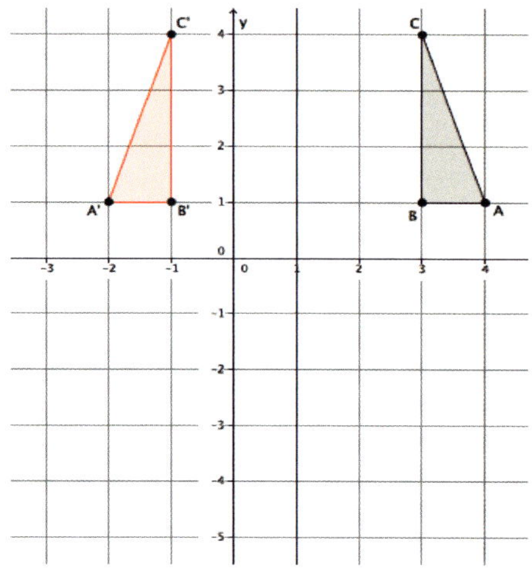

b. Reflect triangle $A'B'C'$ across the horizontal line, parallel to the x-axis going through point $(0, -1)$. Label the transformed points of A', B', C' as A'', B'', C'', respectively.

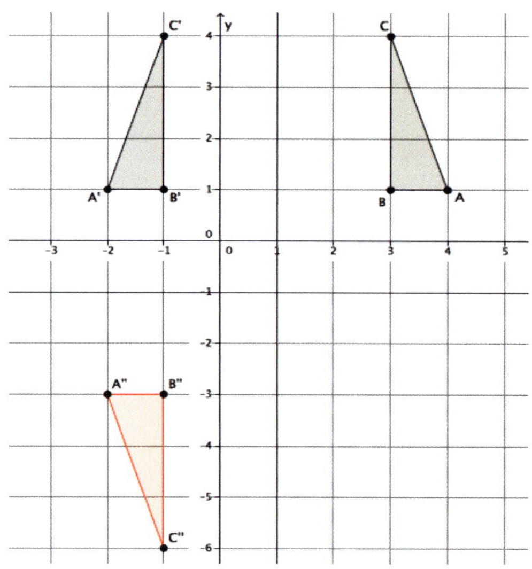

c. Is there a single rigid motion that would map triangle ABC to triangle $A''B''C''$?

Yes, a $180°$ rotation around center $(1, -1)$. The coordinate $(1, -1)$ happens to be the intersection of the two lines of reflection.

Name ______________________________ Date ____________

1.

a. Translate $\triangle XYZ$ along $\overrightarrow{AB}$. Label the image of the triangle with X', Y', and Z'.

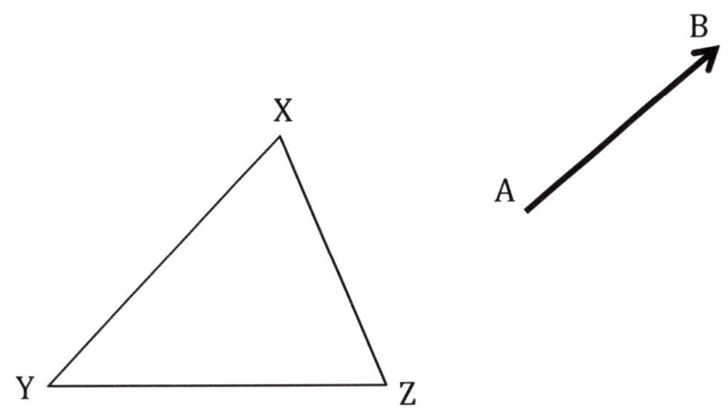

b. Reflect $\triangle XYZ$ across the line of reflection, l. Label the image of the triangle with X', Y', and Z'.

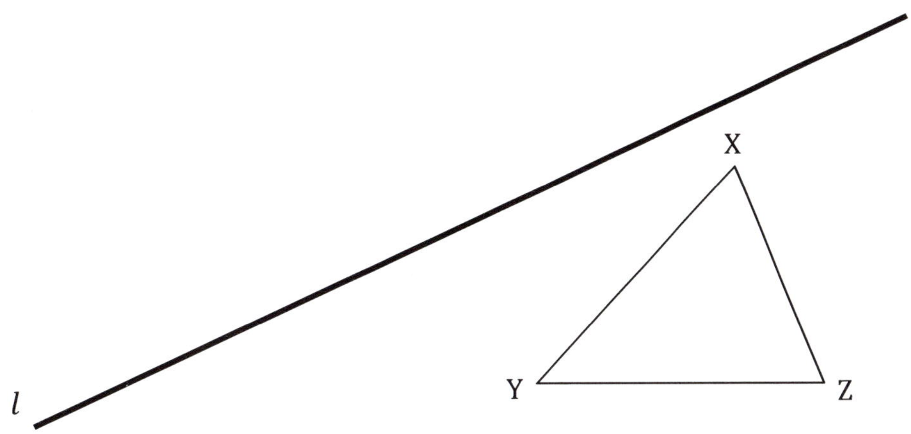

c. Rotate $\triangle XYZ$ around the point $(1,0)$ clockwise $90°$. Label the image of the triangle with X', Y', and Z'.

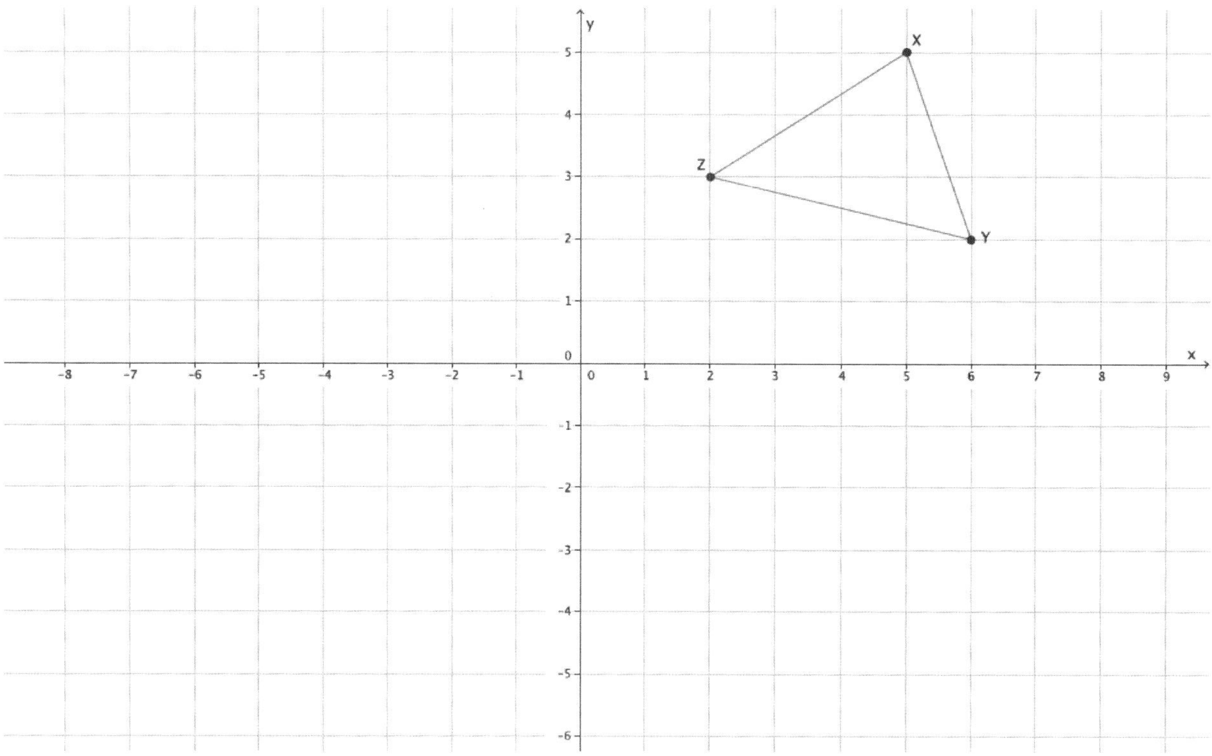

2. Use the picture below to answer the questions.

Figure A has been transformed to Figure B.

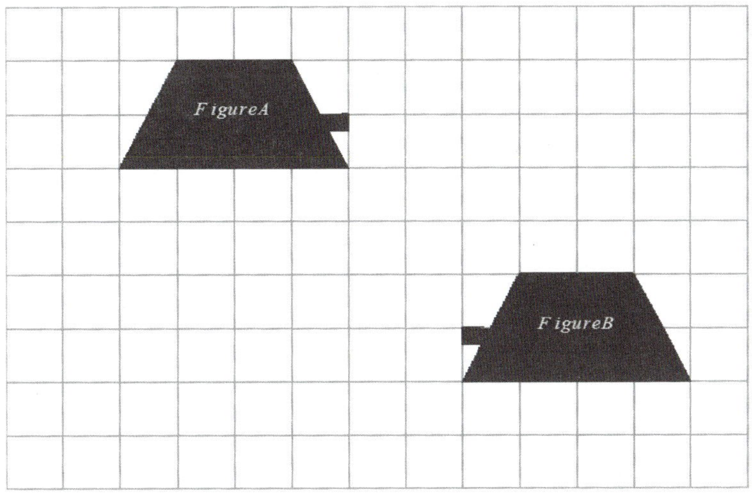

a. Can Figure A be mapped onto Figure B using only translation? Explain. Use drawings as needed in your explanation.

b. Can Figure A be mapped onto Figure B using only reflection? Explain. Use drawings as needed in your explanation.

3. Use the graphs below to answer parts (a) and (b).

   a. Reflect $\triangle XYZ$ over the horizontal line (parallel to the x-axis) through point $(0,1)$. Label the reflected image with $X'Y'Z'$.

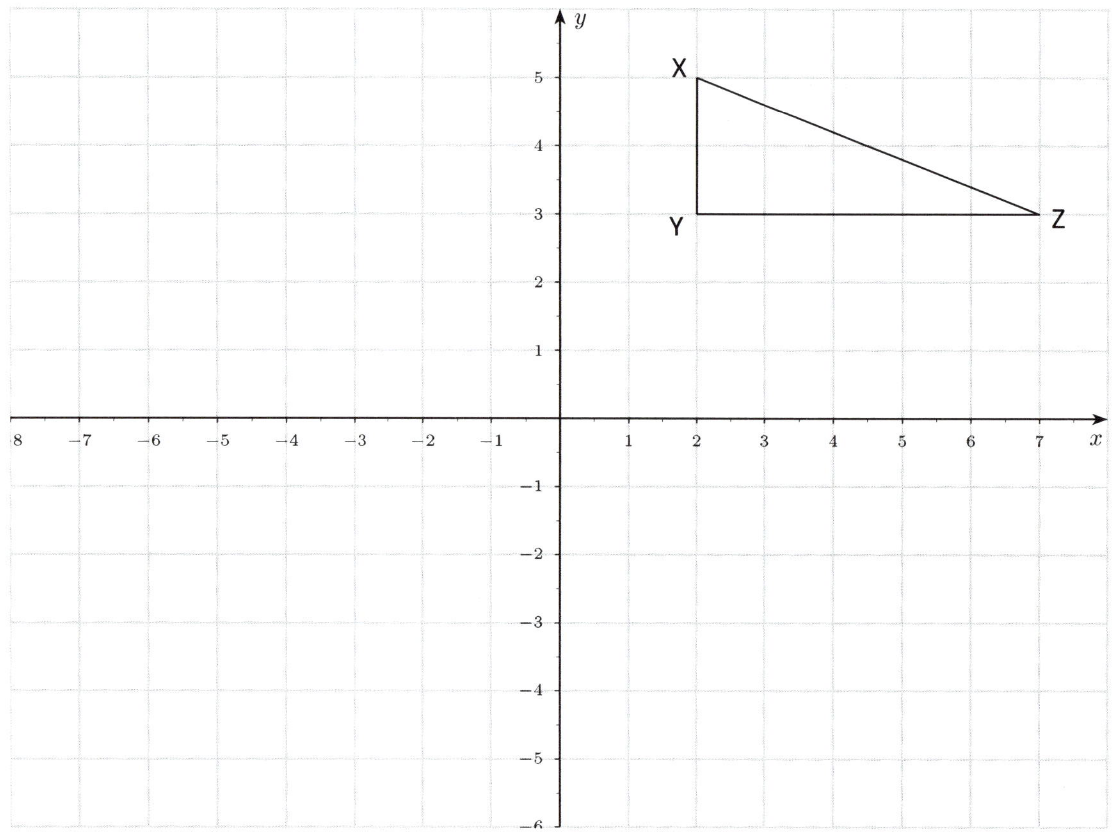

b. One triangle in the diagram below can be mapped onto the other using two reflections. Identify the lines of reflection that would map one onto the other. Can you map one triangle onto the other using just one basic rigid motion? If so, explain.

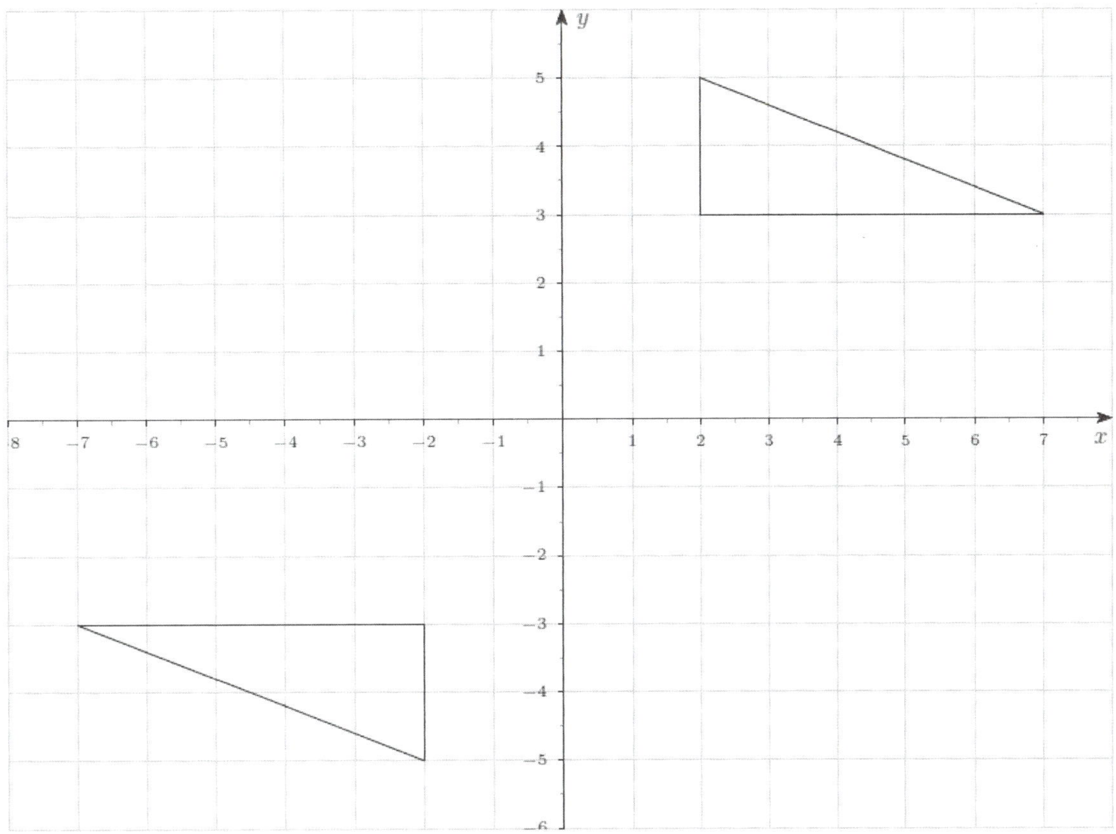

A Progression Toward Mastery					
Assessment Task Item		**STEP 1** Missing or incorrect answer and little evidence of reasoning or application of mathematics to solve the problem.	**STEP 2** Missing or incorrect answer but evidence of some reasoning or application of mathematics to solve the problem.	**STEP 3** A correct answer with some evidence of reasoning or application of mathematics to solve the problem, OR an incorrect answer with substantial evidence of solid reasoning or application of mathematics to solve the problem.	**STEP 4** A correct answer supported by substantial evidence of solid reasoning or application of mathematics to solve the problem.
1	a 8.G.A.1	Student is unable to respond to the question or leaves item blank. Student enlarges or shrinks image. Student reflects or rotates image.	Student translates along a vector. Student uses a different vector than what was given. Student shortens or lengthens given vector. Student does not label image or labels image incorrectly.	Student translates correctly along vector. Student does not label image or labels image incorrectly.	Student translates correctly along vector and labels image correctly.
	b 8.G.A.1	Student is unable to respond to the question or leaves item blank. Student enlarges or shrinks image. Student translates or rotates the image.	Student reflects across line but may reflect across a different line than what is given. Student does not label image or labels image incorrectly. The orientation of the image may be incorrect.	Student reflects correctly across line. Student does not label image or labels image incorrectly.	Student reflects correctly across line and labels image correctly.
	c 8.G.A.1	Student is unable to respond to the question or leaves item blank. Student translates the triangle to the correct quadrant. Student reflects the triangle to the correct quadrant.	Student rotates about the point $(1,0)$. Student rotates the triangle counterclockwise $90°$. Student rotates more or less than $90°$. Student does not label image or labels image incorrectly.	Student rotates about the point $(1,0)$ clockwise $90°$. Student does not label image or labels image incorrectly.	Student rotates about the point $(1,0)$ clockwise $90°$ and labels image correctly.

EUREKA MATH™

2	a 8.G.A.1	Student answers with yes or no only. Student is unable to give any explanation (pictorially or written).	Student answers with yes or no. Student shows some reasoning (pictorially or written) to solve the problem. Student shows no application of mathematics to solve the problem.	Student answers correctly with no. Student uses a pictorial explanation only as evidence of reasoning. Some evidence of mathematical reasoning is evident in explanation. Student does not use mathematical vocabulary in explanation.	Student answers correctly with no and uses mathematical vocabulary in explanation. Student may use pictorial explanation to enhance mathematical explanation.
	b 8.G.A.1	Student answers with yes or no only. Student is unable to give any explanation (pictorially or written).	Student answers with yes or no. Student shows some reasoning (pictorially or written) to solve the problem. Student shows no application of mathematics to solve the problem.	Student answers correctly with no. Student uses a pictorial explanation only as evidence of reasoning. Some evidence of mathematical reasoning is evident in explanation. Student does not use mathematical vocabulary in explanation.	Student answers correctly with no and uses mathematical vocabulary in explanation. Student may use pictorial explanation to enhance mathematical explanation.
3	a 8.G.A.1	Student is unable to respond to the question or leaves item blank. Student shows no reasoning or application of mathematics to solve the problem.	Student reflects triangle across any line other than the line $y = 1$. The orientation of the triangle may or may not be correct. Student may or may not label the triangle correctly.	Student reflects triangle across the line $y = 1$. The orientation of the triangle is correct. Student may or may not label the triangle correctly.	Student reflects triangle across the line $y = 1$, and the orientation of the triangle is correct. Student labels the triangle correctly.
	b 8.G.A.1	Student is unable to respond to the question or leaves item blank. Student answers with yes or no only. Student may or may not identify the lines of reflection. No evidence of mathematical reasoning is used in written explanation.	Student answers with yes or no. Student may or may not identify the lines of reflection. Student identifies a rotation as the rigid motion. Student may or may not identify the degree of rotation or the center of rotation. Some evidence of mathematical reasoning is used in written explanation.	Student answers correctly with yes. Student identifies the lines of reflection. Student identifies a rotation as the rigid motion. Student identifies the degree of rotation. Student may or may not identify the center of rotation. Some evidence of mathematical reasoning is used in written explanation.	Student answers correctly with yes. Student correctly identifies the lines of reflection as $y = 0$ and $x = 0$. Student identifies a rotation as the rigid motion. Student identifies the degree of rotation as 180. Student identifies the center of rotation as the origin and substantial evidence of mathematical reasoning is used in written explanation.

Name ______________________________ Date ______________

1.

a. Translate $\triangle XYZ$ along $\overrightarrow{AB}$. Label the image of the triangle with X', Y', and Z'.

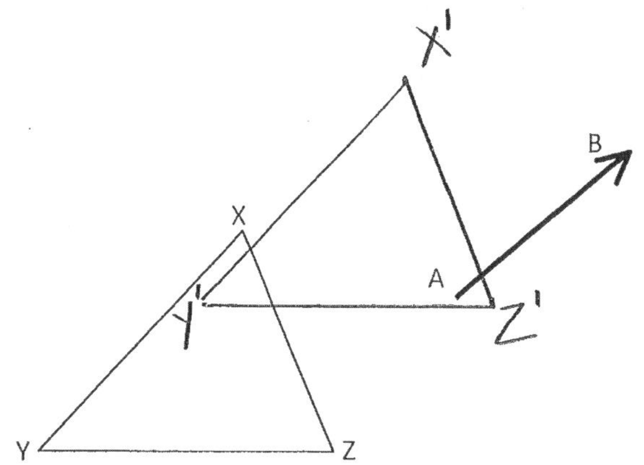

b. Reflect $\triangle XYZ$ across the line of reflection, l. Label the image of the triangle with X', Y', and Z'.

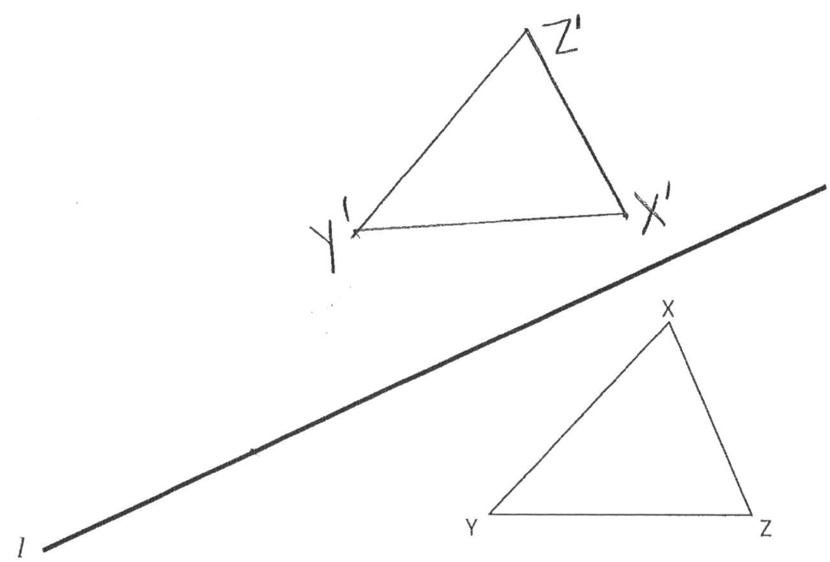

c. Rotate $\triangle XYZ$ around the point $(1,0)$ clockwise $90°$. Label the image of the triangle with X', Y', and Z'.

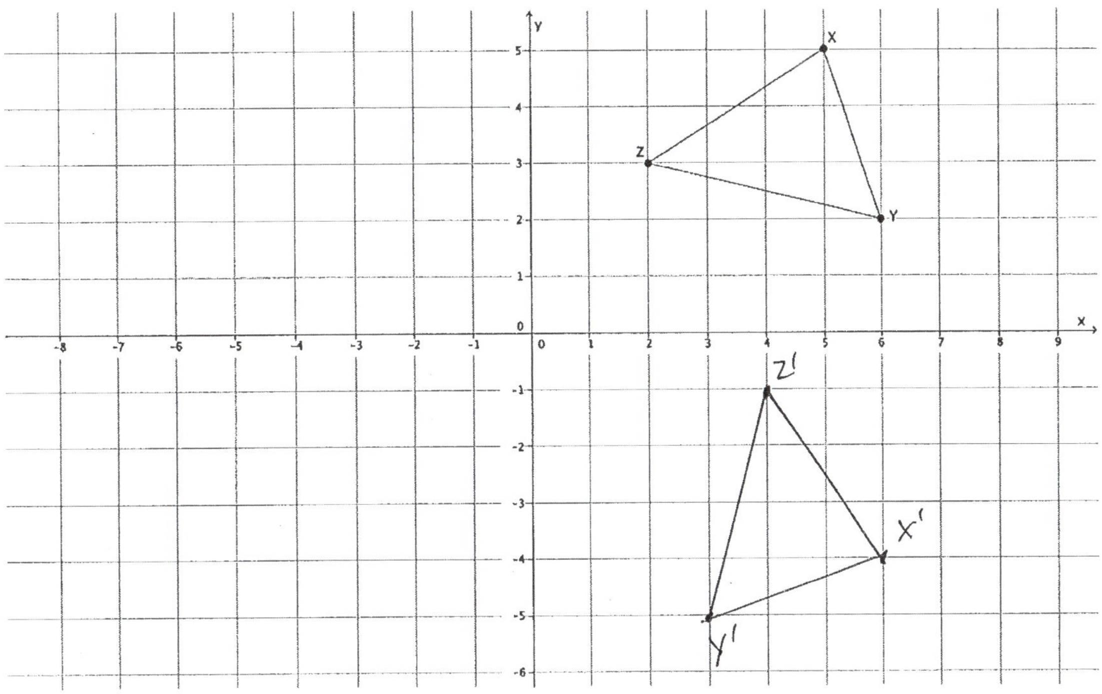

2. Use the picture below to answer the questions.

Figure A has been transformed to Figure B.

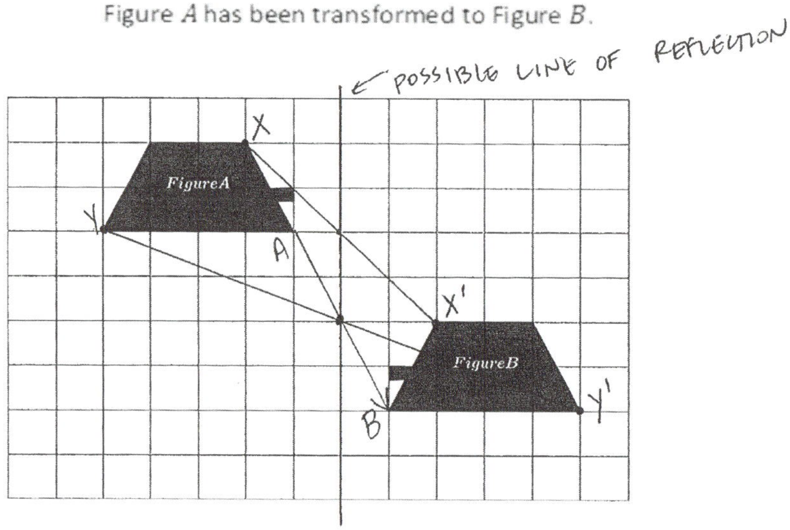

a. Can Figure A be mapped onto Figure B using only translation? Explain. Use drawings as needed in your explanation.

NO, IF I TRANSLATE ALONG VECTOR $\overrightarrow{AB}$ I CAN GET THE LOWER POINT OF FIGURE A TO MAP ONTO THE LOWER LEFT POINT OF FIGURE B (ONE PAIR OF CORRESPONDING POINTS) BUT NO OTHER POINTS OF THE FIGURES COINCIDE.)

b. Can Figure A be mapped onto Figure B using only reflection? Explain. Use drawings as needed in your explanation.

NO, WHEN I CONNECT A POINT OF FIGURE A TO ITS IMAGE ON FIGURE B, THE LINE OF REFLECTION SHOULD BISECT THE SEGMENT. WHEN I CONNECT MIDPOINTS OF $\overline{XX'}$ & $\overline{YY'}$ I GET A POSSIBLE LINE OF REFLECTION, BUT WHEN I CHECK, FIGURE A DOES NOT MAP ONTO FIGURE B.

3. Use the graphs below to answer parts (a) and (b).

   a. Reflect $\triangle XYZ$ over the horizontal line (parallel to the x-axis) through point $(0,1)$. Label the reflected image with $X'Y'Z'$.

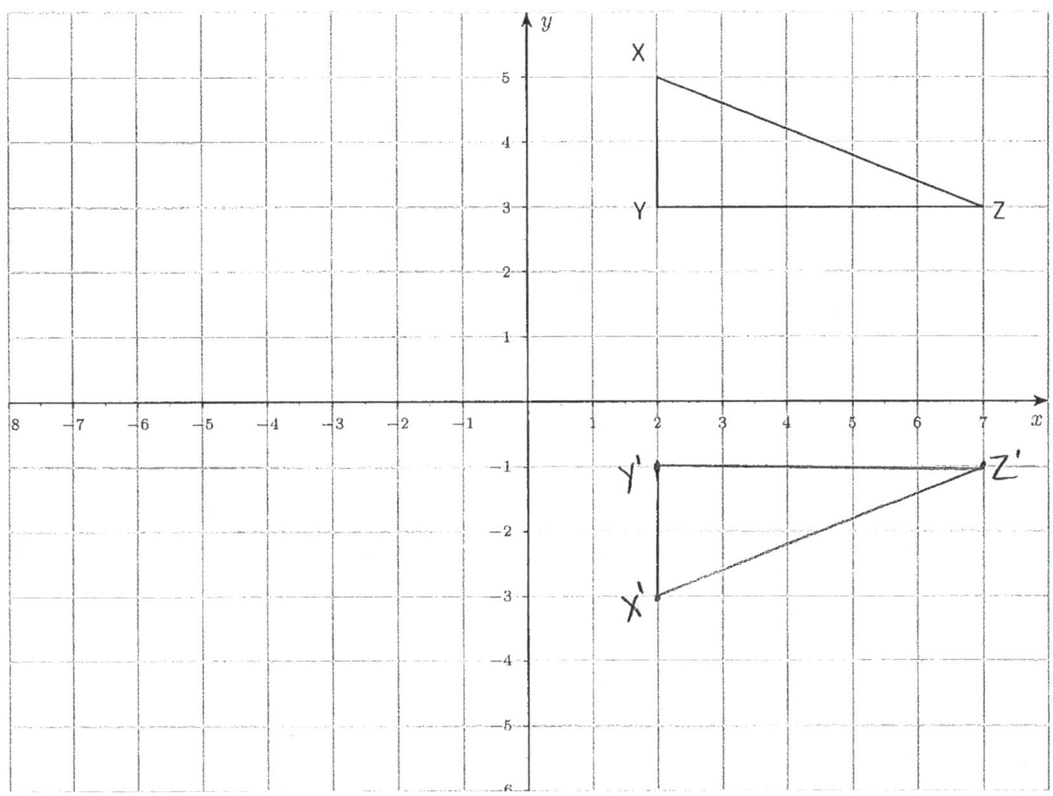

b. One triangle in the diagram below can be mapped onto the other using two reflections. Identify the lines of reflection that would map one onto the other. Can you map one triangle onto the other using just one basic rigid motion? If so, explain.

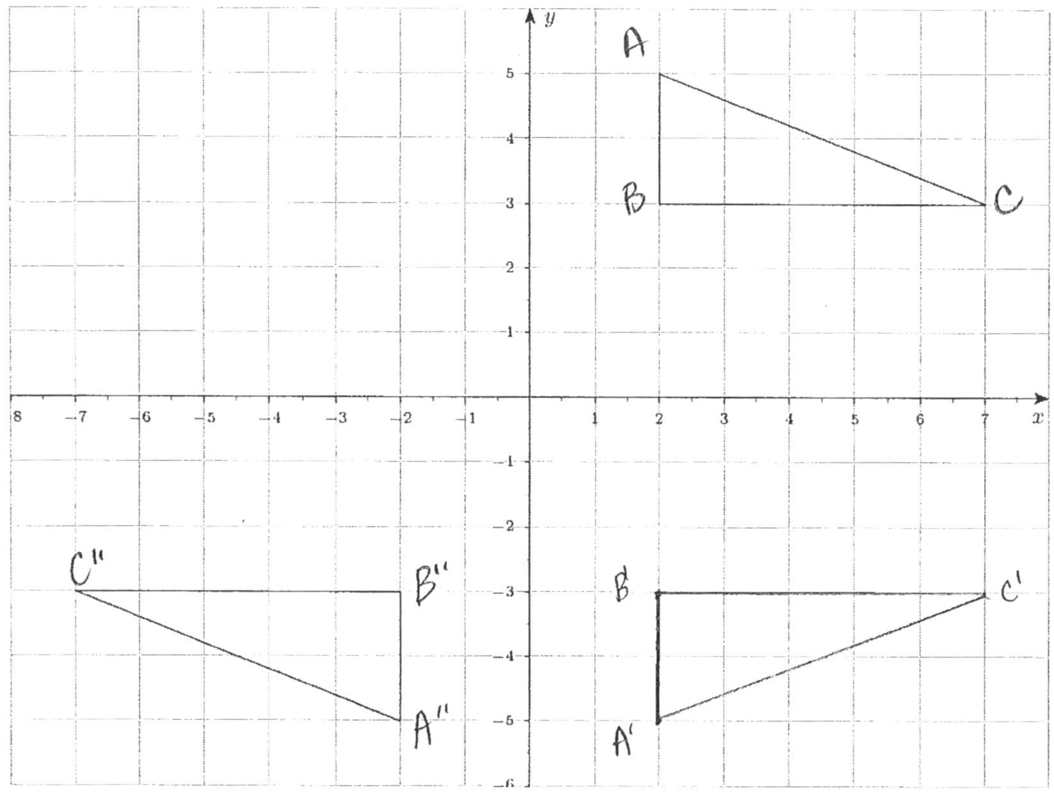

A REFLECTION ACROSS THE X-AXIS MAPS △ABC TO △A'B'C' AND A REFLECTION ACROSS THE Y-AXIS MAPS △A'B'C' TO △A"B"C".

SINCE AB || A"B", BC || B"C", AND AC || A"C" AND THE LENGTHS AB = A"B", BC = B"C", AC = A"C", THEN A 180° ROTATION ABOUT THE ORIGIN WILL MAP △ABC TO △A"B"C".

Topic C

Congruence and Angle Relationships

8.G.A.2, 8.G.A.5

Focus Standards:	8.G.A.2	Understand that a two-dimensional figure is congruent to another if the second can be obtained from the first by a sequence of rotations, reflections, and translations; given two congruent figures, describe a sequence that exhibits the congruence between them.
	8.G.A.5	Use informal arguments to establish facts about the angle sum and exterior angle of triangles, about the angles created when parallel lines are cut by a transversal, and the angle-angle criterion for similarity of triangles. *For example, arrange three copies of the same triangle so that the sum of the three angles appears to form a line, and give an argument in terms of transversals why this is so.*
Instructional Days:	4	
Lesson 11:	Definition of Congruence and Some Basic Properties (S)[1]	
Lesson 12:	Angles Associated with Parallel Lines (E)	
Lesson 13:	Angle Sum of a Triangle (E)	
Lesson 14:	More on the Angles of a Triangle (S)	

Topic C finishes the work of **8.G.A.2** by introducing the concept of congruence as mapping one figure onto another using a sequence of rigid motions. Lesson 11 defines congruence in terms of a sequence of the basic rigid motions (i.e., translations, reflections, and rotations). Students learn the fundamental assumptions that are made about the basic rigid motions that serve as the basis of all geometric investigations.

The concept of congruence and basic rigid motions are used to determine which angles of parallel lines are equal in measure. In Lesson 12, students show why corresponding angles are congruent using translation and why alternate interior angles are congruent using rotation. In Lessons 13 and 14, the knowledge of rigid motions and angle relationships is put to use to develop informal arguments to show that the sum of the degrees of interior angles of a triangle is $180°$. Students are presented with three such arguments as the importance of the theorem justifies the multiple perspectives. Students also take note of a related fact about the exterior angles of triangles.

---

[1]Lesson Structure Key: **P**-Problem Set Lesson, **M**-Modeling Cycle Lesson, **E**-Exploration Lesson, **S**-Socratic Lesson

Lesson 11: Definition of Congruence and Some Basic Properties

Student Outcomes

- Students know the definition of congruence and related notation, that is, $\cong$. Students know that to prove two figures are congruent, there must be a sequence of rigid motions that maps one figure onto the other.
- Students know that the basic properties of congruence are similar to the properties for all three rigid motions (translations, rotations, and reflections).

Classwork

Example 1 (5 minutes)

MP.6

- Sequencing basic rigid motions has been practiced throughout the lessons of Topic B in this module because, in general, the sequence of (a finite number of) basic rigid motions is called a congruence. A geometric figure S is said to be congruent to another geometric figure S' if there is a sequence of rigid motions that maps S to S', that is, $Congruence(S) = S'$. The notation related to congruence is the symbol $\cong$. When two figures are congruent, like S and S', we can write $S \cong S'$.
- We want to describe the sequence of rigid motions that demonstrates the two triangles shown below are congruent, that is, $\triangle ABC \cong \triangle A'B'C'$.

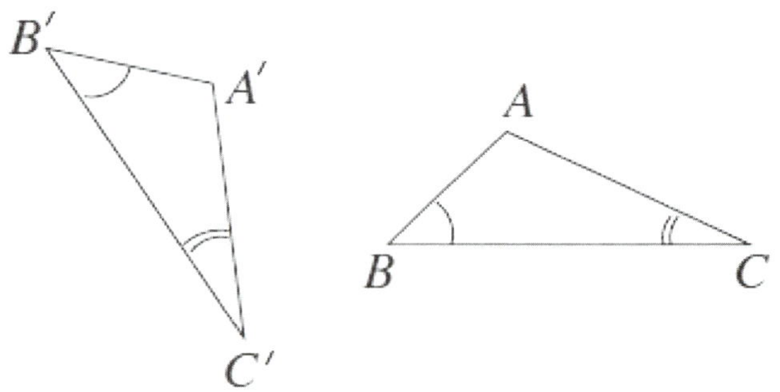

Note to Teacher:

Demonstrate, or have students demonstrate, the rigid motions as they work through the sequence.

- What rigid motion will bring the two triangles together? That is, which motion would bring together at least one pair of corresponding points (vertices)? Be specific.
  - *Translate $\triangle A'B'C'$ along vector $\overrightarrow{A'A}$.*
- What rigid motion would bring together one pair of sides? Be specific.
  - *Rotate d degrees around center A.*
- After these two rigid motions, we have shown that $\triangle ABC \cong \triangle A'B'C'$ through the sequence of a translation followed by a rotation. Notice that only two rigid motions were needed for this sequence. A sequence to demonstrate congruence can be made up of any combination of the basic rigid motions using all three or even just one.

- The concept of *congruence* between two geometric figures is one of the cornerstones of geometry. Congruence is now realized as "a sequence of basic rigid motions that maps one figure onto another."
- Recall the first question raised in this module, "Why move things around?" Now, a complete answer can be given in terms of congruence.

> *Note to Teacher:*
>
> The preceding definition of congruence is meant to replace the existing "same size and same shape" definition.

Example 2 (10 minutes)

- It is said that S is congruent to S' if there is a congruence so that $Congruence(S) = S'$. This leaves open the possibility that, although S is congruent to S', the figure S' may not be congruent to S.

Ask students:

- If there is *a* $Congruence_1$ so that $Congruence_1(S) = S'$, do we know that there will also be a $Congruence_2$ so that $Congruence_2(S') = S$?

Make sure students understand the question.

- Can you say for certain that if they begin by mapping figure 1 onto figure 2, they can also map figure 2 onto figure 1?
  - *Students will likely say yes, but without proof, further work is necessary.*
- Let $Congruence$ be the transformation that first translates by vector $\overrightarrow{MN}$ followed by a reflection across line L. Let S be the figure on the left below, and let S' be the figure on the right below. Then, the equation

$$Congruence(S) = S'$$

says that if we trace S in red on a transparency, then translate the transparency along $\overrightarrow{MN}$, and flip it across L, we get the figure to coincide completely with S'.

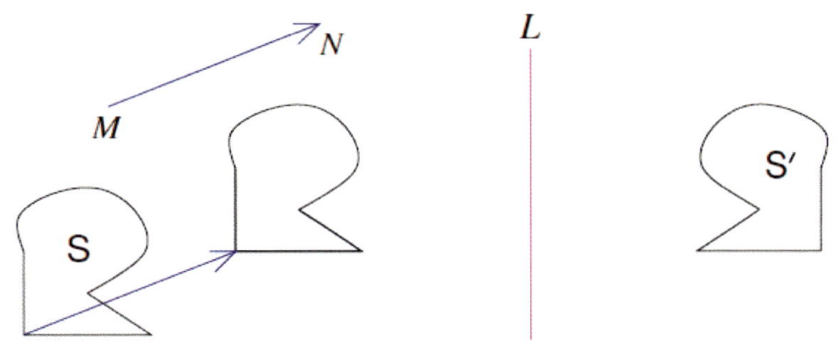

- Now, keeping in mind what we know about how to undo transformations in general, it is obvious how to get a congruence to map S' to S. Namely, tracing the figure S' in red, flipping the transparency across L so the red figure arrives at the figure in the middle, and then translating the figure along vector $\overrightarrow{NM}$. (Note the change in direction of the vector from $\overrightarrow{MN}$.) The red figure now coincides completely with S. The sequence of the reflection across L followed by the translation along vector $\overrightarrow{NM}$ achieves the congruence.

- The general argument is that if there is a $Congruence_1$ so that $Congruence_1(S) = S'$, then there will also be a $Congruence_2$ so that $Congruence_2(S') = S$ is similar. The only additional comment to complete the picture is that, in addition to
    1. The sequence required to show that $Congruence_2$ followed by $Congruence_1$ is equal to $Congruence_1$ followed by $Congruence_2$,
    2. A reflection is undone by a reflection across the same line.

  We also have to draw upon the sequence of rotations that maps a figure onto itself to be certain that each of the three basic rigid motions can be undone by another basic rigid motion.
- In summary, if a figure S is congruent to another figure S', then S' is also congruent to S. In symbols $S \cong S'$. It does not matter whether S or S'comes first.

Exercise 1 (10 minutes)

Students work on Exercise 1 in pairs. Students will likely need some guidance with part (a) of Exercise 1. Provide support, and then allow them to work with a partner to complete parts (b) and (c).

Exercise 1

a. **Describe the sequence of basic rigid motions that shows $S_1 \cong S_2$.**

Let there be the translation along vector $\overrightarrow{AB}$. Let there be a rotation around point B, d_1 degrees. Let there be a reflection across the longest side of the figure so that S_1 maps onto S_2. Then, the $Translation(S_1)$ followed by the rotation followed by the reflection is S_2.

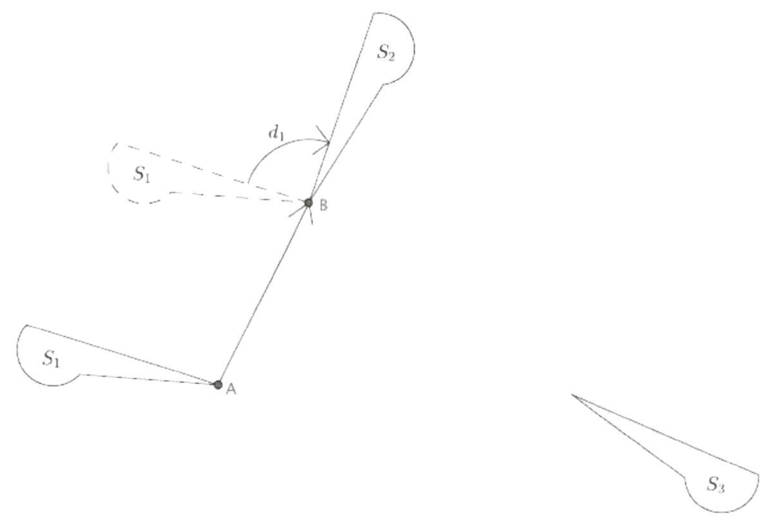

b. **Describe the sequence of basic rigid motions that shows $S_2 \cong S_3$.**

Let there be a translation along vector $\overrightarrow{BC}$. Let there be a rotation around point C, d_2 degrees so that S_2 maps onto S_3. Then, the $Translation(S_2)$ followed by the rotation is S_3.

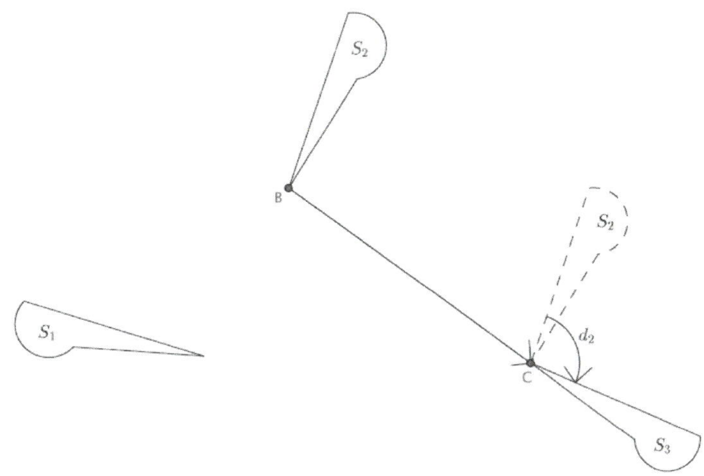

c. **Describe a sequence of basic rigid motions that shows $S_1 \cong S_3$.**

Sample student response: Let there be a translation along vector $\overrightarrow{AC}$. Let there be a rotation around point C, d_3 degrees. Let there be the reflection across the longest side of the figure so that S_1 maps onto S_3. Then, the $Translation(S_1)$ followed by the $Rotation$ followed by the $Reflection$ is S_3. Because we found a congruence that maps S_1 to S_2, that is, $S_1 \cong S_2$, and another congruence that maps S_2 to S_3, that is, $S_2 \cong S_3$, then we know for certain that $S_1 \cong S_3$.

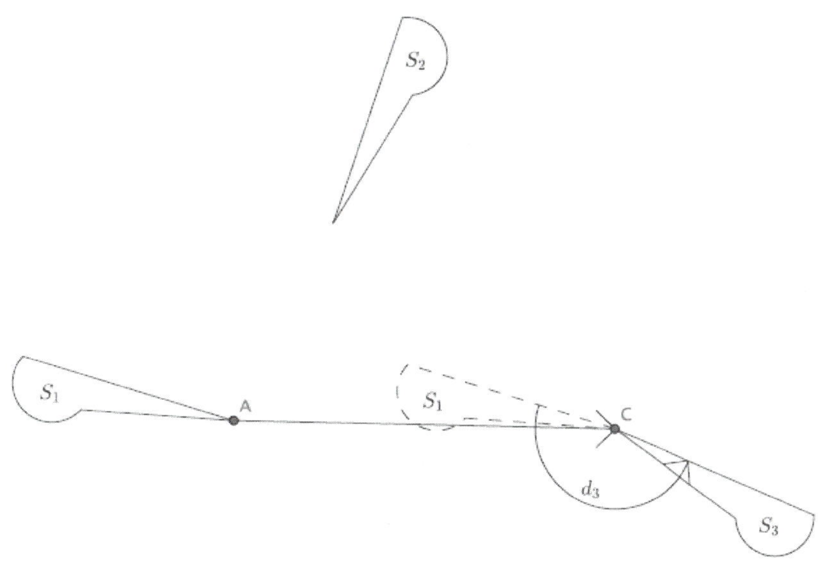

Discussion and Exercise 2 (10 minutes)

Ask students if they really need to do all of the work they did in part (c) of Exercise 1.

Students should say no. The reason they do not need to do all of that work is because they already know that translations, rotations, and reflections preserve angle measures and lengths of segments. For that reason, if they know that $S_1 \cong S_2$ and $S_2 \cong S_3$, then $S_1 \cong S_3$.

Ask students to help summarize the basic properties of all three basic rigid motions.

Elicit from students the following three statements:

MP.8

- *A basic rigid motion maps a line to a line, a ray to a ray, a segment to a segment, and an angle to an angle.*
- *A basic rigid motion preserves lengths of segments.*
- *A basic rigid motion preserves measures of angles.*

Ask students if they believe these same facts are true for *sequences* of basic rigid motions.

- Specifically, under a sequence of a translation followed by a rotation: If there is a translation along a vector $\overrightarrow{AB}$, and there is a rotation of d degrees around a center O, will a figure that is sequenced remain rigid? That is, will lengths and angles be preserved? Will lines remain lines or segments remain segments?
  - *Students should say that yes, sequences of rigid motions also have the same basic properties of rigid motions in general.*

If students are unconvinced, have them complete Exercise 2; then, discuss again.

- Given that sequences enjoy the same basic properties of basic rigid motions, we can state three basic properties of congruences:

MP.8

(Congruence 1) A congruence maps a line to a line, a ray to a ray, a segment to a segment, and an angle to an angle.

(Congruence 2) A congruence preserves lengths of segments.

(Congruence 3) A congruence preserves measures of angles.

Exercise 2

Perform the sequence of a translation followed by a rotation of Figure XYZ, where T is a translation along a vector $\overrightarrow{AB}$, and R is a rotation of d degrees (you choose d) around a center O. Label the transformed figure $X'Y'Z'$. Is $XYZ \cong X'Y'Z'$?

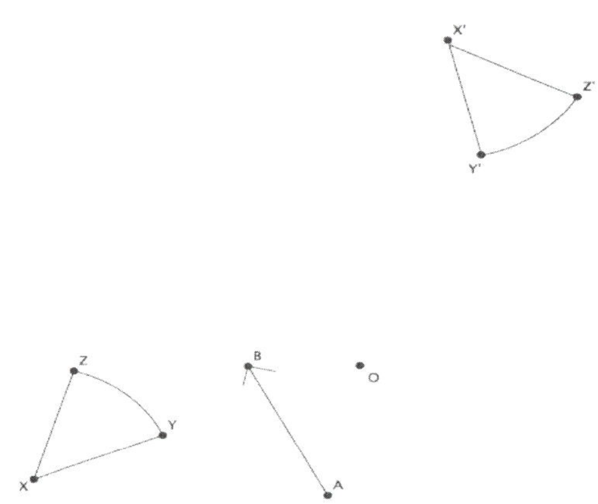

After this exercise, students should be convinced that a sequence of rigid motions maintains the basic properties of individual basic rigid motions. They should clearly see that the figure XYZ that they traced in red is exactly the same, that is, congruent, to the transformed figure $X'Y'Z'$.

Closing (5 minutes)

Summarize, or have students summarize, the lesson.

- We now have a definition for congruence, that is, a sequence of basic rigid motions.
- We now have new notation for congruence, $\cong$.
- The properties that apply to the individual basic rigid motions also apply to congruences.

> **Lesson Summary**
>
> **Given that sequences enjoy the same basic properties of basic rigid motions, we can state three basic properties of congruences:**
>
> **(Congruence 1) A congruence maps a line to a line, a ray to a ray, a segment to a segment, and an angle to an angle.**
>
> **(Congruence 2) A congruence preserves lengths of segments.**
>
> **(Congruence 3) A congruence preserves measures of angles.**
>
> **The notation used for congruence is $\cong$.**

Exit Ticket (5 minutes)

Name ______________________________ Date ____________

Lesson 11: Definition of Congruence and Some Basic Properties

Exit Ticket

1. Is $\triangle ABC \cong \triangle A'B'C'$? If so, describe a sequence of rigid motions that proves they are congruent. If not, explain how you know.

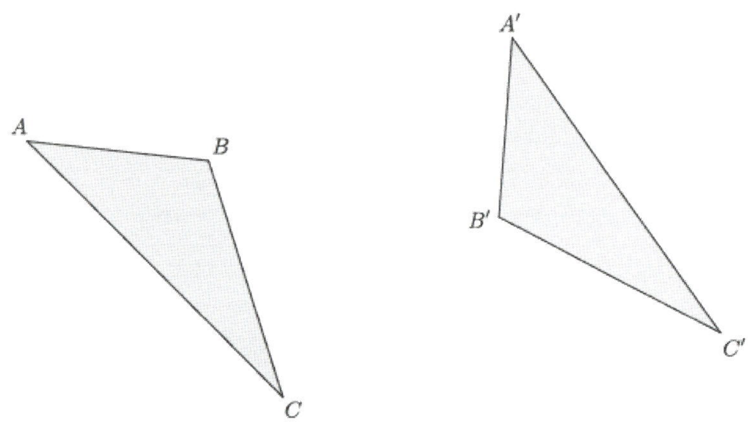

2. Is $\triangle ABC \cong \triangle A'B'C'$? If so, describe a sequence of rigid motions that proves they are congruent. If not, explain how you know.

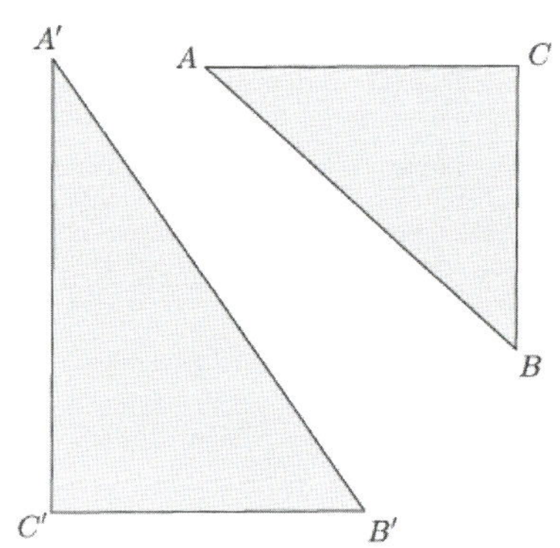

Exit Ticket Sample Solutions

1. Is $\triangle ABC \cong \triangle A'B'C'$? If so, describe a sequence of rigid motions that proves they are congruent. If not, explain how you know.

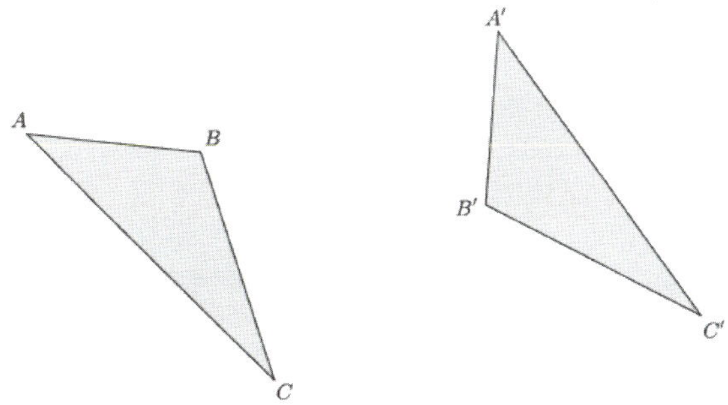

Sample student response: Yes, $\triangle ABC \cong \triangle A'B'C'$.

Translate $\triangle A'B'C'$ along vector $\overrightarrow{A'A}$.

Rotate $\triangle A'B'C'$ around center A d degrees until side $A'C'$ coincides with side AC.

Then, reflect across line AC.

2. Is $\triangle ABC \cong \triangle A'B'C'$? If so, describe a sequence of rigid motions that proves they are congruent. If not, explain how you know.

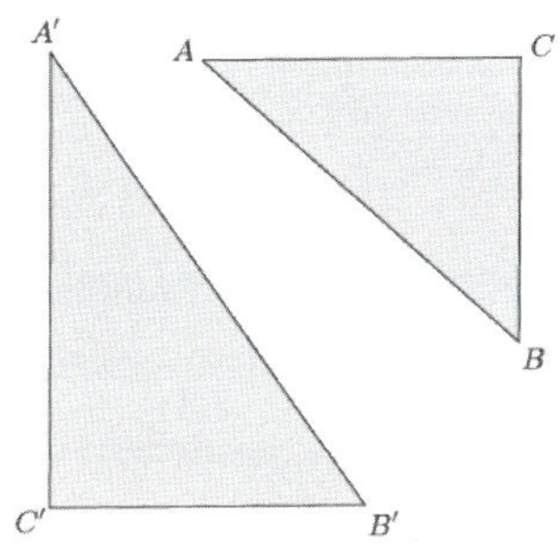

Sample student response: No, $\triangle ABC$ is not congruent to $\triangle A'B'C'$. Though I could translate and rotate to get some of the parts from each triangle to coincide, there is no rigid motion that would map side $A'C'$ to AC or side $A'B'$ to side AB because they are different lengths. Basic rigid motions preserve length, so no sequence would map $\triangle A'B'C'$ onto $\triangle ABC$.

Problem Set Sample Solutions

Students practice describing sequences of rigid motions that produce a congruence.

1. Given two right triangles with lengths shown below, is there one basic rigid motion that maps one to the other? Explain.

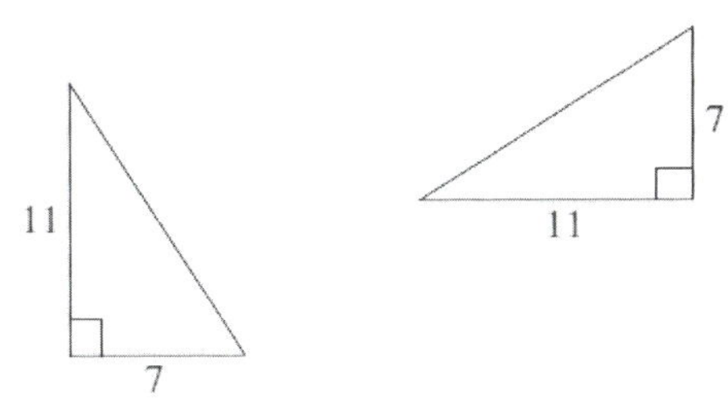

Yes, a rotation of d degrees around some center would map one triangle onto the other. The rotation would map the right angle to the right angle; the sides of length 7 and length 11 would then coincide.

2. **Are the two right triangles shown below congruent? If so, describe a congruence that would map one triangle onto the other.**

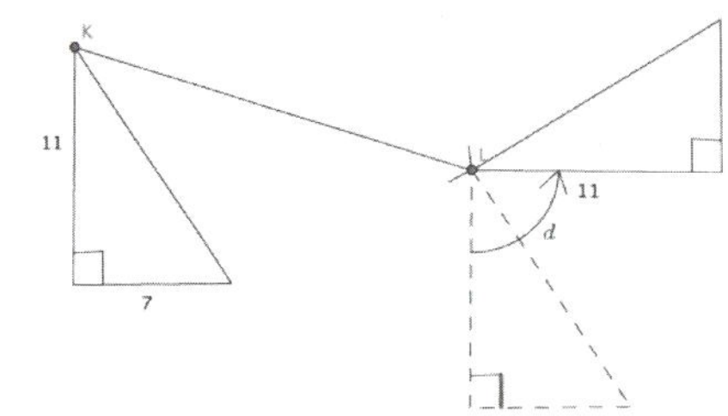

Sample student response: Yes, they are congruent. Let there be the translation along vector $\overrightarrow{KL}$. Let there be the rotation around point L, d degrees. Then, the translation followed by the rotation will map the triangle on the left to the triangle on the right.

3. **Given two rays, $\overrightarrow{OA}$ and $\overrightarrow{O'A'}$:**

   a. **Describe a congruence that maps $\overrightarrow{OA}$ to $\overrightarrow{O'A'}$.**

   Sample student response: Let there be the translation along vector $\overrightarrow{OO'}$. Let there be the rotation around point O' d degrees. Then, the $\mathit{Translation}(\overrightarrow{OA})$ followed by the $\mathit{Rotation}$ is $\overrightarrow{O'A'}$.

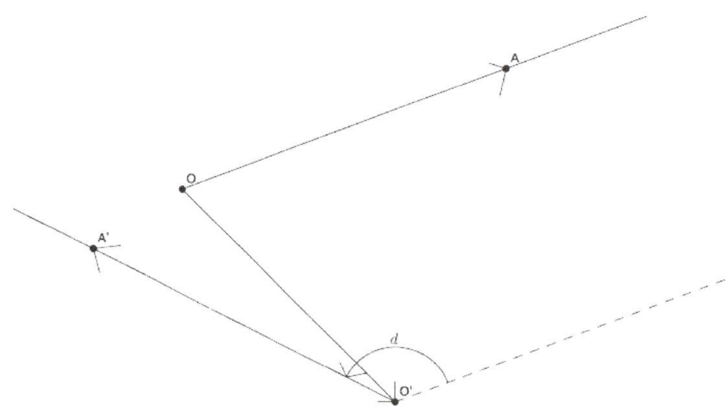

   b. **Describe a congruence that maps $\overrightarrow{O'A'}$ to $\overrightarrow{OA}$.**

   Sample student response: Let there be the translation along vector $\overrightarrow{O'O}$. Let there be the rotation around point O d_1 degrees. Then, the $\mathit{Translation}$ $(\overrightarrow{O'A'})$ followed by the $\mathit{Rotation}$ is $\overrightarrow{OA}$.

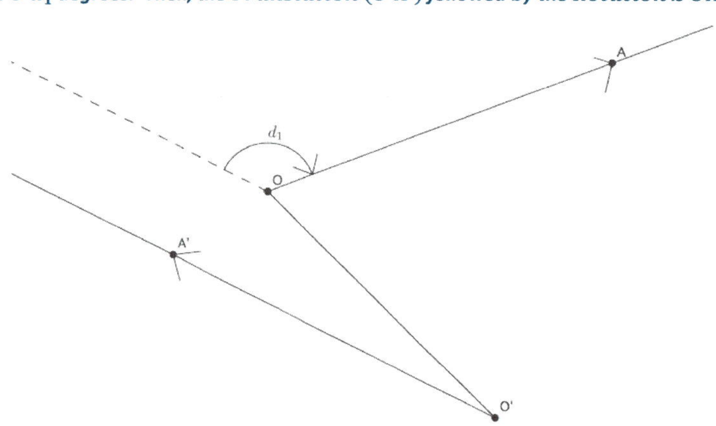

Lesson 12: Angles Associated with Parallel Lines

Student Outcomes

- Students know that corresponding angles, alternate interior angles, and alternate exterior angles of parallel lines are equal. Students know that when these pairs of angles are equal, then lines are parallel.
- Students know that corresponding angles of parallel lines are equal because of properties related to translation. Students know that alternate interior angles of parallel lines are equal because of properties related to rotation.
- Students present informal arguments to draw conclusions about angles formed when parallel lines are cut by a transversal.

Classwork

Exploratory Challenge 1 (7 minutes)

Students complete the Exploratory Challenge individually or in pairs. Students investigate the properties of angles formed by two nonparallel lines cut by a transversal.

Exploratory Challenge 1

In the figure below, L_1 is not parallel to L_2, and m is a transversal. Use a protractor to measure angles 1–8. Which, if any, are equal in measure? Explain why. (Use your transparency if needed.)

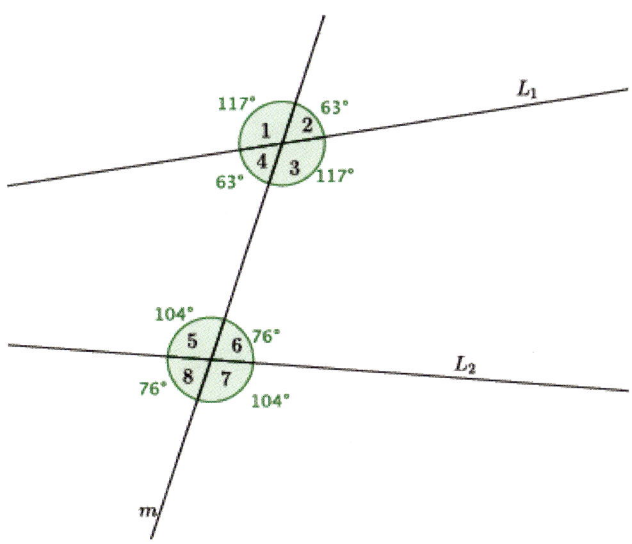

The following angle measures are equal: $\angle 1 = \angle 3$, $\angle 2 = \angle 4$, $\angle 5 = \angle 7$, and $\angle 6 = \angle 8$. The pairs of angles listed are equal because they are vertical angles. Vertical angles are always equal because a rotation of $180°$ around the vertex of the angle will map it to its opposite angle.

Discussion (5 minutes)

Discuss what students noticed about the angles in the first diagram with nonparallel lines. Ask students the following questions during the discussion.

- What did you notice about the pairs of angles in the first diagram when the lines, L_1 and L_2, were not parallel?
  - $\angle 1 = \angle 3$, $\angle 2 = \angle 4$, $\angle 5 = \angle 7$, *and* $\angle 6 = \angle 8$. *Vertical angles were equal in measure.*
- Why are vertical angles equal in measure?
  - *We can rotate the angle around its vertex* $180°$*, and it maps onto its opposite angle. Since rotations are angle-preserving, it means that the angles are equal in measure.*
- Angles that are on the same side of the transversal in corresponding positions (above each of L_1 and L_2 or below each of L_1 and L_2) are called *corresponding angles*. Name a pair of corresponding angles in the diagram. (Note to teacher: Have students name all pairs of corresponding angles from the diagram one pair at a time.)
  - $\angle 1$ *and* $\angle 5$, $\angle 4$ *and* $\angle 8$, $\angle 2$ *and* $\angle 6$, $\angle 3$ *and* $\angle 7$
- When angles are on opposite sides of the transversal and between (inside) the lines L_1 and L_2, they are called *alternate interior angles*. Name a pair of alternate interior angles. (Note to teacher: Have students name both pairs of alternate interior angles from the diagram one pair at a time.)
  - $\angle 4$ *and* $\angle 6$, $\angle 3$ *and* $\angle 5$
- When angles are on opposite sides of the transversal and outside of the parallel lines (above L_1 and below L_2), they are called *alternate exterior angles*. Name a pair of alternate exterior angles. (Note to teacher: Have students name both pairs of alternate exterior angles from the diagram one pair at a time.)
  - $\angle 1$ *and* $\angle 7$, $\angle 2$ *and* $\angle 8$

Exploratory Challenge 2 (9 minutes)

Students complete the Exploratory Challenge individually or in pairs. Students investigate the properties of angles formed by two parallel lines cut by a transversal.

Exploratory Challenge 2

In the figure below, $L_1 \parallel L_2$, and m is a transversal. Use a protractor to measure angles 1–8. List the angles that are equal in measure.

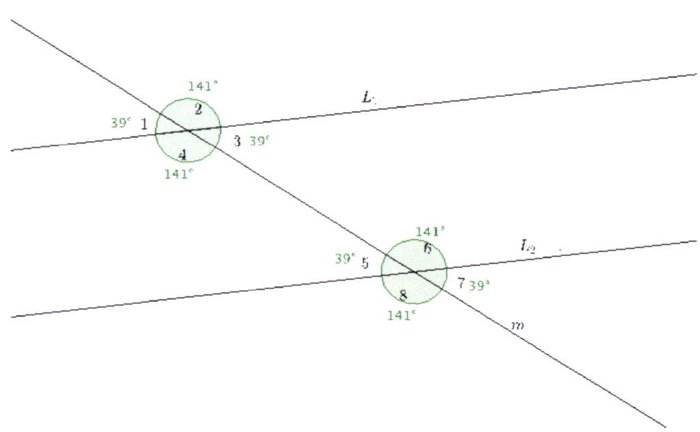

$\angle 1 = \angle 3 = \angle 5 = \angle 7$ ***and*** $\angle 2 = \angle 4 = \angle 6 = \angle 8$

a. **What did you notice about the measures of $\angle 1$ and $\angle 5$? Why do you think this is so? (Use your transparency if needed.)**

$\angle 1$ and $\angle 5$ are equal in measure. We can translate $\angle 1$ along a vector on line m so that the vertex of $\angle 1$ maps onto the vertex of $\angle 5$. Translations are angle-preserving, so the two angles will coincide.

b. **What did you notice about the measures of $\angle 3$ and $\angle 7$? Why do you think this is so? (Use your transparency if needed.) Are there any other pairs of angles with this same relationship? If so, list them.**

$\angle 3$ and $\angle 7$ are equal in measure. We can translate $\angle 3$ along a vector on line m so that the vertex of $\angle 3$ maps onto the vertex of $\angle 7$. Translations are angle-preserving, so the two angles will coincide. Other pairs of angles with this same relationship are $\angle 4$ and $\angle 8$ and $\angle 2$ and $\angle 6$.

c. **What did you notice about the measures of $\angle 4$ and $\angle 6$? Why do you think this is so? (Use your transparency if needed.) Is there another pair of angles with this same relationship?**

The measures of $\angle 4$ and $\angle 6$ are equal. A rotation of $180°$ around a center would map $\angle 4$ to $\angle 6$. Rotations are angle-preserving, so we know that $\angle 4$ and $\angle 6$ are equal. $\angle 3$ and $\angle 5$ have the same relationship.

Discussion (15 minutes)

Discuss what students noticed about the angles in the second diagram with parallel lines. Ask students the following questions during the discussion.

- We are going to discuss what you observed when you measured and compared angles in Exploratory Challenge 2. What we will do is make some informal arguments to prove some of the things you noticed. Each time you answer "why," "how do you know," or "explain," you are making an informal argument.
- Were the vertical angles in Exploratory Challenge 2 equal like they were in Exploratory Challenge 1? Why?
  - *Yes, for the same reason they were equal in the first diagram, rotation.*
- What other angles were equal in the second diagram when the lines L_1 and L_2 were parallel?
  - $\angle 1 = \angle 3 = \angle 5 = \angle 7$ *and* $\angle 2 = \angle 4 = \angle 6 = \angle 8$
- Let's look at just $\angle 1$ and $\angle 5$. What kind of angles are these, and how do you know?
  - *They are corresponding angles because they are on the same side of the transversal and in corresponding positions (i.e., above each of L_1 and L_2 or below each of L_1 and L_2).*
- We have already said that these two angles are equal in measure. Who can explain why this is so?
  - *Translation along a vector in line m will map $\angle 1$ onto $\angle 5$. Translations preserve degrees of angles, so the two angles are equal in measure.*

Insert and label a point at the vertex of each angle, points A and B, for example. Draw the vector between the two points. Trace one of the angles on a transparency, and demonstrate to students that a translation along a vector of exactly length AB would map one angle onto the other. Further, if there is a translation along vector $\overrightarrow{AB}$, then $Translation(L_1)$ maps to L_2. The vector $\overrightarrow{AB}$ on line m is also translated but remains on line m. Therefore, all of the angles formed by the intersection of L_1 and m are translated along $\overrightarrow{AB}$ according to the translation. That is why the corresponding angles are equal in measure, that is, $Translation(\angle 1)$ maps to $\angle 5$.

- What did you notice about $\angle 3$ and $\angle 7$?
  - *These two angles were also equal in measure for the same reason as $\angle 1$ and $\angle 5$.*

- What other pairs of corresponding angles are in the diagram?
  - *$\angle 4$ and $\angle 8$ and $\angle 2$ and$\angle 6$*
- In Exploratory Challenge 1, the pairs of corresponding angles we named were not equal in measure. Given the information provided about each diagram, can you think of why this is so?
  - *In the first diagram, the lines L_1 and L_2 were not parallel. A translation of one of the angles would not map onto the other angle.*
- Are $\angle 4$ and $\angle 6$ corresponding angles? If not, why not?
  - *No, they are not on the same side of the transversal in corresponding locations.*
- What kind of angles are $\angle 4$ and $\angle 6$? How do you know?
  - *$\angle 4$ and $\angle 6$ are alternate interior angles because they are on opposite sides of the transversal and inside the lines L_1 and L_2.*
- We have already said that $\angle 4$ and $\angle 6$ are equal in measure. Why do you think this is so?
  - *You can use rotation to map one of the angles onto the other. Rotations are angle-preserving.*

Mark the midpoint, point M for example, of the segment between the vertices of the two angles. Trace one of the angles on a transparency, and demonstrate to students that a $180°$ rotation around the point M would map one of the angles onto the other. Further, we know that a $180°$ rotation of a line around a point not on the line maps to a parallel line. If we let there be the rotation of $180°$ around point M, then $Rotation(L_1)$ maps to L_2. The rotation of point M maps to itself under rotation, and the line containing M, line m, will also map to itself under rotation. For that reason, the angles formed at the intersection of L_1 and line m, under rotation, will map to angles at the intersection of L_2 and line m but on the other side of the transversal, specifically, $Rotation(\angle 4)$ maps to $\angle 6$.

- Name another pair of alternate interior angles.
  - *$\angle 3$ and $\angle 5$*
- In Exploratory Challenge 1, the pairs of alternate interior angles we named were not equal in measure. Given the information provided about each diagram, can you think of why this is so?
  - *In the first diagram, the lines L_1 and L_2 were not parallel. A rotation around a point would not map one angle onto the other angle.*
- Are $\angle 1$ and $\angle 7$ corresponding angles? If not, why not?
  - *No, they are not on the same side of the transversal, so they cannot be corresponding angles.*
- Are $\angle 1$ and $\angle 7$ alternate interior angles? If not, why not?
  - *No, they are not inside of the lines L_1 and L_2, so they cannot be interior angles.*
- What kind of angles are $\angle 1$ and $\angle 7$?
  - *They are alternate exterior angles because they are on opposite sides of the transversal and above L_1 and below L_2.*
- Name another pair of alternate exterior angles.
  - *$\angle 2$ and $\angle 8$*
- These pairs of alternate exterior angles were not equal in measure in Exploratory Challenge 1. Given the information provided about each diagram, can you think of why this is so?
  - *In the first diagram, the lines L_1 and L_2 were not parallel. A rotation of one of the angles would not map onto the other angle.*

- If you know that pairs of corresponding angles, alternate interior angles, and alternate exterior angles are congruent, what do you think is true about the lines?
    - *The lines are parallel when pairs of corresponding angles, alternate interior angles, and alternate exterior angles are congruent.*

State the following theorem and its converse:

- **THEOREM**: When parallel lines are cut by a transversal, then the pairs of corresponding angles are congruent, the pairs of alternate interior angles are congruent, and the pairs of alternate exterior angles are congruent.
- The converse of the theorem states that if you know that corresponding angles are congruent, then you can be sure that the lines cut by a transversal are parallel. How could you phrase the converse of the theorem with respect to other types of angles we have learned?
    - *When alternate interior angles are congruent, then the lines cut by a transversal are parallel.*
    - *When alternate exterior angles are congruent, then the lines cut by a transversal are parallel.*

Closing (4 minutes)

Provide students two minutes to fill in the example portion of the lesson summary in the student materials.

Summarize, or have students summarize, the lesson.

- When a pair of parallel lines are cut by a transversal, then any corresponding angles, any alternate interior angles, and any alternate exterior angles are equal in measure.
- The reason that specific pairs of angles are equal is because of the properties we learned about the basic rigid motions, specifically that they are angle-preserving.
- When a pair of nonparallel lines are cut by a transversal, any corresponding angles, any alternate interior angles, and any alternate exterior angles are not equal in measure.

Lesson Summary

Angles that are on the same side of the transversal in corresponding positions (above each of L_1 and L_2 or below each of L_1 and L_2) are called *corresponding angles*. For example, $\angle 2$ and $\angle 4$ are corresponding angles.

When angles are on opposite sides of the transversal and between (inside) the lines L_1 and L_2, they are called *alternate interior angles*. For example, $\angle 3$ and $\angle 7$ are alternate interior angles.

When angles are on opposite sides of the transversal and outside of the parallel lines (above L_1 and below L_2), they are called *alternate exterior angles*. For example, $\angle 1$ and $\angle 5$ are alternate exterior angles.

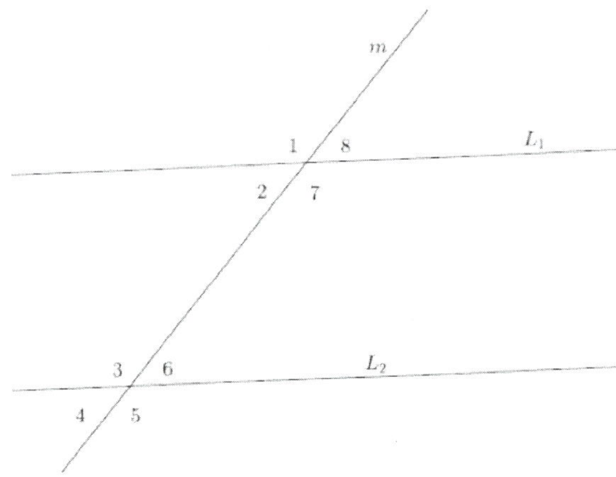

When parallel lines are cut by a transversal, any corresponding angles, any alternate interior angles, and any alternate exterior angles are equal in measure. If the lines are not parallel, then the angles are not equal in measure.

Exit Ticket (5 minutes)

Name ______________________________ Date ______________

Lesson 12: Angles Associated with Parallel Lines

Exit Ticket

Use the diagram to answer Questions 1 and 2. In the diagram, lines L_1 and L_2 are intersected by transversal m, forming angles 1–8, as shown.

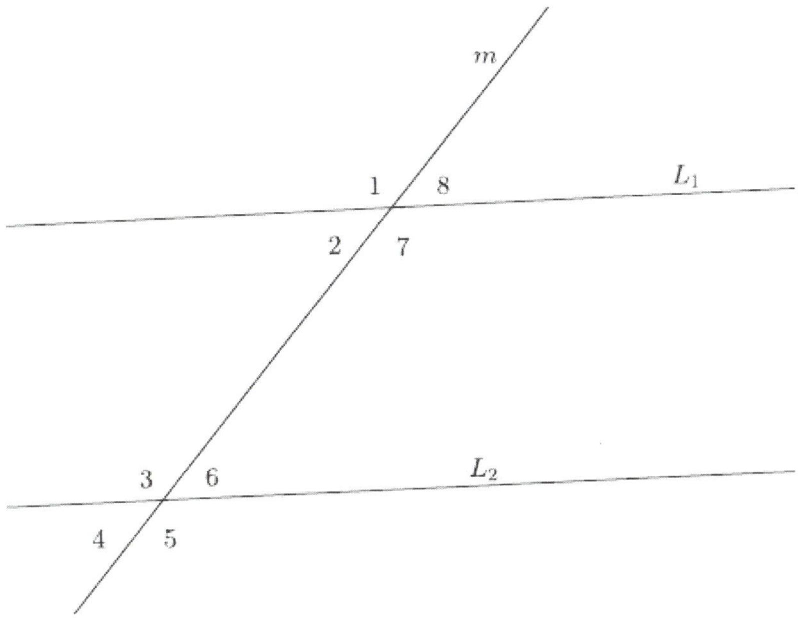

1. If $L_1 \parallel L_2$, what do you know about $\angle 2$ and $\angle 6$? Use informal arguments to support your claim.

2. If $L_1 \parallel L_2$, what do you know about $\angle 1$ and $\angle 3$? Use informal arguments to support your claim.

Exit Ticket Sample Solutions

Use the diagram to answer Problems 1 and 2. In the diagram, lines L_1 and L_2 are intersected by transversal m, forming angles 1–8, as shown.

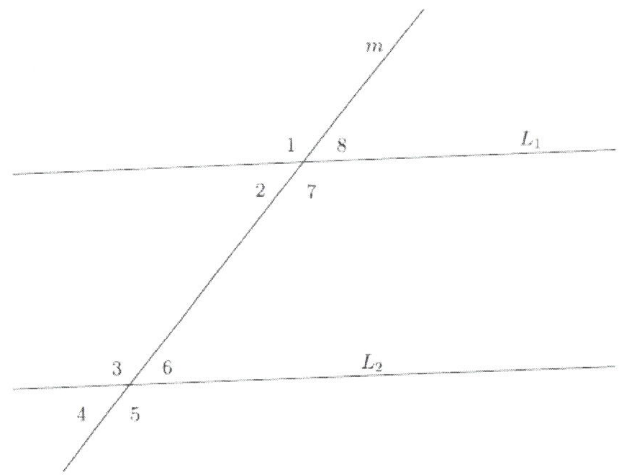

1. **If $L_1 \parallel L_2$, what do you know about $\angle 2$ and $\angle 6$? Use informal arguments to support your claim.**

   They are alternate interior angles because they are on opposite sides of the transversal and inside of lines L_1 and L_2. Also, the angles are equal in measure because the lines L_1 and L_2 are parallel. If we rotated angle 2 $180°$ around the midpoint of the segment between the parallel lines, then it would map onto angle 6.

2. **If $L_1 \parallel L_2$, what do you know about $\angle 1$ and $\angle 3$? Use informal arguments to support your claim.**

   They are corresponding angles because they are on the same side of the transversal and above each of lines L_1 and L_2. Also, the angles are equal in measure because the lines L_1 and L_2 are parallel. If we translated angle 1 along a vector (the same length as the segment between the parallel lines), then it would map onto angle 3.

Problem Set Sample Solutions

Students practice identifying corresponding, alternate interior, and alternate exterior angles from a diagram.

Use the diagram below to do Problems 1–10.

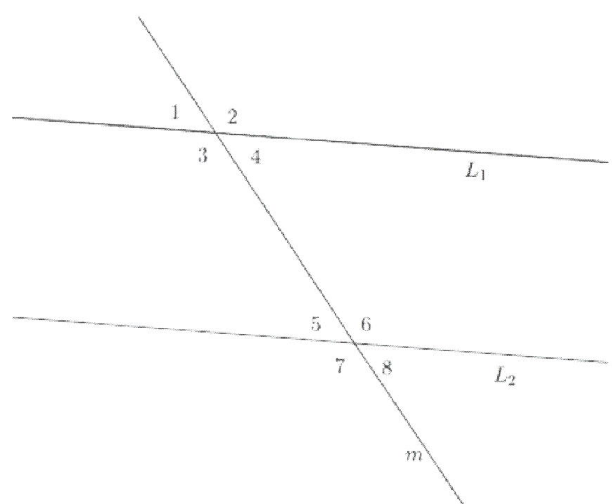

1. **Identify all pairs of corresponding angles. Are the pairs of corresponding angles equal in measure? How do you know?**

   $\angle 1$ and $\angle 5$, $\angle 4$ and $\angle 8$, $\angle 2$ and $\angle 6$, $\angle 3$ and $\angle 7$

   There is no information provided about the lines in the diagram being parallel. For that reason, we do not know if the pairs of corresponding angles are equal in measure. If we knew the lines were parallel, we could use translation to map one angle onto another.

2. **Identify all pairs of alternate interior angles. Are the pairs of alternate interior angles equal in measure? How do you know?**

   $\angle 4$ and $\angle 5$, $\angle 3$ and $\angle 6$

   There is no information provided about the lines in the diagram being parallel. For that reason, we do not know if the pairs of alternate interior angles are equal in measure. If the lines were parallel, we could use rotation to show that the pairs of angles would map onto one another, proving they are equal in measure.

3. **Use an informal argument to describe why $\angle 1$ and $\angle 8$ are equal in measure if $L_1 \parallel L_2$.**

   The reason that angle 1 and angle 8 are equal in measure when the lines are parallel is because you can rotate around the midpoint of the segment between the parallel lines. A rotation would then map angle 1 onto angle 8, showing that they are congruent and equal in measure.

4. **Assuming $L_1 \parallel L_2$, if the measure of $\angle 4$ is $73°$, what is the measure of $\angle 8$? How do you know?**

   The measure of $\angle 8$ is $73°$. This must be true because they are corresponding angles of parallel lines.

5. **Assuming $L_1 \parallel L_2$, if the measure of $\angle 3$ is $107°$ degrees, what is the measure of $\angle 6$? How do you know?**

   The measure of $\angle 6$ is $107°$. This must be true because they are alternate interior angles of parallel lines.

6. **Assuming $L_1 \parallel L_2$, if the measure of $\angle 2$ is $107°$, what is the measure of $\angle 7$? How do you know?**

   The measure of $\angle 7$ is $107°$. This must be true because they are alternate exterior angles of parallel lines.

7. **Would your answers to Problems 4–6 be the same if you had not been informed that $L_1 \parallel L_2$? Why or why not?**

   No. The fact that the lines are parallel is the reason we can state that specific pairs of angles are equal in measure. We can use basic rigid motions to prove that angles associated with parallel lines have the property of being equal in measure when they are corresponding, alternate interior, or alternate exterior angles. If the lines are not parallel, then we could still classify the angles, but we would not know anything about their measures.

8. **Use an informal argument to describe why $\angle 1$ and $\angle 5$ are equal in measure if $L_1 \parallel L_2$.**

   The reason that angle 1 and angle 5 are equal in measure when the lines are parallel is because you can translate along a vector equal in length of the segment between the parallel lines; then, angle 1 would map onto angle 5.

9. **Use an informal argument to describe why $\angle 4$ and $\angle 5$ are equal in measure if $L_1 \parallel L_2$.**

   The reason that angle 4 and angle 5 are equal in measure when the lines are parallel is because when you rotate angle 4 around the midpoint of the segment between the parallel lines, angle 4 will map onto angle 5.

10. **Assume that L_1 is not parallel to L_2. Explain why $\angle 3 \neq \angle 7$.**

    If the lines are not parallel, then all we know about angle 3 and angle 7 is that they are corresponding angles. If the lines are parallel, we could use translation to map one angle onto the other to show that they are equal in measure. However, we are to assume that the lines are not parallel, which means that their corresponding angles will not be equal in measure.

Lesson 13: Angle Sum of a Triangle

Student Outcomes

- Students know the angle sum theorem for triangles; the sum of the interior angles of a triangle is always $180°$.
- Students present informal arguments to draw conclusions about the angle sum of a triangle.

Classwork

Concept Development (3 minutes)

- The angle sum theorem for triangles states that the sum of the measures of the interior angles of a triangle is always $180°$ ($\angle$ sum of $\triangle$).
- It does not matter what kind of triangle it is (i.e., acute, obtuse, right); when the measure of the three angles are added, the sum is always $180°$.

Concept Development

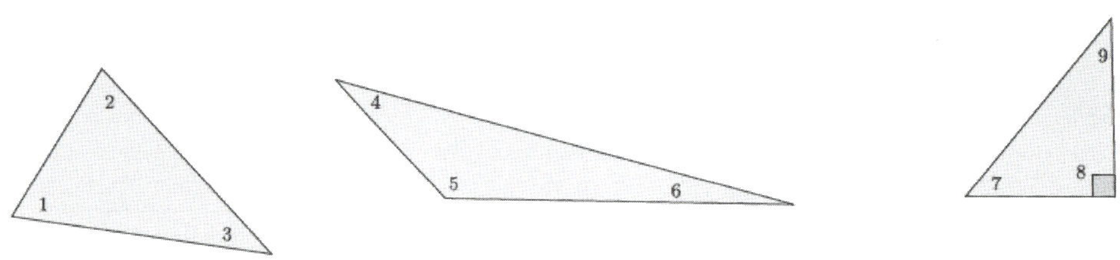

$$m\angle 1 + m\angle 2 + m\angle 3 = m\angle 4 + m\angle 5 + m\angle 6 = m\angle 7 + m\angle 8 + m\angle 9 = 180°$$

Note that the sum of the measures of angles 7 and 9 must equal $90°$ because of the known right angle in the right triangle.

We want to prove that the angle sum of any triangle is $180°$. To do so, we use some facts that we already know about geometry:

- A straight angle is $180°$ in measure.
- Corresponding angles of parallel lines are equal in measure (corr. $\angle$'s, $\overline{AB} \parallel \overline{CD}$).
- Alternate interior angles of parallel lines are equal in measure (alt. $\angle$'s, $\overline{AB} \parallel \overline{CD}$).

Exploratory Challenge 1 (13 minutes)

Provide students 10 minutes of work time. Once the 10 minutes have passed, review the solutions with students before moving on to Exploratory Challenge 2.

Exploratory Challenge 1

Let triangle ABC be given. On the ray from B to C, take a point D so that C is between B and D. Through point C, draw a segment parallel to $\overline{AB}$, as shown. Extend the segments AB and CE. Line AC is the transversal that intersects the parallel lines.

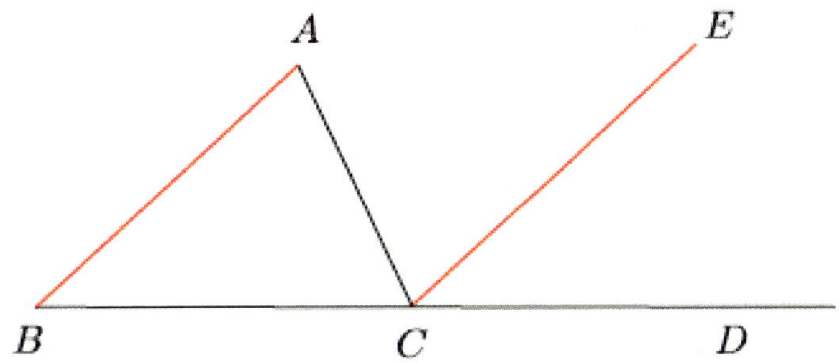

a. Name the three interior angles of triangle ABC.

$\angle ABC$, $\angle BAC$, $\angle BCA$

MP.3

b. Name the straight angle.

$\angle BCD$

Our goal is to show that the measures of the three interior angles of triangle ABC are equal to the measures of the angles that make up the straight angle. We already know that a straight angle is $180°$ in measure. If we can show that the interior angles of the triangle are the same as the angles of the straight angle, then we will have proven that the sum of the measures of the interior angles of the triangle have a sum of $180°$.

c. What kinds of angles are $\angle ABC$ and $\angle ECD$? What does that mean about their measures?

$\angle ABC$ and $\angle ECD$ are corresponding angles. Corresponding angles of parallel lines are equal in measure (corr. $\angle$'s, $\overline{AB} \parallel \overline{CE}$).

d. What kinds of angles are $\angle BAC$ and $\angle ECA$? What does that mean about their measures?

$\angle BAC$ and $\angle ECA$ are alternate interior angles. Alternate interior angles of parallel lines are equal in measure (alt. $\angle$'s, $\overline{AB} \parallel \overline{CE}$).

e. We know that $m\angle BCD = m\angle BCA + m\angle ECA + m\angle ECD = 180°$. Use substitution to show that the measures of the three interior angles of the triangle have a sum of $180°$.

$m\angle BCD = m\angle BCA + m\angle BAC + m\angle ABC = 180°$ ($\angle$ sum of $\triangle$)

Exploratory Challenge 2 (20 minutes)

Provide students 15 minutes of work time. Once the 15 minutes have passed, review the solutions with students.

Exploratory Challenge 2

The figure below shows parallel lines L_1 and L_2. Let m and n be transversals that intersect L_1 at points B and C, respectively, and L_2 at point F, as shown. Let A be a point on L_1 to the left of B, D be a point on L_1 to the right of C, G be a point on L_2 to the left of F, and E be a point on L_2 to the right of F.

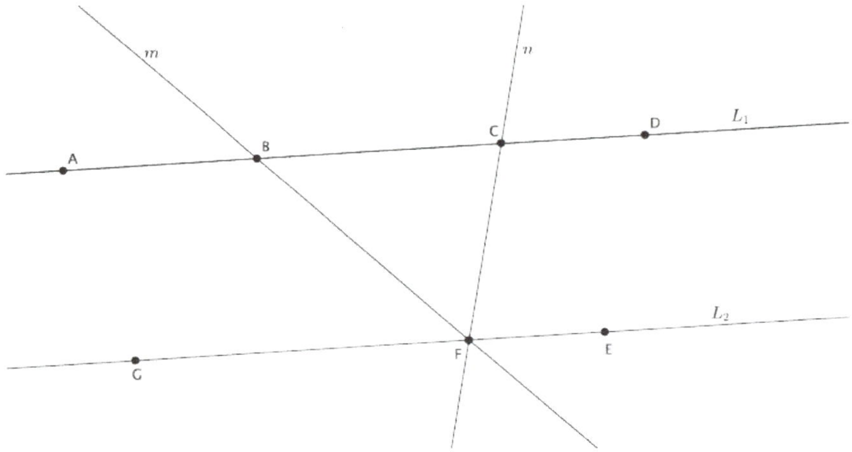

a. **Name the triangle in the figure.**

$\triangle\ BCF$

b. **Name a straight angle that will be useful in proving that the sum of the measures of the interior angles of the triangle is $180°$.**

$\angle GFE$

As before, our goal is to show that the sum of the measures of the interior angles of the triangle are equal to the measure of the straight angle. Use what you learned from Exploratory Challenge 1 to show that the measures of the interior angles of a triangle have a sum of $180°$.

c. **Write your proof below.**

MP.3

The straight angle $\angle GFE$ is comprised of $\angle GFB$, $\angle BFC$, and $\angle EFC$. Alternate interior angles of parallel lines are equal in measure (alt. $\angle$'s, $\overline{AD} \parallel \overline{CE}$). For that reason, $\angle BCF = \angle EFC$ and $\angle CBF = \angle GFB$. Since $\angle GFE$ is a straight angle, it is equal to $180°$. Then, $\angle GFE = \angle GFB + \angle BFC + \angle EFC = 180°$. By substitution, $\angle GFE = \angle CBF + \angle BFC + \angle BCF = 180°$. Therefore, the sum of the measures of the interior angles of a triangle is $180°$ ($\angle$ sum of $\triangle$).

Closing (4 minutes)

Summarize, or have students summarize, the lesson.

- All triangles have interior angles whose measures sum to $180°$.
- We can prove that the sum of the measures of the interior angles of a triangle are equal to the measure of a straight angle using what we know about alternate interior angles and corresponding angles of parallel lines.

G8-M2-B1-v1.3.0_05/04/15

Lesson Summary

All triangles have a sum of measures of the interior angles equal to 180°.

The proof that a triangle has a sum of measures of the interior angles equal to 180° is dependent upon the knowledge of straight angles and angle relationships of parallel lines cut by a transversal.

Exit Ticket (5 minutes)

Name ______________________________ Date ______________

Lesson 13: Angle Sum of a Triangle

Exit Ticket

1. If $L_1 \parallel L_2$, and $L_3 \parallel L_4$, what is the measure of $\angle 1$? Explain how you arrived at your answer.

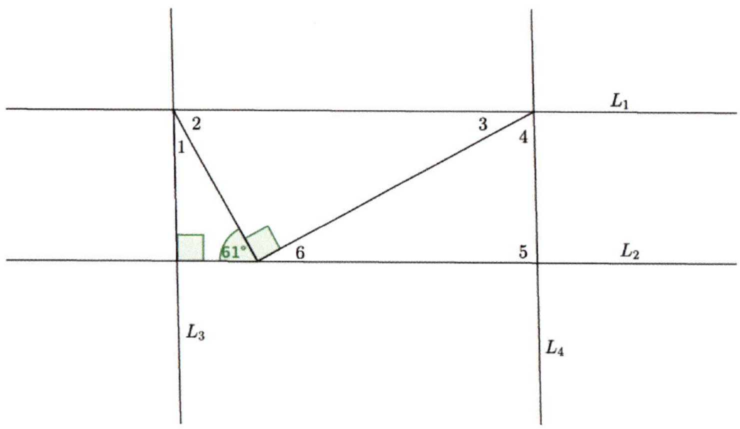

2. Given that line AB is parallel to line CE, present an informal argument to prove that the measures of the interior angles of triangle ABC have a sum of $180°$.

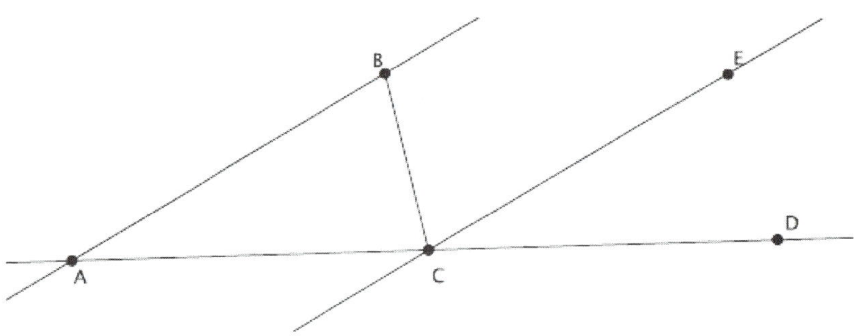

Exit Ticket Sample Solutions

1. If $L_1 \parallel L_2$, and $L_3 \parallel L_4$, what is the measure of $\angle 1$? Explain how you arrived at your answer.

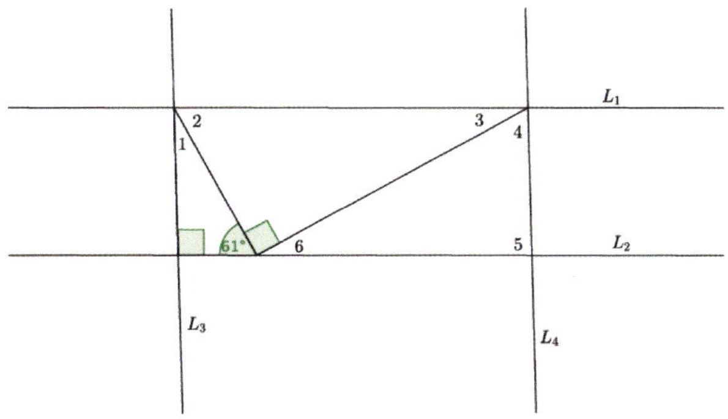

The measure of angle 1 is $29°$. I know that the angle sum of triangles is $180°$. I already know that two of the angles of the triangle are $90°$ and $61°$.

2. Given that line AB is parallel to line CE, present an informal argument to prove that the measures of the interior angles of triangle ABC have a sum of $180°$.

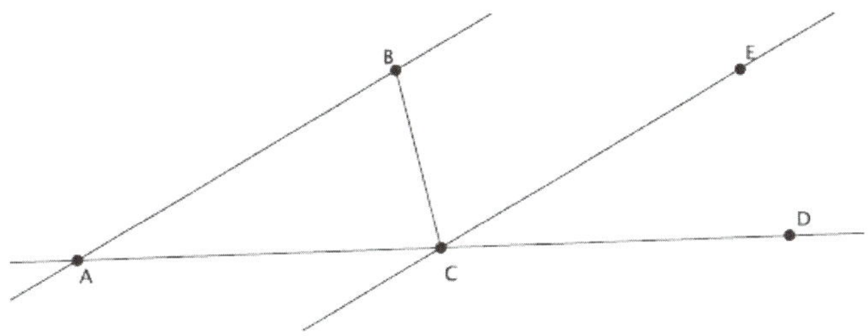

Since $\overleftrightarrow{AB}$ is parallel to $\overleftrightarrow{CE}$, the corresponding angles $\angle BAC$ and $\angle ECD$ are equal in measure. Similarly, $\angle ABC$ and $\angle ECB$ are equal in measure because they are alternate interior angles. Since $\angle ACD$ is a straight angle (i.e., equal to $180°$ in measure), substitution shows that triangle ABC has a sum of $180°$. Specifically, the straight angle is made up of $\angle ACB$, $\angle ECB$, and $\angle ECD$. $\angle ACB$ is one of the interior angles of the triangle and one of the angles of the straight angle. We know that $\angle ABC$ has the same measure as $\angle ECB$ and that $\angle BAC$ has the same measure as $\angle ECD$. Therefore, the sum of the measures of the interior angles will be the same as the sum of the measures of the angles of the straight angle, which is $180°$.

Problem Set Sample Solutions

Students practice presenting informal arguments about the sum of the angles of a triangle using the theorem to find the measures of missing angles.

1. **In the diagram below, line AB is parallel to line CD, that is, $L_{AB} \parallel L_{CD}$. The measure of $\angle ABC$ is $28°$ and the measure of $\angle EDC$ is $42°$. Find the measure of $\angle CED$. Explain why you are correct by presenting an informal argument that uses the angle sum of a triangle.**

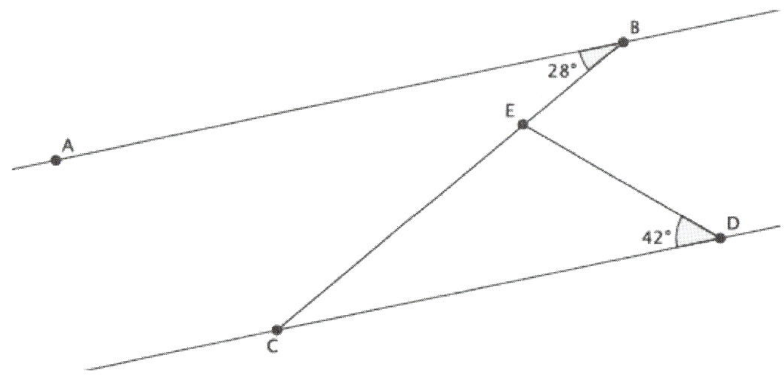

The measure of $\angle CED$ is $110°$. This is the correct measure for the angle because $\angle ABC$ and $\angle DCE$ are alternate interior angles of parallel lines. That means that the angles are congruent and have the same measure. Since the angle sum of a triangle is $180°$, then $\angle CED = 180° - (28° + 42°)$ and $\angle CED$ is $110°$.

2. **In the diagram below, line AB is parallel to line CD, that is, $L_{AB} \parallel L_{CD}$. The measure of $\angle ABE$ is $38°$, and the measure of $\angle EDC$ is $16°$. Find the measure of $\angle BED$. Explain why you are correct by presenting an informal argument that uses the angle sum of a triangle. (Hint: Find the measure of $\angle CED$ first, and then use that measure to find the measure of $\angle BED$.)**

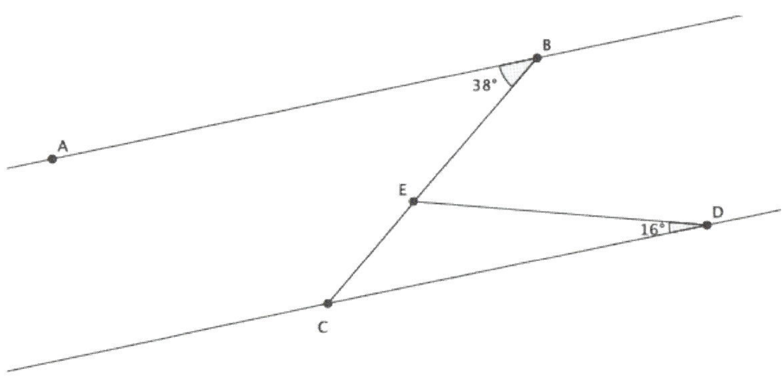

The measure of $\angle BED$ is $54°$. This is the correct measure for the angle because $\angle ABC$ and $\angle DCE$ are alternate interior angles of parallel lines. That means that the angles are congruent and have the same measure. Since the angle sum of a triangle is $180°$, then $\angle CED = 180° - (38° + 16°) = 126°$. The straight angle $\angle BEC$ is made up of $\angle CED$ and $\angle BED$. Since we know straight angles measure $180°$ and $\angle CED = 126°$, then $\angle BED$ is $54°$.

3. In the diagram below, line AB is parallel to line CD, that is, $L_{AB} \parallel L_{CD}$. The measure of $\angle ABE$ is $56°$, and the measure of $\angle EDC$ is $22°$. Find the measure of $\angle BED$. Explain why you are correct by presenting an informal argument that uses the angle sum of a triangle. (Hint: Extend the segment BE so that it intersects line CD.)

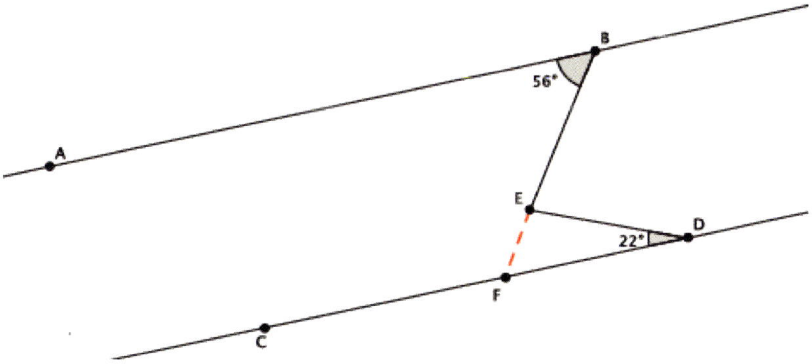

The measure of $\angle BED$ is $78°$. This is the correct measure for the angle because $\angle ABE$ and $\angle DFE$ are alternate interior angles of parallel lines. That means that the angles are congruent and have the same measure. Since the angle sum of a triangle is $180°$, then $\angle FED = 180° - (56° + 22°) = 102°$. The straight angle $\angle BEF$ is made up of $\angle FED$ and $\angle BED$. Since straight angles measure $180°$ and $\angle FED$ is $102°$, then $\angle BED$ is $78°$.

4. What is the measure of $\angle ACB$?

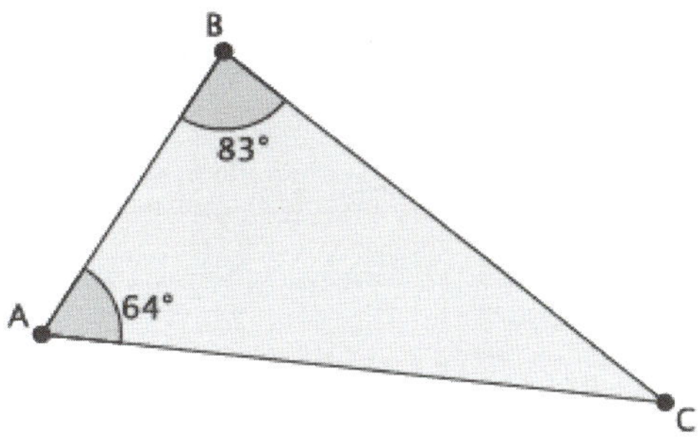

The measure of $\angle ACB$ is $180° - (83° + 64°)$, which is equal to $33°$.

5. What is the measure of $\angle EFD$?

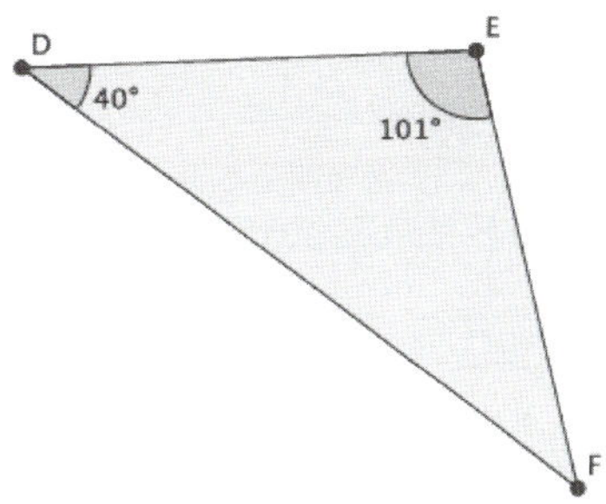

The measure of $\angle EFD$ is $180° - (101° + 40°)$, which is equal to $39°$.

6. What is the measure of $\angle HIG$?

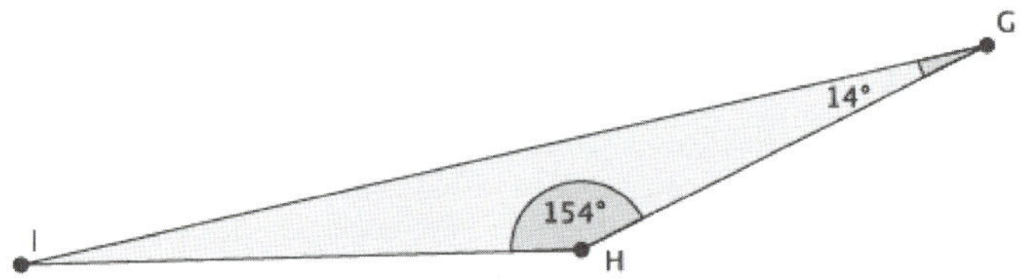

The measure of $\angle HIG$ *is* $180° - (154° + 14°)$, *which is equal to* $12°$.

7. What is the measure of $\angle ABC$?

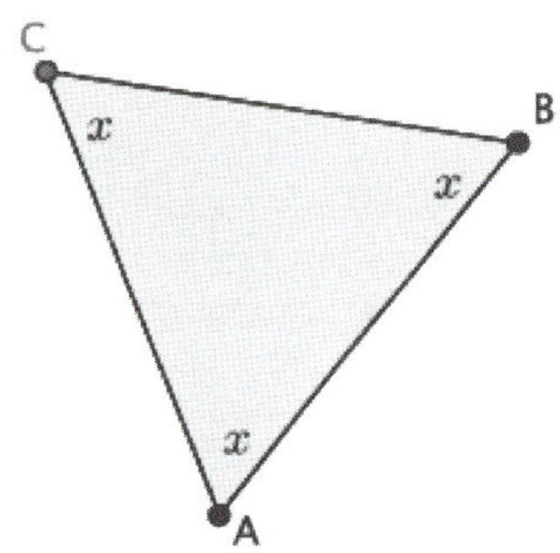

The measure of $\angle ABC$ *is* $60°$ *because* $60° + 60° + 60°$, *which is equal to* $180°$.

8. Triangle DEF is a right triangle. What is the measure of $\angle EFD$?

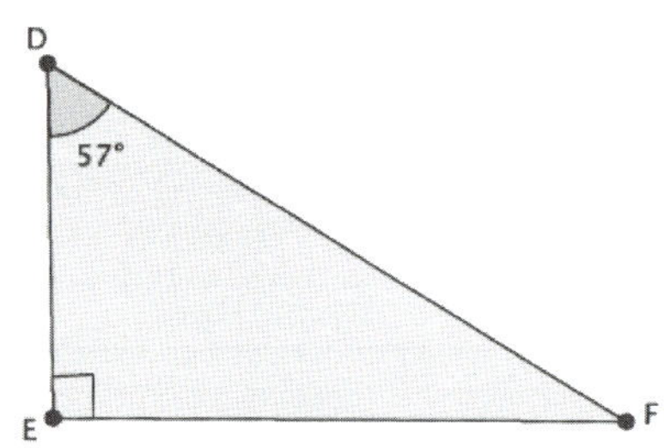

The measure of $\angle EFD$ *is* $90° - 57°$, *which is equal to* $33°$.

9. In the diagram below, Lines L_1 and L_2 are parallel. Transversals r and s intersect both lines at the points shown below. Determine the measure of $\angle JMK$. Explain how you know you are correct.

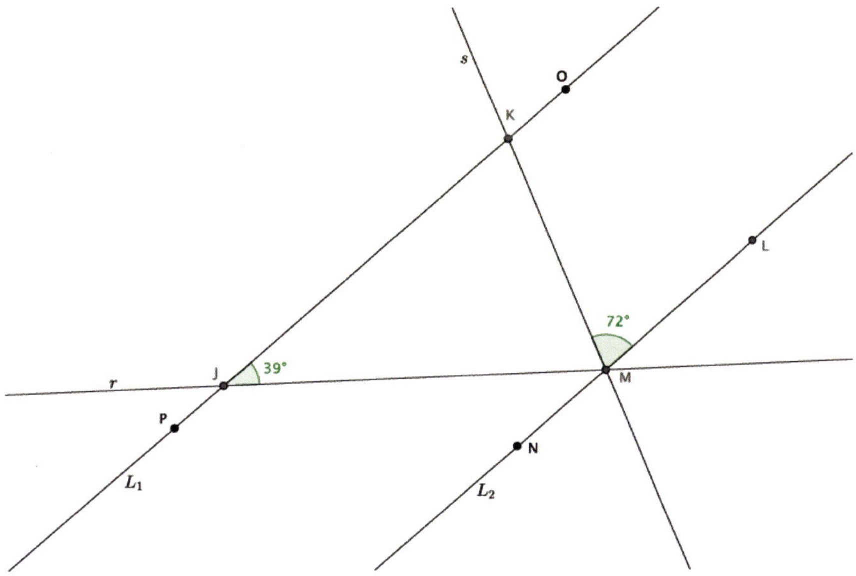

The Lines L_1 and L_2 are parallel, which means that the alternate interior angles formed by the transversals are equal. Specifically, $\angle LMK = \angle JKM = 72°$. Since $\triangle JKM$ has a sum of interior angles equal to $180°$, $\angle KJM + \angle JMK + \angle JKM = 180°$. By substitution, we have $39° + \angle JMK + 72° = 180°$; therefore, $\angle JMK$ is $69°$.

Lesson 14: More on the Angles of a Triangle

Student Outcomes

- Students know a third informal proof of the angle sum theorem.
- Students know how to find missing interior and exterior angle measures of triangles and present informal arguments to prove their answer is correct.

Lesson Notes

Students see one final informal proof of the angle sum of a triangle before moving on to working with exterior angles of triangles.

Classwork

Discussion (7 minutes)

Let's look at one final proof that the sum of the degrees of the interior angles of a triangle is $180°$.

- Start with a rectangle. What properties do rectangles have?
    - *All four angles are right angles; opposite sides are equal in length.*

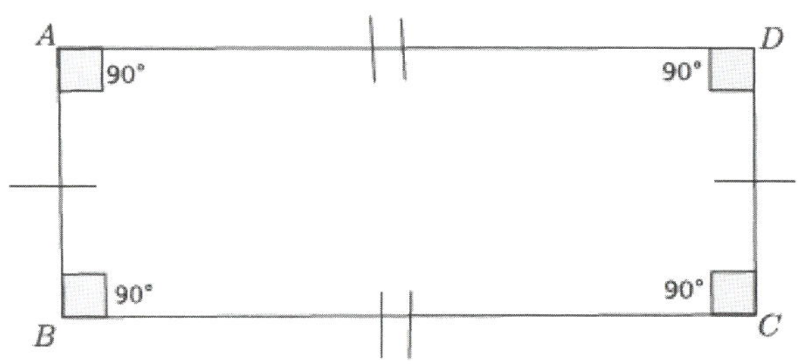

- If we draw a diagonal that connects A to C (or we could choose to connect B to D), what shapes are formed?
    - *We get two triangles.*

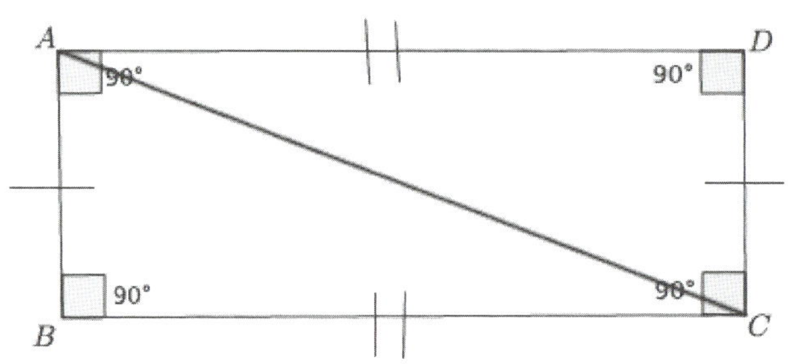

- What do we know about these triangles, and how do we know it?
  - *The triangles are congruent. We can trace one of the triangles and, through a sequence of basic rigid motions, map it onto the other triangle.*
- Our goal is to show that the angle sum of a triangle is $180°$. We know that when we draw a diagonal through a rectangle, we get two congruent triangles. How can we put this information together to show that the sum of angles in a triangle is $180°$?
  - *The rectangle has four right angles, which means that the sum of the angles of the rectangle is $4(90°)$ or $360°$. Since the diagonal divides the rectangle into two congruent triangles, each triangle will have exactly half the total degrees of the rectangle. Since $360° \div 2$ is $180°$, each triangle has a sum of angles equal to $180°$.*

Discussion (7 minutes)

Now, let's look at what is called the *exterior angle of a triangle*. An exterior angle is formed when one of the sides of the triangle is extended. The interior angles are inside the triangle, so the exterior angle is outside of the triangle along the extended side. In triangle ABC, the exterior angles are $\angle CBD$, $\angle ECA$, and $\angle BAF$.

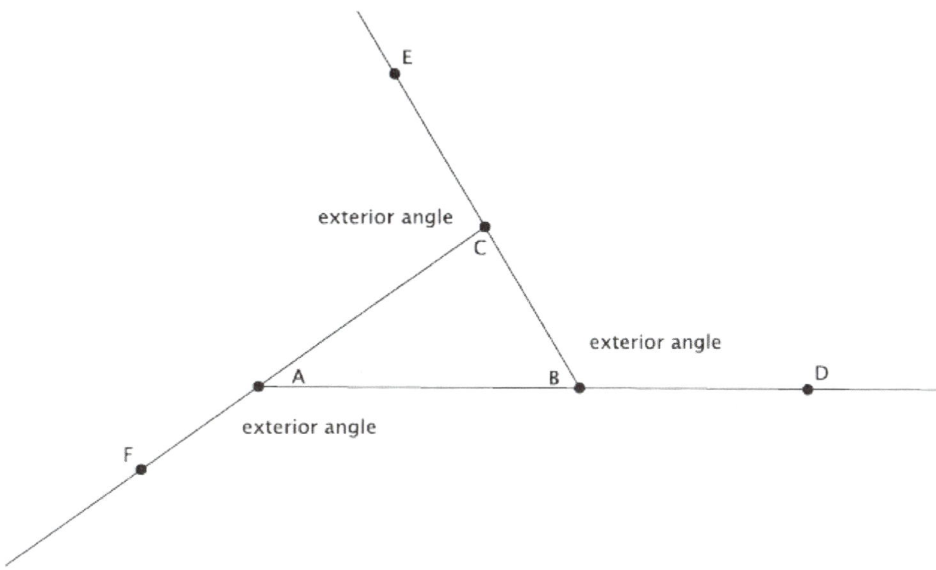

- What do we know about the sum of interior angles of a triangle? Name the angles.
  - *The sum of the interior angles $\angle ABC$, $\angle BCA$, and $\angle CAB$ of the triangle is $180°$.*
- What do we know about the degree of a straight angle?
  - *A straight angle has a measure of $180°$.*
- Let's look specifically at straight angle $\angle ABD$. Name the angles that make up this straight angle.
  - *$\angle ABC$ and $\angle CBD$*

- Because the triangle and the straight angle both have measures of $180°$, we can write them as equal to one another. That is, since
$$\angle ABC + \angle BCA + \angle CAB = 180°,$$
and
$$\angle ABC + \angle CBD = 180°,$$
then,
$$\angle ABC + \angle BCA + \angle CAB = \angle ABC + \angle CBD.$$
- Which angle is common to both the triangle and the straight angle?
  - $\angle ABC$
- If we subtract the measure of $\angle ABC$ from both the triangle and the straight angle, we get
$$\angle ABC - \angle ABC + \angle BCA + \angle CAB = \angle ABC - \angle ABC + \angle CBD$$
$$\angle BCA + \angle CAB = \angle CBD.$$
- What kind of angle is $\angle CBD$?
  - *It is the exterior angle of the triangle.*
- We call angles $\angle BCA$ and $\angle CAB$ the remote interior angles because they are the farthest, or most remote, from the exterior angle $\angle CBD$. Each of the remote angles shares one side with the angle adjacent to the exterior angle. The equation $\angle BCA + \angle CAB = \angle CBD$ means that the sum of measures of the remote interior angles is equal to the measure of the exterior angle of the triangle.

Exercises 1–4 (8 minutes)

Students work in pairs to identify the remote interior angles and corresponding exterior angle of the triangle in Exercises 1–3. After most students have finished Exercises 1–3, provide the correct answers before they move on to the next exercise. In Exercise 4, students recreate the reasoning of Example 1 for another exterior angle of the triangle.

Scaffolding:

Keep the work of Example 1 visible while students work on Exercises 1–4.

Exercises 1–4

Use the diagram below to complete Exercises 1–4.

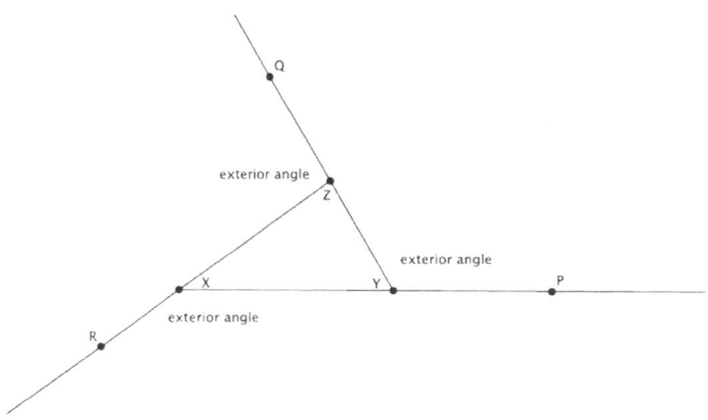

1. **Name an exterior angle and the related remote interior angles.**

   The exterior angle is $\angle ZYP$, and the related remote interior angles are $\angle YZX$ and $\angle ZXY$.

2. **Name a second exterior angle and the related remote interior angles.**

   The exterior angle is $\angle XZQ$, and the related remote interior angles are $\angle ZYX$ and $\angle ZXY$.

3. **Name a third exterior angle and the related remote interior angles.**

   The exterior angle is $\angle RXY$, and the related remote interior angles are $\angle ZYX$ and $\angle XZY$.

4. **Show that the measure of an exterior angle is equal to the sum of the measures of the related remote interior angles.**

   Triangle XYZ has interior angles $\angle XYZ$, $\angle YZX$, and $\angle ZXY$. The sum of those angles is $180°$. The straight angle $\angle XYP$ also has a measure of $180°$ and is made up of angles $\angle XYZ$ and $\angle ZYP$. Since the triangle and the straight angle have the same number of degrees, we can write the sum of their respective angles as an equality:

   $$\angle XYZ + \angle YZX + \angle ZXY = \angle XYZ + ZYP.$$

   Both the triangle and the straight angle share $\angle XYZ$. We can subtract the measure of that angle from the triangle and the straight angle. Then, we have

   $$\angle YZX + \angle ZXY = \angle ZYP,$$

   where the angle $\angle ZYP$ is the exterior angle, and the angles $\angle YZX$ and $\angle ZXY$ are the related remote interior angles of the triangle. Therefore, the sum of the measures of the remote interior angles of a triangle are equal to the measure of the exterior angle.

Example 1 (2 minutes)

Ask students what they need to do to find the measure of angle x. Then, have them work on personal white boards and show their answer.

Example 1

Find the measure of angle x.

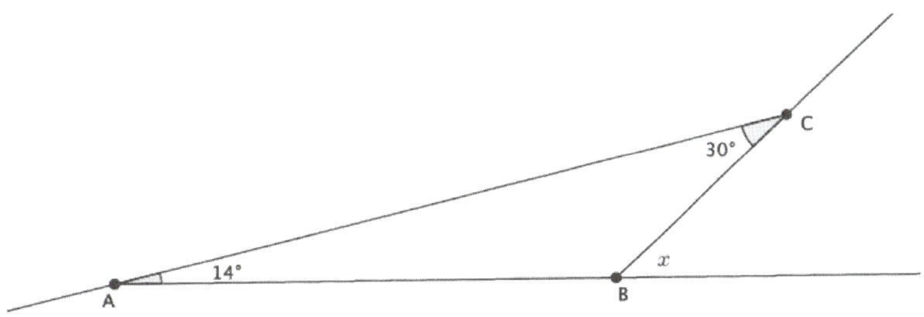

We need to find the sum of the measures of the remote interior angles to find the measure of the exterior angle x: $14 + 30 = 44$. Therefore, the measure of $\angle x$ is $44°$.

- Present an informal argument that proves you are correct.
  - *We know that triangles have a sum of interior angles that is equal to $180°$. We also know that straight angles are $180°$. $\angle ABC$ must be $136°$, which means that $\angle x$ is $44°$.*

Example 2 (2 minutes)

Ask students what they need to do to find the measure of angle x. Then, have them work on personal white boards and show their answer.

Example 2

Find the measure of angle x.

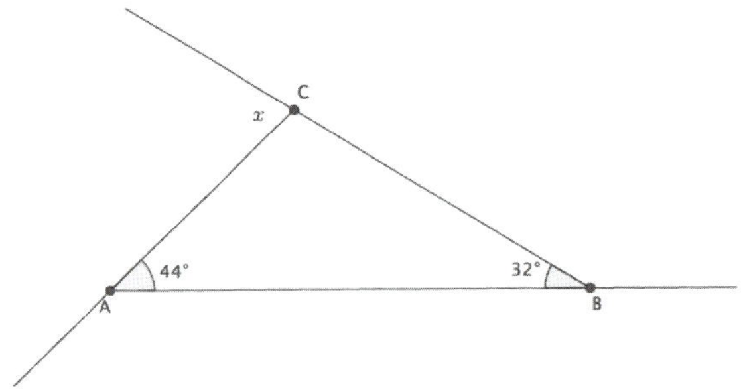

We need to find the sum of the measures of the remote interior angles to find the measure of the exterior angle x: $44 + 32 = 76$. Therefore, the measure of $\angle x$ is $76°$.

- Present an informal argument that proves you are correct.
  - *We know that triangles have a sum of interior angles that is equal to $180°$. We also know that straight angles are $180°$. $\angle ACB$ must be $104°$, which means that $\angle x$ is $76°$.*

Example 3 (2 minutes)

Ask students what they need to do to find the measure of angle x. Then, have them work on personal white boards and show their answers. Make sure students see that this is not like the last two examples. They must pay attention to the information that is provided and not expect to always do the same procedure.

Example 3

Find the measure of angle x.

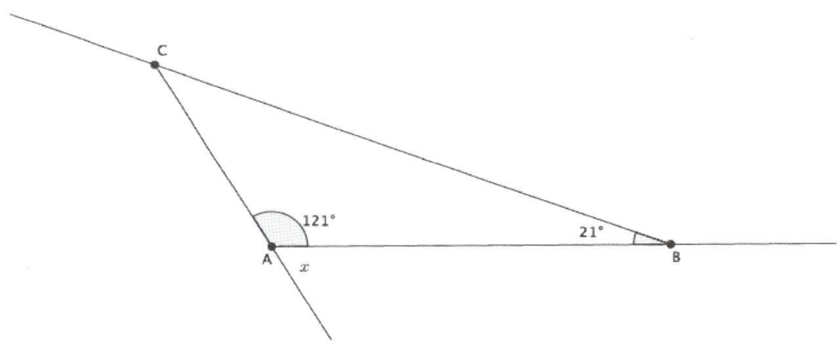

$180 - 121 = 59$. Therefore, the measure of $\angle x$ is $59°$.

Students should notice that they are not given the two remote interior angles associated with the exterior angle x. For that reason, they must use what they know about straight angles (or supplementary angles) to find the measure of angle x.

Example 4 (2 minutes)

Ask students what they need to do to find the measure of angle x. Then, have them work on personal white boards and show their answers. Make sure students see that this is not like the last three examples. They must pay attention to the information that is provided and not expect to always do the same procedure.

Example 4

Find the measure of angle x.

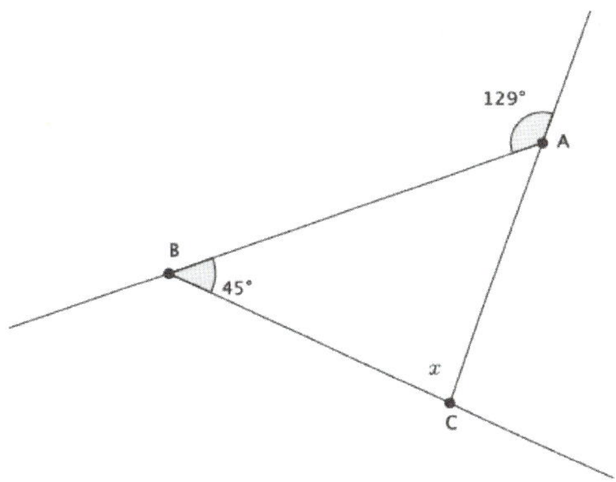

$129 - 45 = 84$. Therefore, the measure of $\angle x$ is $84°$.

Students should notice that they are given just one of the remote interior angle measures and the exterior angle measure. For that reason, they need to subtract 45 from the exterior angle to find the measure of angle x.

Exercises 5–10 (6 minutes)

Students complete Exercises 5–10 independently. Check solutions once most students have finished.

Exercise 5–10

5. **Find the measure of angle x. Present an informal argument showing that your answer is correct.**

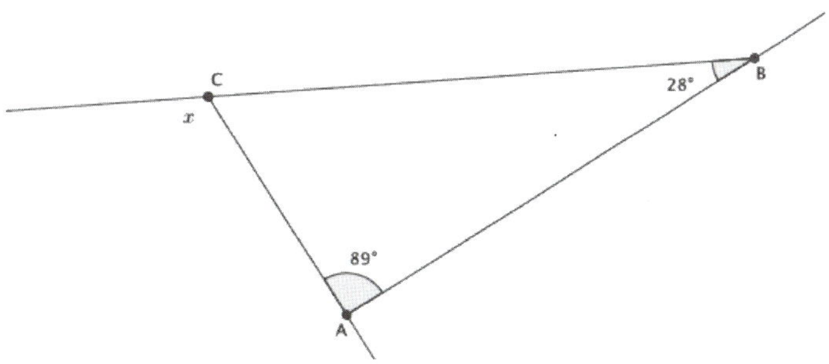

Since $89 + 28 = 117$, the measure of angle x is $117°$. We know that triangles have a sum of interior angles that is equal to $180°$. We also know that straight angles are $180°$. $\angle ACB$ must be $63°$, which means that $\angle x$ is $117°$.

6. Find the measure of angle x. Present an informal argument showing that your answer is correct.

Since $59 + 52 = 111$, the measure of angle x is $111°$. We know that triangles have a sum of interior angles that is equal to $180°$. We also know that straight angles are $180°$. $\angle CAB$ must be $69°$, which means that $\angle x$ is $111°$.

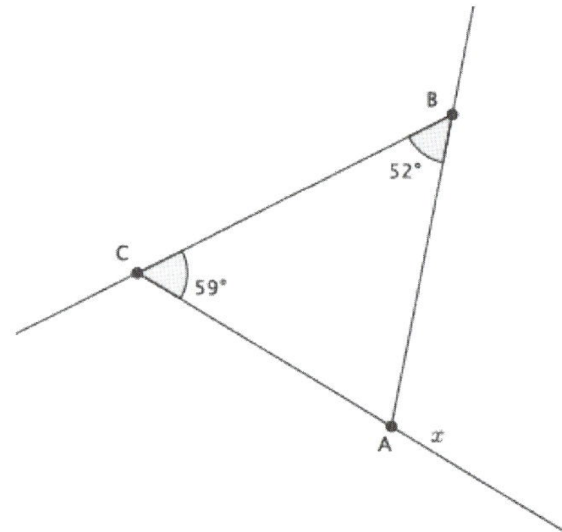

7. Find the measure of angle x. Present an informal argument showing that your answer is correct.

Since $180 - 79 = 101$, the measure of angle x is $101°$. We know that straight angles are $180°$, and the straight angle in the diagram is made up of $\angle ABC$ and $\angle x$. $\angle ABC$ is $79°$, which means that $\angle x$ is $101°$.

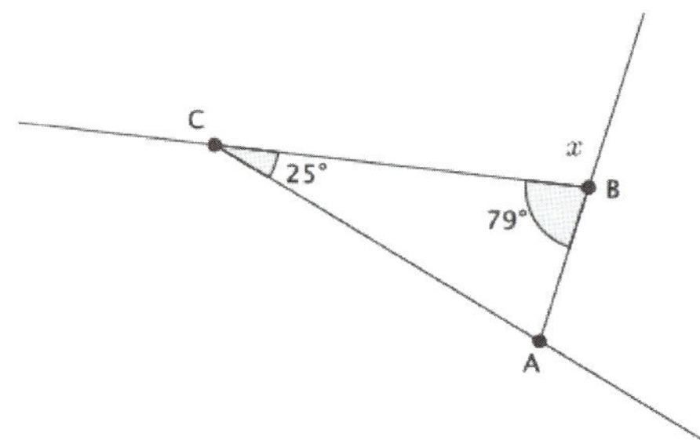

8. Find the measure of angle x. Present an informal argument showing that your answer is correct.

Since $71 + 74 = 145$, the measure of angle x is $145°$. We know that triangles have a sum of interior angles that is equal to $180°$. We also know that straight angles are $180°$. $\angle ACB$ must be $35°$, which means that $\angle x$ is $145°$.

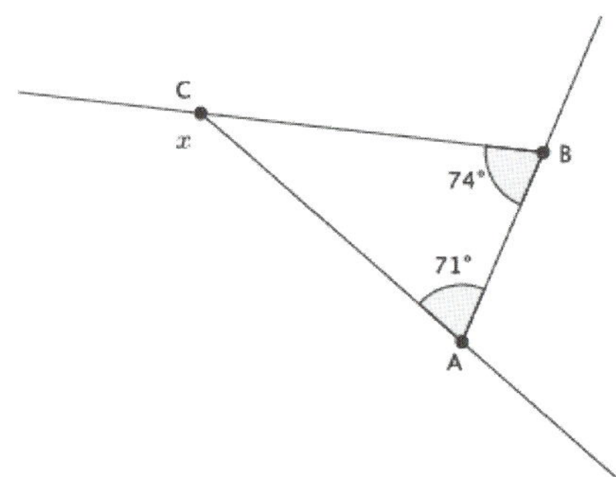

9. Find the measure of angle x. Present an informal argument showing that your answer is correct.

Since $107 + 32 = 139$, the measure of angle x is $139°$. We know that triangles have a sum of interior angles that is equal to $180°$. We also know that straight angles are $180°$. $\angle CBA$ must be $41°$, which means that x is $139°$.

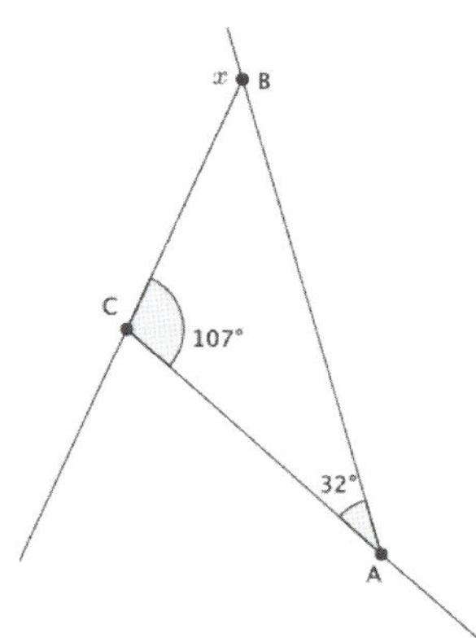

G8-M2-B1-v1.3.0_05/04/15

10. **Find the measure of angle x. Present an informal argument showing that your answer is correct.**

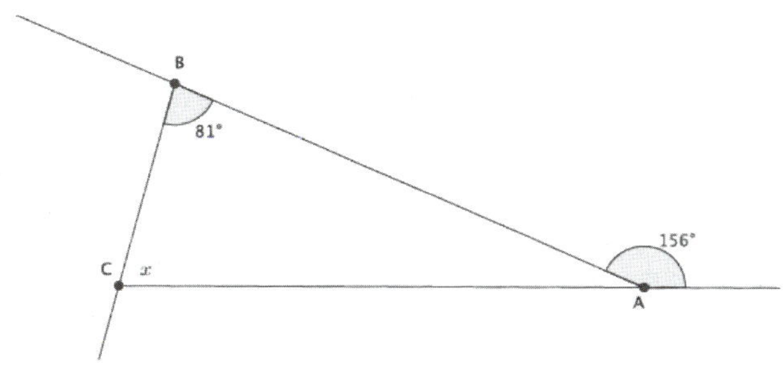

Since $156 - 81 = 75$, the measure of angle x is $75°$. We know that triangles have a sum of interior angles that is equal to $180°$. We also know that straight angles are $180°$. $\angle BAC$ must be $24°$ because it is part of the straight angle. Then, $\angle x = 180° - (81° + 24°)$, which means $\angle x$ is $75°$.

Closing (4 minutes)

Summarize, or have students summarize, the lesson.

- We learned another proof as to why the sum of the measures of the interior angles of a triangle are equal to $180°$ with respect to a triangle being exactly half of a rectangle.
- We learned the definitions of exterior angles and remote interior angles.
- The sum of the measures of the remote interior angles of a triangle is equal to the measure of the related exterior angle.

Lesson Summary

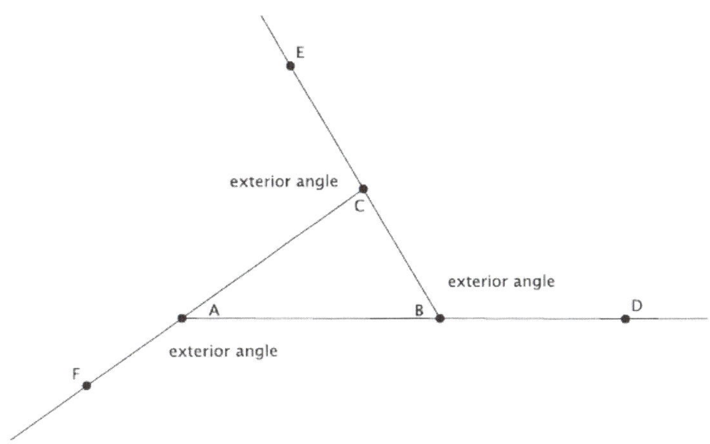

The sum of the measures of the remote interior angles of a triangle is equal to the measure of the related exterior angle. For example, $\angle CAB + \angle ABC = \angle ACE$.

Exit Ticket (5 minutes)

Name ______________________________ Date ______________

Lesson 14: More on the Angles of a Triangle

Exit Ticket

1. Find the measure of angle p. Present an informal argument showing that your answer is correct.

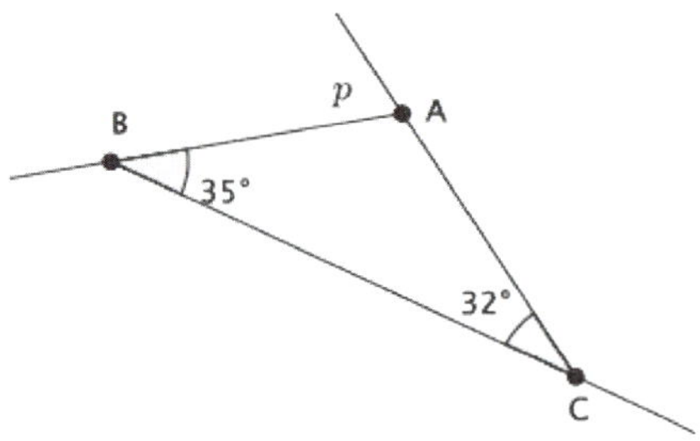

2. Find the measure of angle q. Present an informal argument showing that your answer is correct.

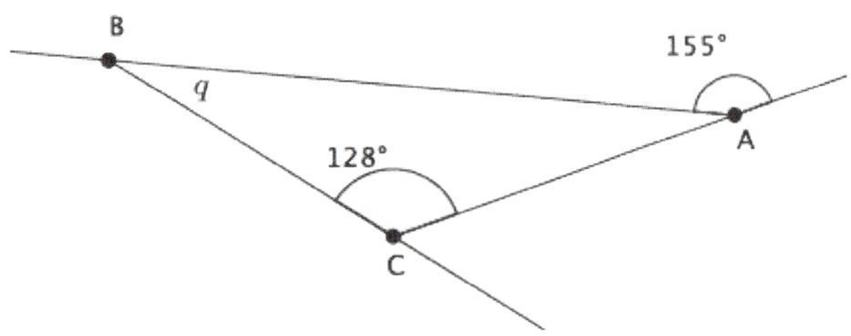

3. Find the measure of angle r. Present an informal argument showing that your answer is correct.

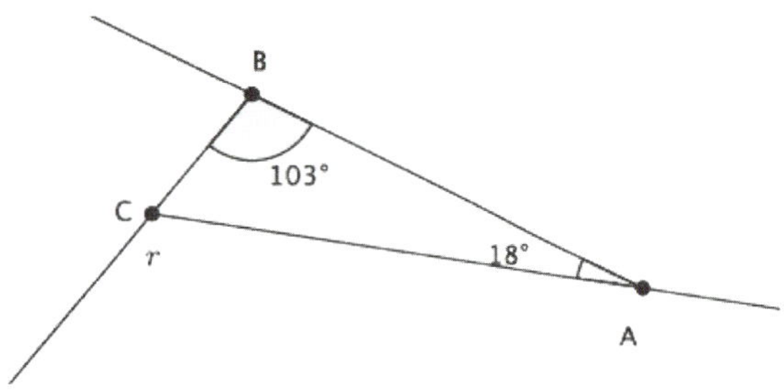

Exit Ticket Sample Solutions

1. **Find the measure of angle p. Present an informal argument showing that your answer is correct.**

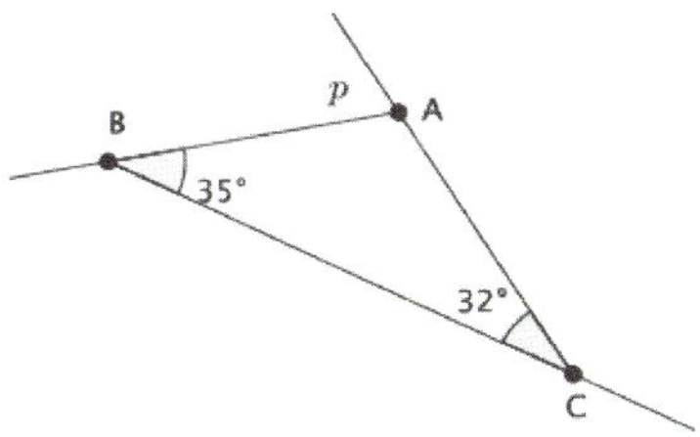

The measure of angle p is $67°$. We know that triangles have a sum of interior angles that is equal to $180°$. We also know that straight angles are $180°$. $\angle BAC$ must be $113°$, which means that $\angle p$ is $67°$.

2. **Find the measure of angle q. Present an informal argument showing that your answer is correct.**

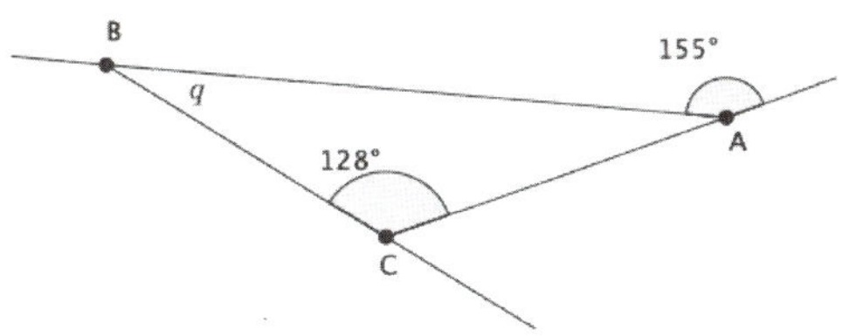

The measure of angle q is $27°$. We know that triangles have a sum of interior angles that is equal to $180°$. We also know that straight angles are $180°$. $\angle CAB$ must be $25°$, which means that $\angle q$ is $27°$.

3. **Find the measure of angle r. Present an informal argument showing that your answer is correct.**

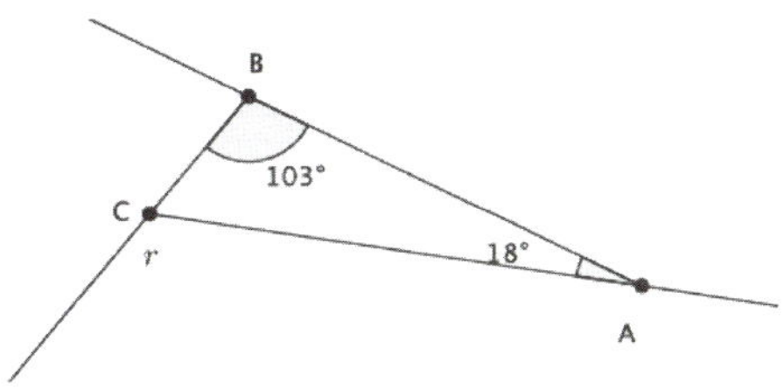

The measure of angle r is $121°$. We know that triangles have a sum of interior angles that is equal to $180°$. We also know that straight angles are $180°$. $\angle BCA$ must be $59°$, which means that $\angle r$ is $121°$.

Problem Set Sample Solutions

Students practice finding missing angle measures of triangles.

For each of the problems below, use the diagram to find the missing angle measure. Show your work.

1. **Find the measure of angle x. Present an informal argument showing that your answer is correct.**

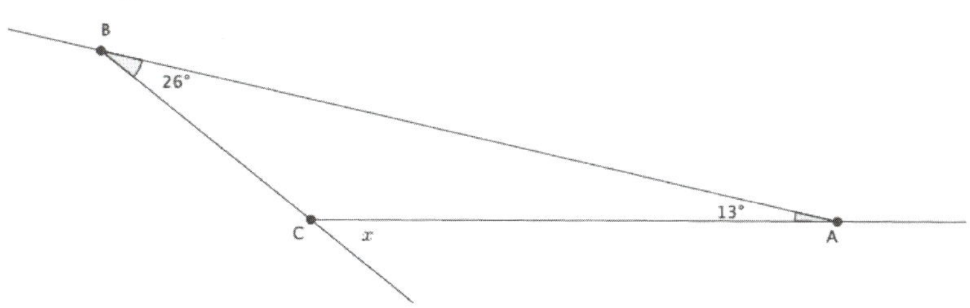

Since $26 + 13 = 39$, the measure of angle x is $39°$. We know that triangles have a sum of interior angles that is equal to $180°$. We also know that straight angles are $180°$. $\angle BCA$ must be $141°$, which means that $\angle x$ is $39°$.

G8-M2-B1-v1.3.0_05/04/15

2. **Find the measure of angle x.**

Since $52 + 44 = 96$, the measure of angle x is $96°$.

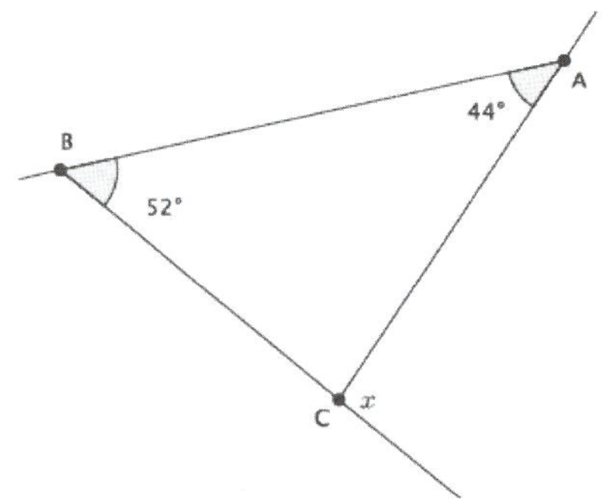

3. **Find the measure of angle x. Present an informal argument showing that your answer is correct.**

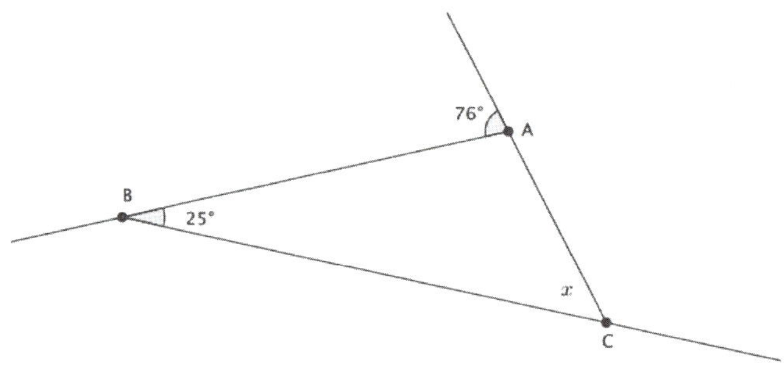

Since $76 - 25 = 51$, the measure of $\angle x$ is $51°$. We know that triangles have a sum of interior angles that is equal to $180°$. We also know that straight angles are $180°$. $\angle BAC$ must be $104°$ because it is part of the straight angle. Then, $x = 180° - (104° + 25°)$, which means $\angle x$ is $51°$.

4. **Find the measure of angle x.**

Since $27 + 52 = 79$, the measure of angle x is $79°$.

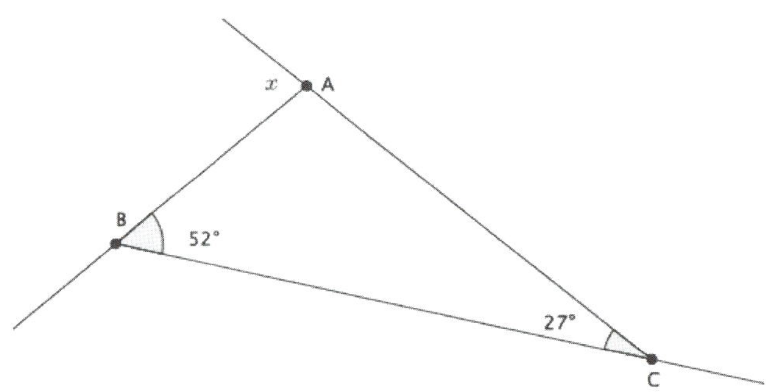

5. **Find the measure of angle x.**

Since $180 - 104 = 76$, the measure of angle x is $76°$.

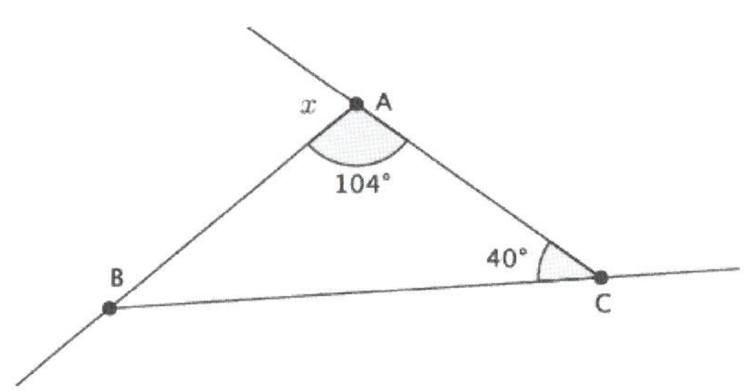

6. **Find the measure of angle x.**

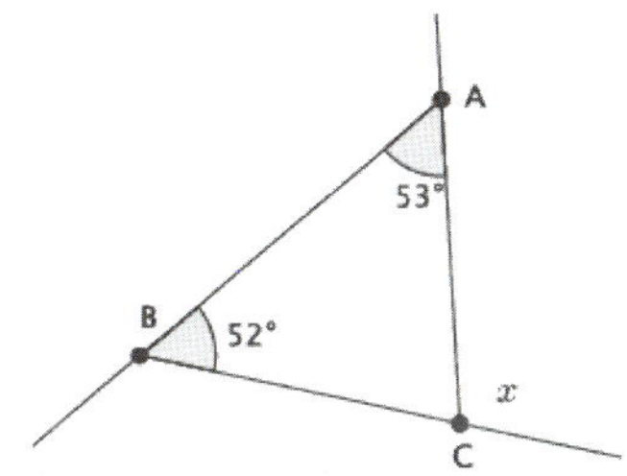

Since $52 + 53 = 105$, the measure of angle x is $105°$.

7. **Find the measure of angle x.**

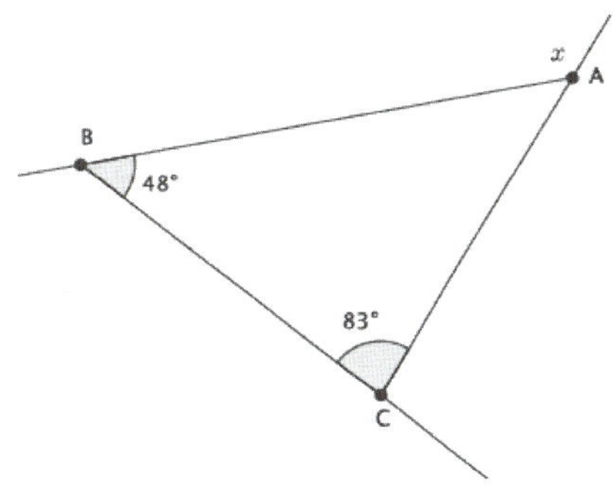

Since $48 + 83 = 131$, the measure of angle x is $131°$.

8. **Find the measure of angle x.**

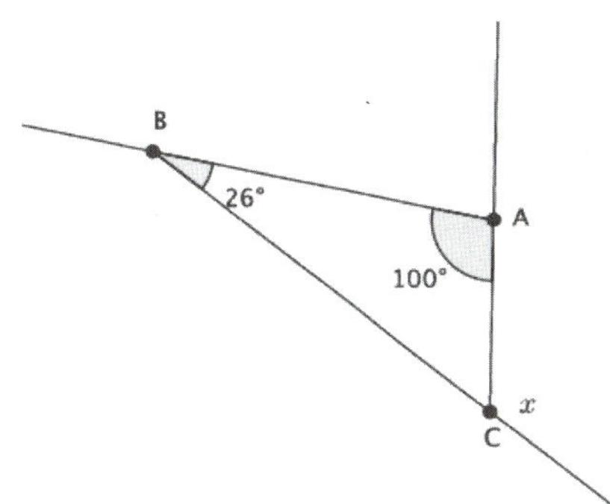

Since $100 + 26 = 126$, the measure of angle x is $126°$.

9. **Find the measure of angle x.**

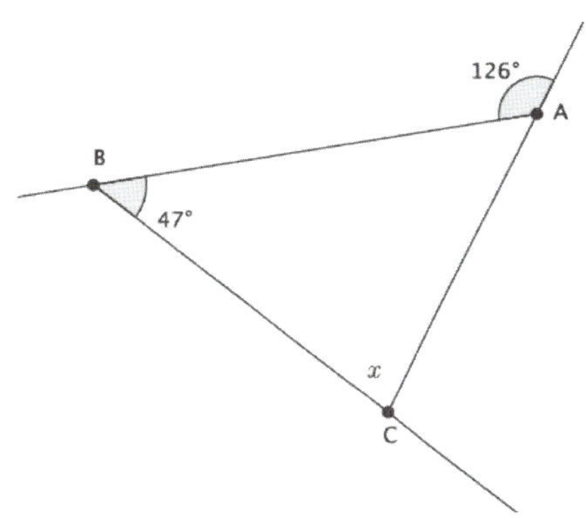

Since $126 - 47 = 79$, the measure of angle x is $79°$.

10. **Write an equation that would allow you to find the measure of angle x. Present an informal argument showing that your answer is correct.**

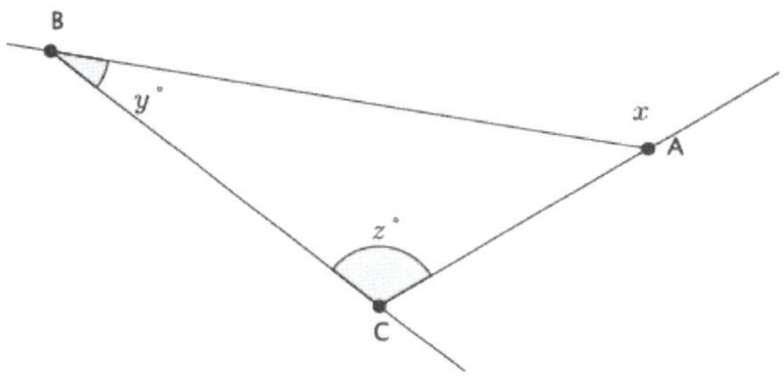

Since $y + z = x$, the measure of angle x is $(y + z)°$. We know that triangles have a sum of interior angles that is equal to $180°$. We also know that straight angles are $180°$.
Then, $\angle y + \angle z + \angle BAC = 180°$, and $\angle x + \angle BAC = 180°$. Since both equations are equal to $180°$, then $\angle y + \angle z + \angle BAC = \angle x + \angle BAC$. Subtract $\angle BAC$ from each side of the equation, and you get $\angle y + \angle z = \angle x$.

Name ________________________________ Date ________________

1. $\triangle ABC \cong \triangle A'B'C'$. Use the picture to answer the question below.

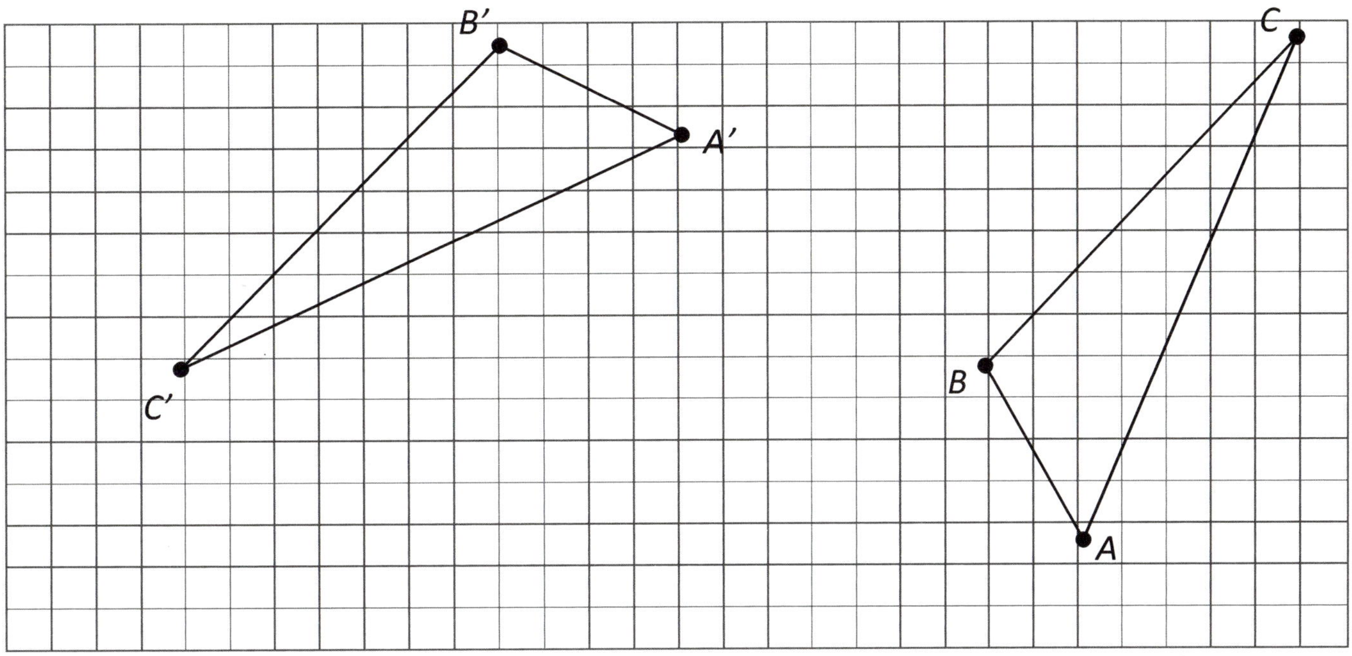

Describe a sequence of rigid motions that would prove a congruence between $\triangle ABC$ and $\triangle A'B'C'$.

2. Use the diagram to answer the question below.

   $k \parallel l$

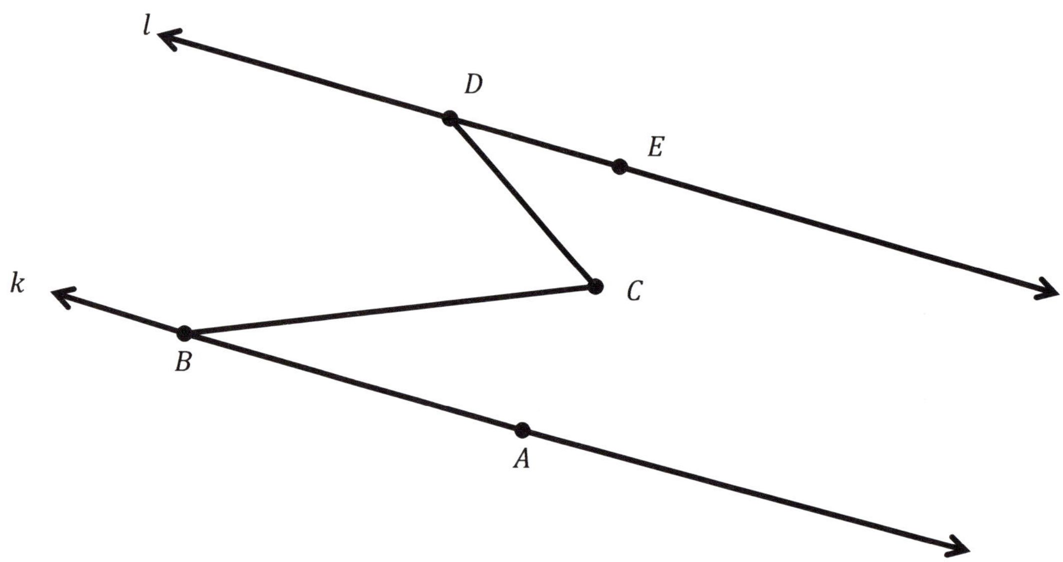

Line k is parallel to line l. $m\angle EDC = 41°$ and $m\angle ABC = 32°$. Find the $m\angle BCD$. Explain in detail how you know you are correct. Add additional lines and points as needed for your explanation.

3. Use the diagram below to answer the questions that follow. Lines L_1 and L_2 are parallel, $L_1 \parallel L_2$. Point N is the midpoint of segment GH.

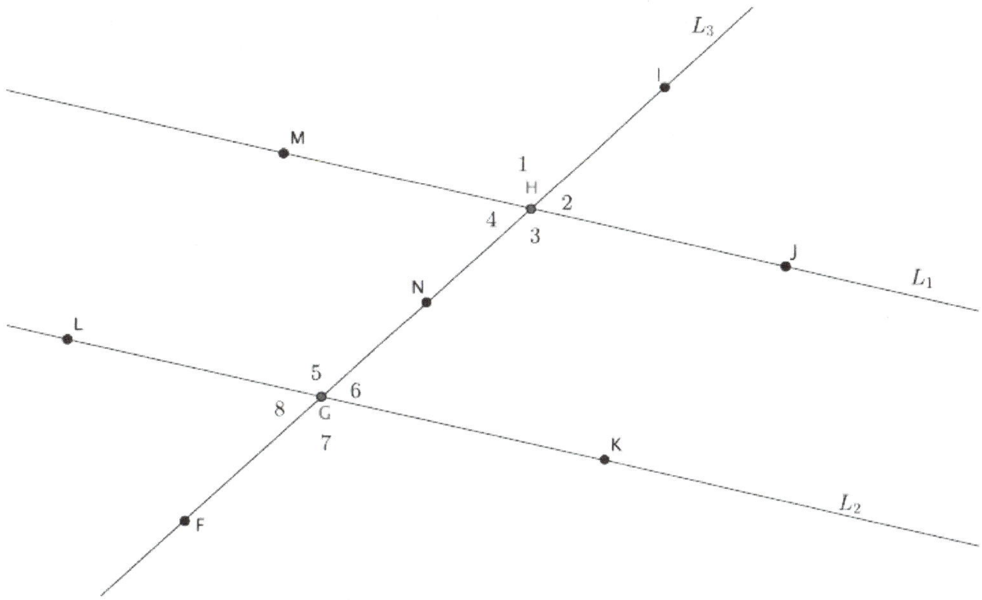

a. If the measure of $\angle IHM$ is $125°$, what is the measure of $\angle IHJ$? $\angle JHN$? $\angle NHM$?

b. What can you say about the relationship between $\angle 4$ and $\angle 6$? Explain using a basic rigid motion. Name another pair of angles with this same relationship.

c. What can you say about the relationship between $\angle 1$ and $\angle 5$? Explain using a basic rigid motion. Name another pair of angles with this same relationship.

A Progression Toward Mastery					
Assessment Task Item		**STEP 1** Missing or incorrect answer and little evidence of reasoning or application of mathematics to solve the problem.	**STEP 2** Missing or incorrect answer but evidence of some reasoning or application of mathematics to solve the problem.	**STEP 3** A correct answer with some evidence of reasoning or application of mathematics to solve the problem, OR an incorrect answer with substantial evidence of solid reasoning or application of mathematics to solve the problem.	**STEP 4** A correct answer supported by substantial evidence of solid reasoning or application of mathematics to solve the problem.
1	8.G.A.2	Student is unable to respond to the question or leaves item blank. Student does not describe a sequence. Student shows no reasoning or application of mathematics to solve the problem.	Student identifies an incorrect sequence of rigid motions. Student uses little or no mathematical vocabulary or notation in sequence. Some evidence of mathematical reasoning is used in sequence.	Student identifies a correct sequence of rigid motions but lacks precision. Student may or may not use mathematical vocabulary or notation sequence. Some evidence of mathematical reasoning is used in sequence.	Student identifies a correct sequence of rigid motions with precision. Student uses mathematical vocabulary and notation in sequence. Substantial evidence of mathematical reasoning is used in sequence.
2	8.G.A.5	Student is unable to respond to the question or leaves item blank. Student shows no reasoning or application of mathematics to solve the problem.	Student calculates the measurement of the angle but makes calculation errors. Student attempts to use auxiliary lines to solve the problem. Student shows little or no reasoning in the written explanation. Student does not use any theorem in the written explanation.	Student calculates the measurement of the angle but makes calculation errors. Student uses auxiliary lines to solve the problem. Student shows some reasoning in the written explanation. Student may or may not use the correct theorem in the written explanation.	Student calculates the measurement of the angle correctly as $73°$. Student uses auxiliary lines to solve the problem. Student shows substantial reasoning in the written explanation including information about congruent angles being equal, straight angles having $180°$, triangle sum being $180°$, and the sum of remote interior angles being equal to the exterior angle of a triangle.

EUREKA MATH™

3	a 8.G.A.5	Student is unable to respond to the questions or leaves item blank. Student shows no reasoning or application of mathematics to solve the problem.	Student makes calculation errors. Student answers part of the question correctly (e.g., $\angle IHM = \angle JHN = 125°$ but omits $\angle IHJ = \angle NHM = 55°$). OR Student answers with all four angles as the same measure.	Student shows some application of mathematics to solve the problem. Student makes calculation errors. Student reverses the answers (i.e., $\angle IHM = \angle JHN = 55°$ or $\angle IHJ = \angle NHM = 125°$).	Student answers correctly with $\angle IHM = \angle JHN = 125°$ and $\angle IHJ = \angle NHM = 55°$ for measures of ALL four angles.
	b 8.G.A.5	Student is unable to respond to the questions or leaves item blank. Student shows no reasoning or application of mathematics to solve the problem. Student does not include a written explanation.	Student answers the name of the angles incorrectly. Student incorrectly identifies the other angles with the same relationship. Student includes a written explanation. Student references a rigid motion, translation, rotation, and reflection. The written explanation is not mathematically based (e.g., "they look the same").	Student may answer the name of the angles incorrectly but correctly identifies the other angles with the same relationship. Student uses some mathematical vocabulary in the written explanation. Student references rotation but may not reference all of the key points in the written explanation.	Student answers correctly by calling the angles alternate interior angles. Student names $\angle 3$ and $\angle 5$ as angles with the same relationship. Student uses mathematical vocabulary in the written explanation. Student references ALL of the following key points: N is the midpoint of HG, rotation of $180°$ around N, and rotation is angle-preserving in the written explanation. The written explanation is thorough and complete.
	c 8.G.A.5	Student is unable to respond to the questions or leaves item blank. Student shows no reasoning or application of mathematics to solve the problem. Student does not include a written explanation.	Student answers the name of the angles incorrectly. Student incorrectly identifies the other angles with the same relationship. Student includes a written explanation. Student references a rigid motion, translation, rotation, reflection. The written explanation is not mathematically based (e.g., "they look the same").	Student identifies the name of the angles incorrectly but does correctly identify the other angles with the same relationship. Student uses some mathematical vocabulary in the written explanation. Student references translation but may not reference all of the key points in the written explanation.	Student answers correctly by calling the angles corresponding angles. Student names $\angle 2$ and $\angle 6$ (or $\angle 3$ and $\angle 7$ or $\angle 4$ and $\angle 8$) as angles with the same relationship. Student uses mathematical vocabulary in the written explanation. Student references ALL of the following key points: translation along vector HG, translation maps parallel lines to parallel lines, and translation is angle-preserving in the written explanation. The written explanation is thorough and complete.

G8-M2-B1-v1.3.0_05/04/15

Name ______________________________ Date ______________

1. $\triangle ABC \cong \triangle A'B'C'$. Use the picture to answer the question below.

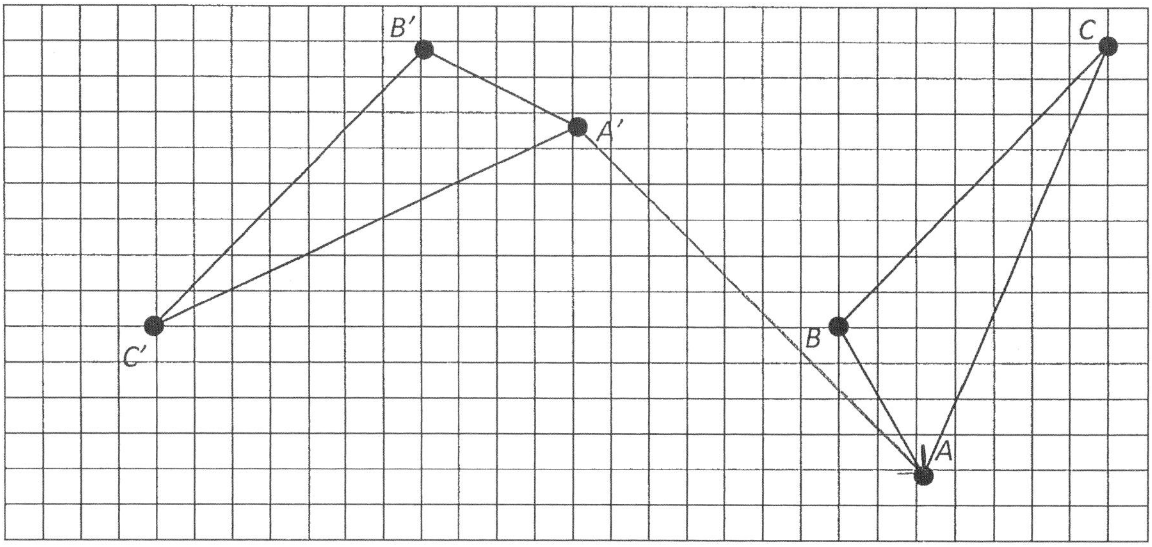

Describe a sequence of rigid motions that would prove a congruence between $\triangle ABC$ and $\triangle A'B'C'$.

LET T BE THE TRANSLATION ALONG $\overrightarrow{A'A}$ SO THAT $T(A')=A$.
LET R BE THE ROTATION AROUND A, d DEGREES SO THAT $R(A'B') = AB$. BY HYPOTHESIS $|AB|=|A'B'|$.
LET Λ BE THE REFLECTION ACROSS L_{AB}. AGAIN BY HYPOTHESIS $|\angle A|=|\angle A|$, $|\angle B|=|\angle B|$, SO THE COMPOSITION $\Lambda \circ R \circ T$ WILL MAP $\Delta A'B'C'$ TO ΔABC, ie., $\Lambda(R(T(\Delta A'B'C'))) = \Delta ABC$

2. Use the diagram to answer the question below.

$k \parallel l$

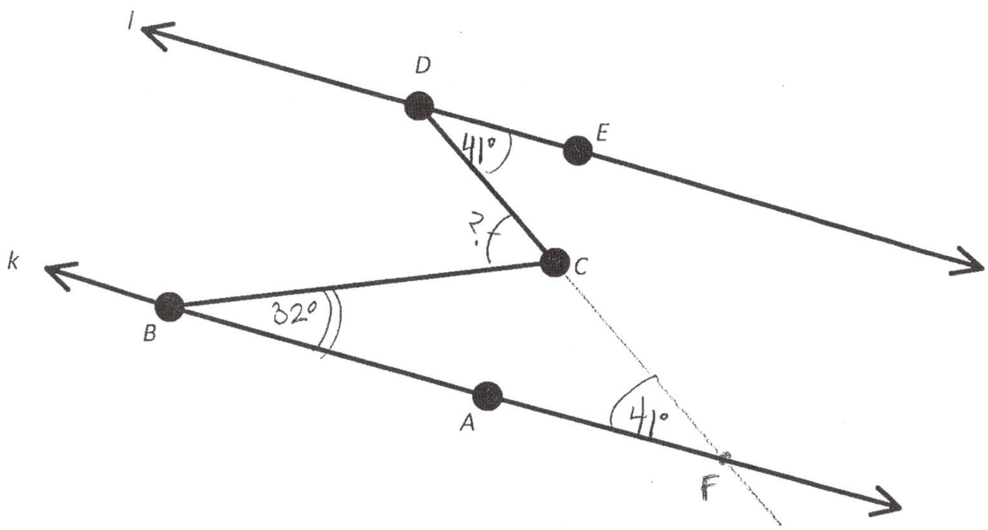

Line k is parallel to line l. $m\angle EDC = 41°$ and $m\angle ABC = 32°$. Find the $m\angle BCD$. Explain in detail how you know you are correct. Add additional lines and points as needed for your explanation.

LET F BE A POINT ON LINE k SO THAT ∠DCF IS A STRAIGHT ANGLE. THEN BECAUSE k∥l, ∠EDC ≅ ∠CFA AND HAVE EQUAL MEASURE. ∠ABC AND ∠CFA ARE THE REMOTE INTERIOR ANGLES OF △BCF WHICH MEANS ∠BCD = ∠ABC + CFA. THEREFORE ∠BCD = 32 + 41 = 73°.

3. Use the diagram below to answer the questions that follow. Lines L_1 and L_2 are parallel, $L_1 \parallel L_2$. Point N is the midpoint of segment GH.

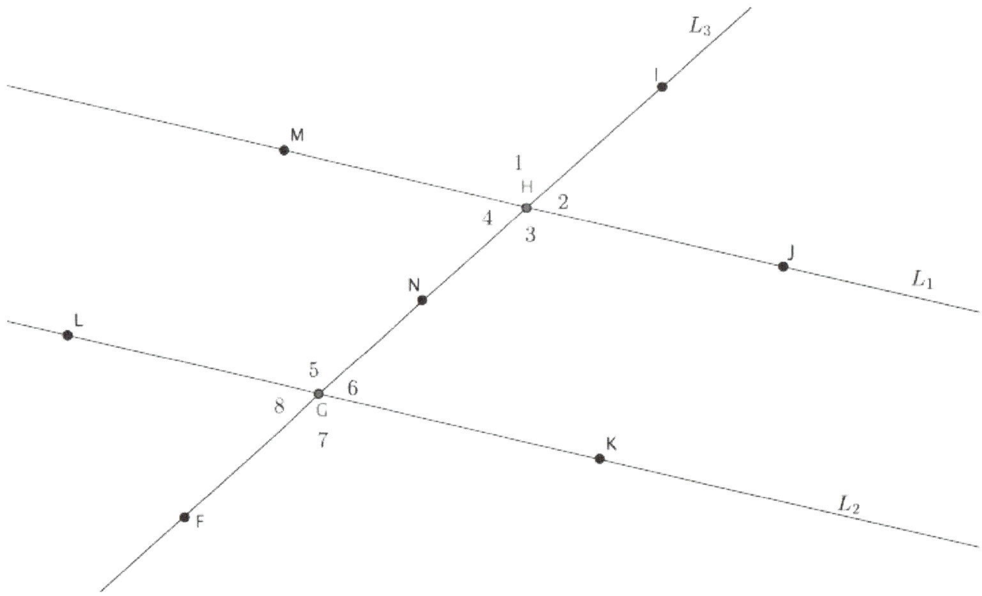

a. If the measure of $\angle IHM$ is $125°$, what is the measure of $\angle IHJ$? $\angle JHN$? $\angle NHM$?

∠IHJ = 55°   ∠JHN = 125°   ∠NHM = 55°

b. What can you say about the relationship between $\angle 4$ and $\angle 6$? Explain using a basic rigid motion. Name another pair of angles with this same relationship.

∠4 & ∠6 ARE ALTERNATE INTERIOR ANGLES THAT ARE EQUAL BECAUSE L₁//L₂. LET R BE A ROTATION OF 180° AROUND POINT N. THEN R(N)=N; R(L3)=L3; AND R(L1)=L2. ROTATIONS ARE DEGREE PRESERVING SO R(∠4)=∠6.

∠3 & ∠5 ARE ALSO ALTERNATE INTERIOR ANGLES THAT ARE EQUAL.

c. What can you say about the relationship between $\angle 1$ and $\angle 5$? Explain using a basic rigid motion. Name another pair of angles with this same relationship.

∠1 & ∠5 ARE CORRESPONDING ANGLES THAT ARE EQUAL BECAUSE L1//L2. LET T BE THE TRANSLATION ALONG VECTOR $\overrightarrow{GH}$. THEN T(L2)=L1 AND T(∠5)=∠1.

∠3 & ∠7 ARE ALSO CORRESPONDING ANGLES THAT ARE EQUAL.

EUREKA MATH™

(Optional) Topic D

The Pythagorean Theorem

8.G.B.6, 8.G.B.7

Focus Standards:	8.G.B.6	Explain a proof of the Pythagorean Theorem and its converse.
	8.G.B.7	Apply the Pythagorean Theorem to determine unknown side lengths in right triangles in real-world and mathematical problems in two and three dimensions.
Instructional Days:	2	
Lesson 15:	Informal Proof of the Pythagorean Theorem (S)[1]	
Lesson 16:	Applications of the Pythagorean Theorem (P)	

In Topic D, students are guided through the square within a square proof of the Pythagorean theorem, which requires students to know that congruent figures also have congruent areas. Once proved, students practice using the Pythagorean theorem and its converse in Lesson 16 to find unknown side lengths in right triangles. Students apply their knowledge of the Pythagorean theorem to real-world problems that involve two-and three-dimensional figures.

[1]Lesson Structure Key: **P**-Problem Set Lesson, **M**-Modeling Cycle Lesson, **E**-Exploration Lesson, **S**-Socratic Lesson

Lesson 15: Informal Proof of the Pythagorean Theorem

Student Outcomes

- Students are introduced to the Pythagorean theorem and are shown an informal proof of the theorem.
- Students use the Pythagorean theorem to find the length of the hypotenuse of a right triangle.

Lesson Notes

Since **8.G.B.6** and **8.G.B.7** are post-test standards, this lesson is designated as an extension lesson for this module. However, the content within this lesson is prerequisite knowledge for Module 7. If this lesson is not used with students as part of the work within Module 2, it must be used with students prior to beginning work on Module 7. Please realize that many mathematicians agree that the Pythagorean theorem is the most important theorem in geometry and has immense implications in much of high school mathematics in general (e.g., when studying quadratics and trigonometry). It is crucial that students see the teacher explain several proofs of the Pythagorean theorem and practice using it before being expected to produce a proof on their own.

Classwork

Concept Development (5 minutes)

The Pythagorean theorem is a famous theorem that is used throughout much of high school mathematics. For that reason, students see several proofs of the theorem throughout the year and have plenty of practice using it. The first thing to know about the Pythagorean theorem is what it states.

PYTHAGOREAN THEOREM: If the lengths of the legs of a right triangle are a and b, and the length of the hypotenuse is c, then $a^2 + b^2 = c^2$.

Given a right triangle ABC with C being the vertex of the right angle, then the sides $\overline{AC}$ and $\overline{BC}$ are called the *legs* of $\triangle ABC$, and $\overline{AB}$ is called the *hypotenuse* of $\triangle ABC$.

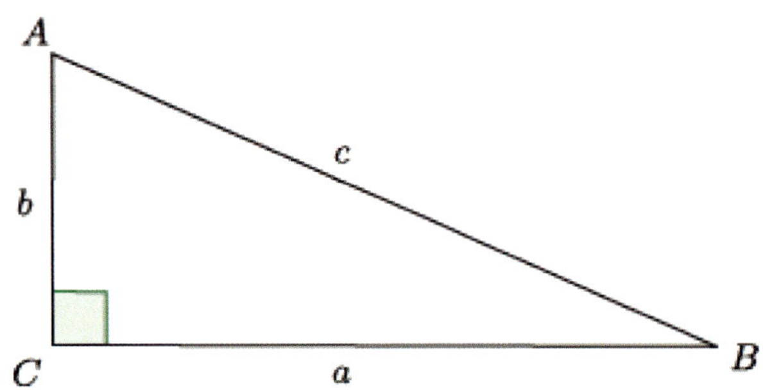

> *Scaffolding:*
>
> Draw arrows, one at a time, to show that each side is the opposite of the given angle.

Take note of the fact that side a is opposite the angle A, side b is opposite the angle B, and side c is opposite the angle C.

Discussion (15 minutes)

The first proof of the Pythagorean theorem requires knowledge of some basic facts about geometry.

1. Congruent triangles have equal areas.
2. All corresponding parts of congruent triangles are congruent.
3. The triangle sum theorem. (∠ sum of △)
   a. For right triangles, the two angles that are not the right angle have a sum of $90°$. (∠ sum of rt. △)

What we look at next is what is called a square within a square. The outside square has side lengths $(a + b)$, and the inside square has side lengths c. Our goal is to show that $a^2 + b^2 = c^2$. To accomplish this goal, we compare the total area of the outside square with the parts it is composed of, that is, the four triangles and the smaller inside square.

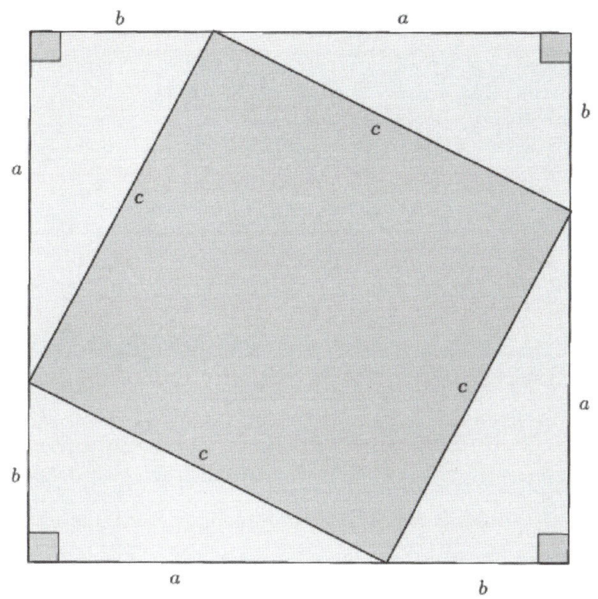

> *Note to Teacher:*
>
> Remind students to use the distributive law to determine the area of the outside square. Also remind them to use what they know about exponential notation to simplify the expression.

Ask students the following questions during the discussion:

- Looking at the outside square only, the square with side lengths $(a + b)$, what is its area?
  - *The area of the outside square is $(a + b)^2$ which after expanding is $a^2 + 2ab + b^2$.*
- Are the four triangles with sides lengths a and b congruent? If so, how do you know?
  - *Yes, the triangles are congruent. From the diagram, we can see that each triangle has a right angle, and each triangle has side lengths of a and b. Our rigid motions preserve those measures, and we can trace one triangle and use rigid motions to prove that they are congruent.*
- What is the area of just one triangle?
  - $\frac{1}{2}ab$
- Does each triangle have the same area? If so, what is the sum of all four of those areas?
  - *Yes, each triangle has the same area because they are congruent. The sum of all four triangles is $4\left(\frac{1}{2}ab\right)$ or after simplifying, $2ab$.*

- We called this entire figure a square within a square, but we want to make sure that the figure in the center is indeed a square. To do so, we need to look at the angles of the triangles. First of all, what do we know about corresponding angles of congruent triangles?
  - *Corresponding angles of congruent triangles are also congruent and equal in measure.*

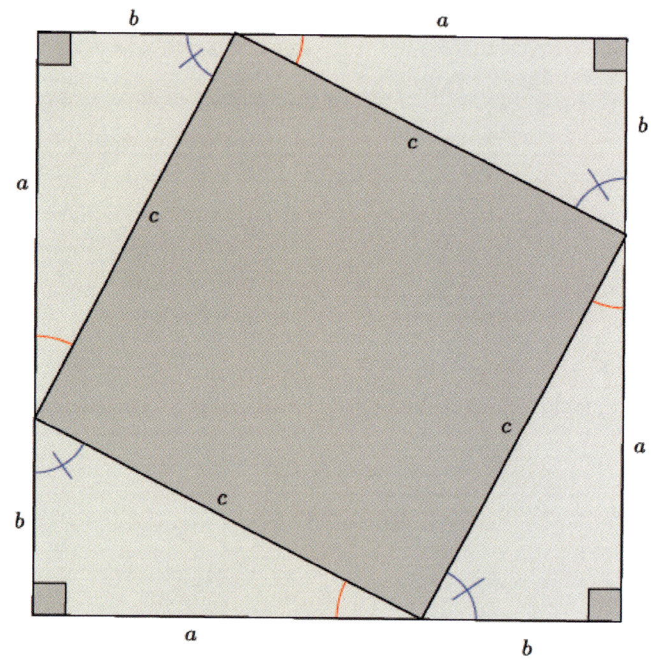

- So we know that the angles marked by the red arcs are equal in measure, and the angles marked with the blue arcs and line are equal in measure. What do we know about the sum of the interior angles of a triangle (∠ sum of △)?
  - *The sum of the measures of the interior angles of a triangle is* $180°$.
- What is the sum of the measures of the two interior angles of a right triangle, not including the right angle (∠ sum of rt. △)? How do you know?
  - *For right triangles, we know that one angle has a measure of* $90°$. *Since the sum of the measures of all three angles must be* $180°$, *we know that the other two angle measures must have a sum of* $90°$.
- Now, look at just one side of the figure. We have an angle with a red arc and an angle with a blue arc. In between them is another angle that we do not know the measure of. All three angles added together make up the straight side of the outside figure. What must be the measure of the unknown angle (the measure of the angle between the red and blue arcs)? How do you know?
  - *Since the angle with the red arc and the angle with the blue arc must have a sum of* $90°$, *and all three angles together must make a straight angle measuring* $180°$ *(∠s on a line), the unknown angle must equal* $90°$.
- That means that the figure with side lengths c must be a square. It is a figure with four equal sides and four right angles. What is the area of this square?
  - *The area of the square must be* c^2.

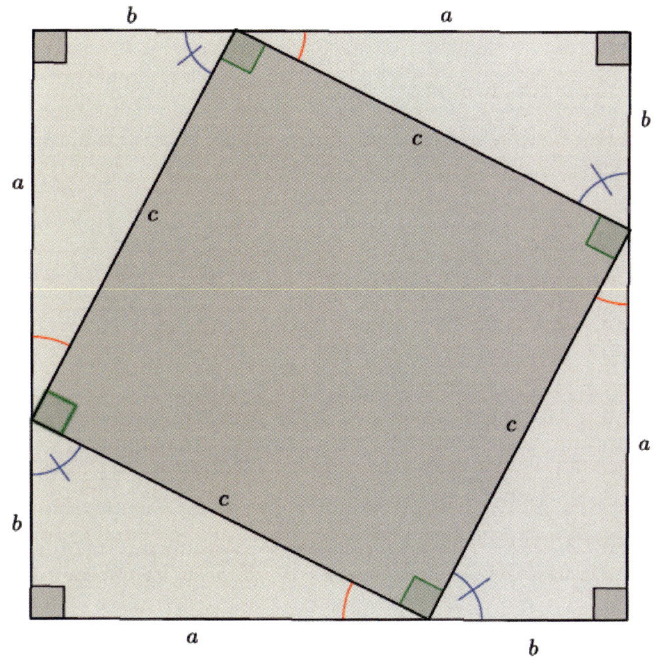

- Recall our goal: To show that $a^2 + b^2 = c^2$. To accomplish this goal, we compare the total area of the outside square with the parts it is composed of, that is, the four triangles and the smaller, inside square. Do we have everything we need to accomplish our goal?
  - *Yes, we know the area of the outside square is $a^2 + 2ab + b^2$; the sum of the areas of the four triangles is $2ab$, and the area of the inside square, c^2.*

Show students the end of the square within a square proof:

Total area of the outside square = area of four triangles + area of inside square

$$a^2 + 2ab + b^2 = 2ab + c^2$$
$$a^2 + 2ab - 2ab + b^2 = 2ab - 2ab + c^2$$
$$a^2 + b^2 = c^2$$

Thus, we have shown the Pythagorean theorem to be true using a square within a square.

Example 1 (2 minutes)

Example 1

Now that we know what the Pythagorean theorem is, let's practice using it to find the length of a hypotenuse of a right triangle.

Determine the length of the hypotenuse of the right triangle.

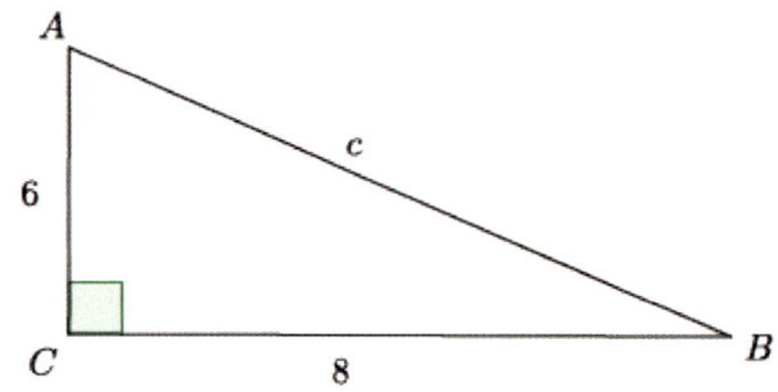

The Pythagorean theorem states that for right triangles $a^2 + b^2 = c^2$, where a and b are the legs, and c is the hypotenuse. Then,

$$\begin{aligned} a^2 + b^2 &= c^2 \\ 6^2 + 8^2 &= c^2 \\ 36 + 64 &= c^2 \\ 100 &= c^2. \end{aligned}$$

Since we know that $100 = 10^2$, we can say that the hypotenuse c is 10.

Example 2 (3 minutes)

Example 2

Determine the length of the hypotenuse of the right triangle.

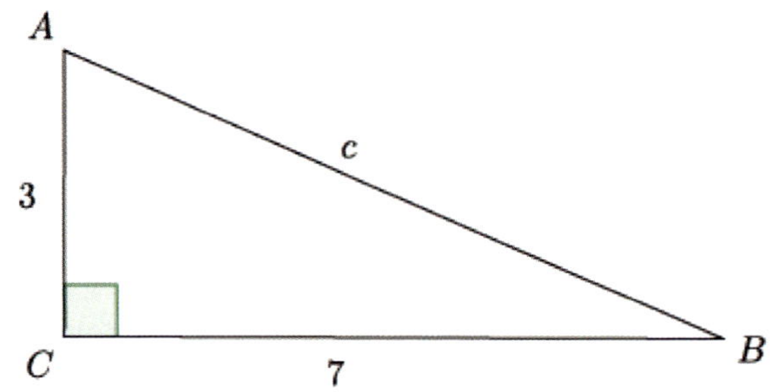

- Based on our work in the last example, what should we do to find the length of the hypotenuse?
  - *Use the Pythagorean theorem, and replace a and b with 3 and 7. Then,*

$$\begin{aligned} a^2 + b^2 &= c^2 \\ 3^2 + 7^2 &= c^2 \\ 9 + 49 &= c^2 \\ 58 &= c^2. \end{aligned}$$

- Since we do not know what number times itself produces 58, for now we can leave our answer as $58 = c^2$. Later this year, we learn how to determine the actual value for c for problems like this one.

Exercises 1–5 (10 minutes)

Exercises 1–5

For each of the exercises, determine the length of the hypotenuse of the right triangle shown. Note: Figures are not drawn to scale.

1.

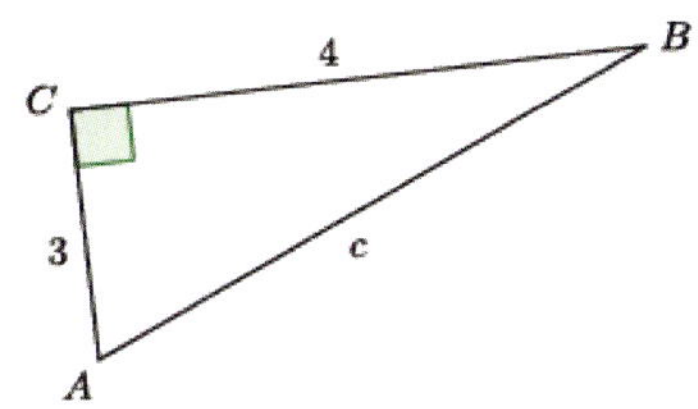

$$\begin{aligned} a^2 + b^2 &= c^2 \\ 3^2 + 4^2 &= c^2 \\ 9 + 16 &= c^2 \\ 25 &= c^2 \\ 5 &= c \end{aligned}$$

2.

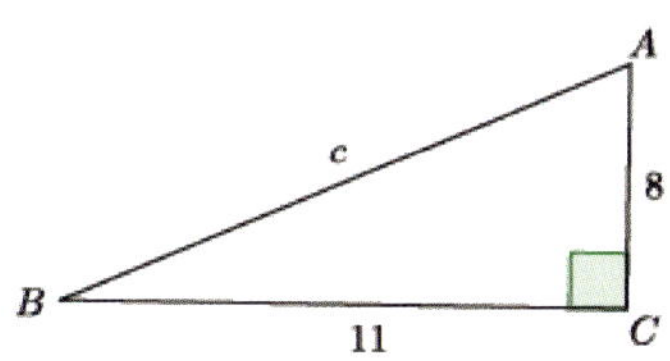

$$\begin{aligned} a^2 + b^2 &= c^2 \\ 8^2 + 11^2 &= c^2 \\ 64 + 121 &= c^2 \\ 185 &= c^2 \end{aligned}$$

3.

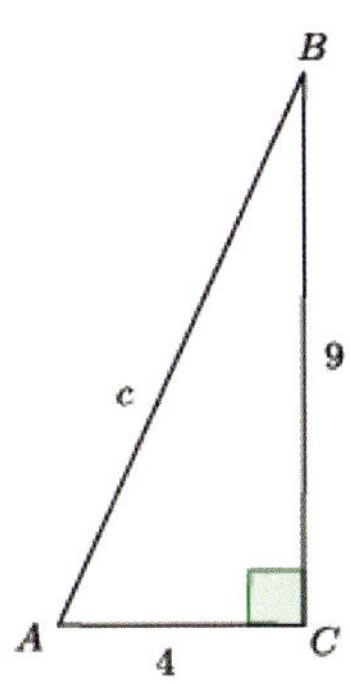

$$\begin{aligned} a^2 + b^2 &= c^2 \\ 4^2 + 9^2 &= c^2 \\ 16 + 81 &= c^2 \\ 97 &= c^2 \end{aligned}$$

4.

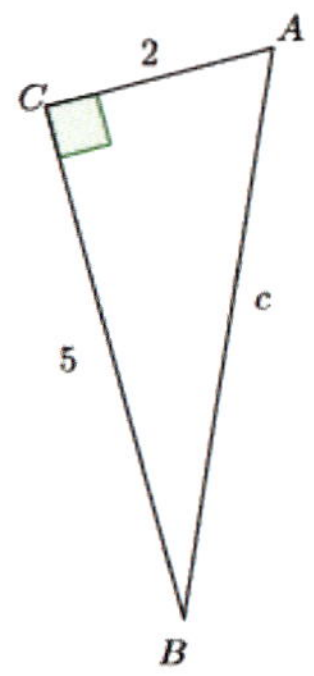

$$a^2 + b^2 = c^2$$
$$2^2 + 5^2 = c^2$$
$$4 + 25 = c^2$$
$$29 = c^2$$

5.

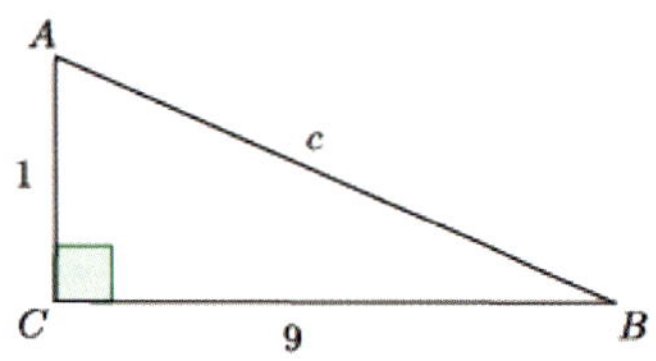

$$a^2 + b^2 = c^2$$
$$1^2 + 9^2 = c^2$$
$$1 + 81 = c^2$$
$$82 = c^2$$

Closing (5 minutes)

Summarize, or have students summarize, the lesson.

- We were shown a proof for the Pythagorean theorem that required us to find the area of four congruent triangles and two squares.
- We learned that right triangles have sides a and b, known as legs, and a side c, known as the hypotenuse.
- We know that for right triangles, $a^2 + b^2 = c^2$.
- We learned how to use the Pythagorean theorem to find the length of the hypotenuse of a right triangle.

Lesson Summary

Given a right triangle ABC with C being the vertex of the right angle, then the sides $\overline{AC}$ and $\overline{BC}$ are called the *legs* of $\triangle ABC$, and $\overline{AB}$ is called the *hypotenuse* of $\triangle ABC$.

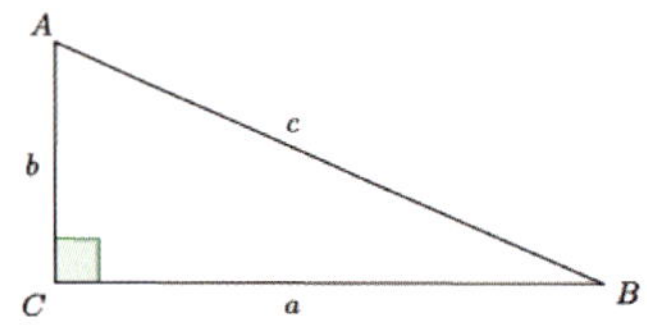

Take note of the fact that side a is opposite the angle A, side b is opposite the angle B, and side c is opposite the angle C.

The Pythagorean theorem states that for any right triangle, $a^2 + b^2 = c^2$.

Exit Ticket (5 minutes)

Name ______________________________ Date ________________

Lesson 15: Informal Proof of the Pythagorean Theorem

Exit Ticket

1. Label the sides of the right triangle with leg, leg, and hypotenuse.

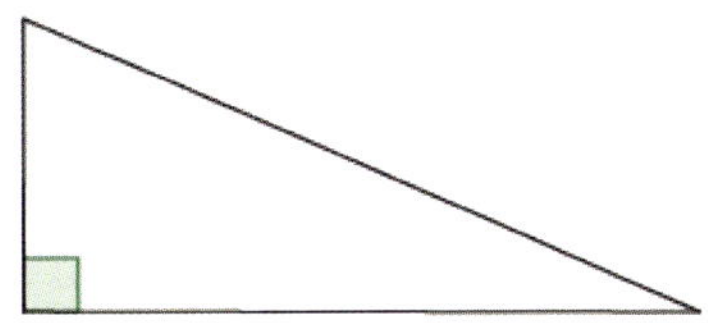

2. Determine the length of c in the triangle shown.

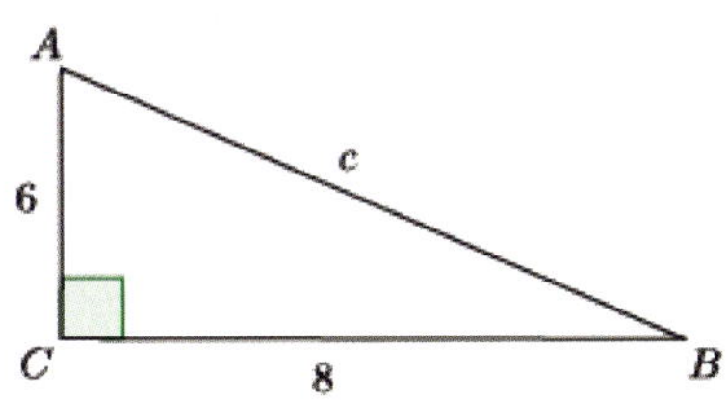

3. Determine the length of c in the triangle shown.

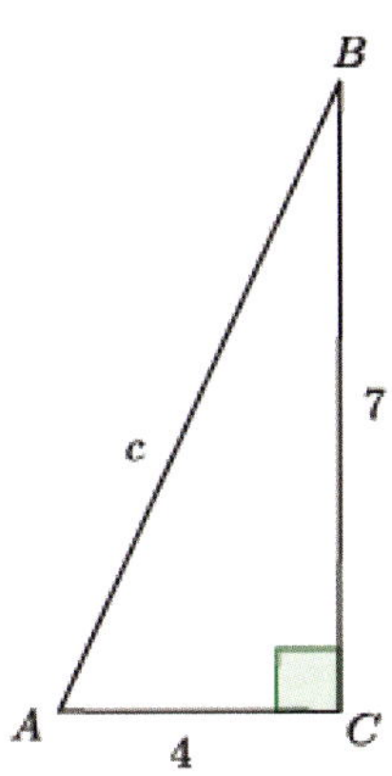

Exit Ticket Sample Solutions

1. **Label the sides of the right triangle with leg, leg, and hypotenuse.**

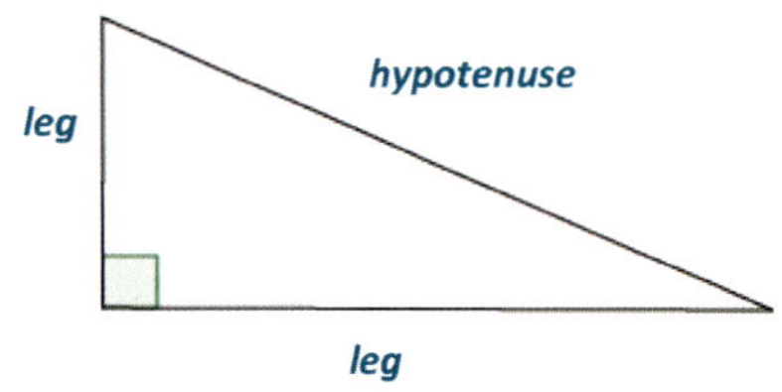

2. **Determine the length of c in the triangle shown.**

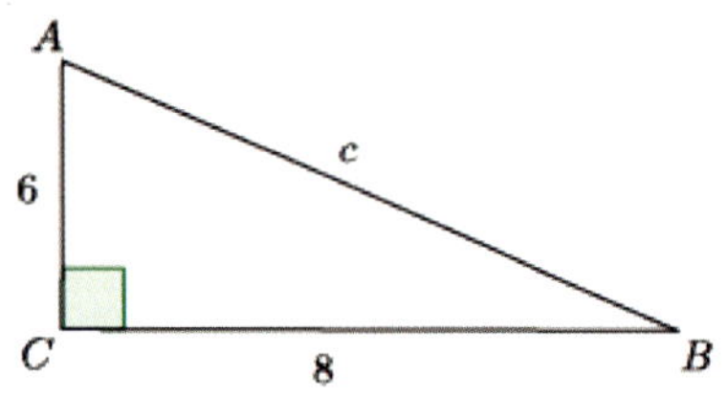

$$a^2 + b^2 = c^2$$
$$6^2 + 8^2 = c^2$$
$$36 + 64 = c^2$$
$$100 = c^2$$
$$10 = c$$

3. **Determine the length of c in the triangle shown.**

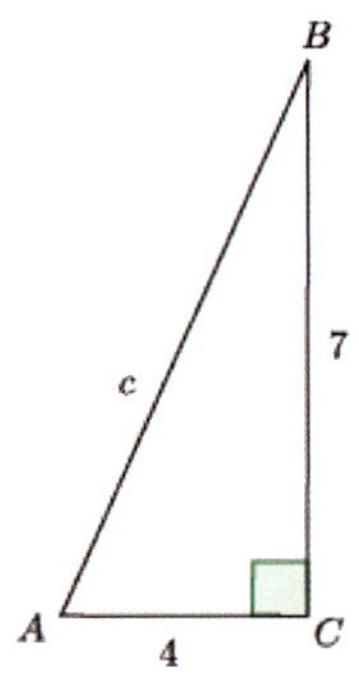

$$a^2 + b^2 = c^2$$
$$4^2 + 7^2 = c^2$$
$$16 + 49 = c^2$$
$$65 = c^2$$

Problem Set Sample Solutions

Students practice using the Pythagorean theorem to find the length of the hypotenuse of a right triangle.

For each of the problems below, determine the length of the hypotenuse of the right triangle shown. Note: Figures are not drawn to scale.

1.

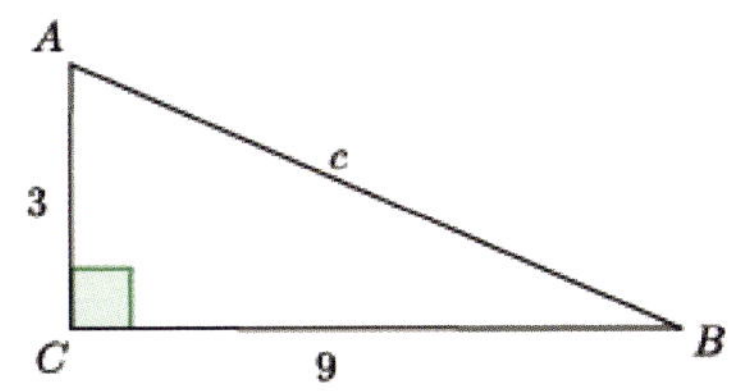

$$a^2 + b^2 = c^2$$
$$3^2 + 9^2 = c^2$$
$$9 + 81 = c^2$$
$$90 = c^2$$

2.

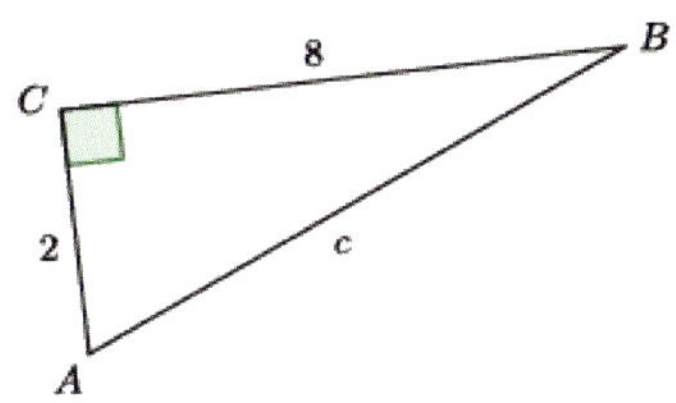

$$a^2 + b^2 = c^2$$
$$8^2 + 2^2 = c^2$$
$$64 + 4 = c^2$$
$$68 = c^2$$

3.

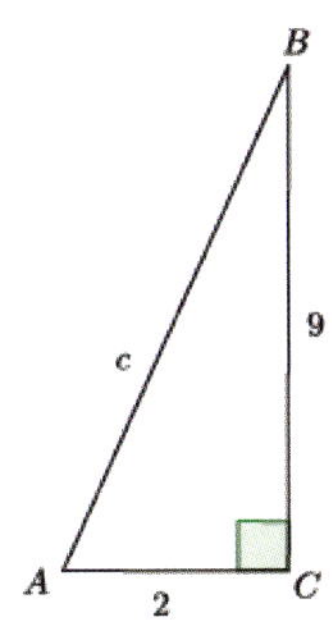

$$a^2 + b^2 = c^2$$
$$9^2 + 2^2 = c^2$$
$$81 + 4 = c^2$$
$$85 = c^2$$

4.

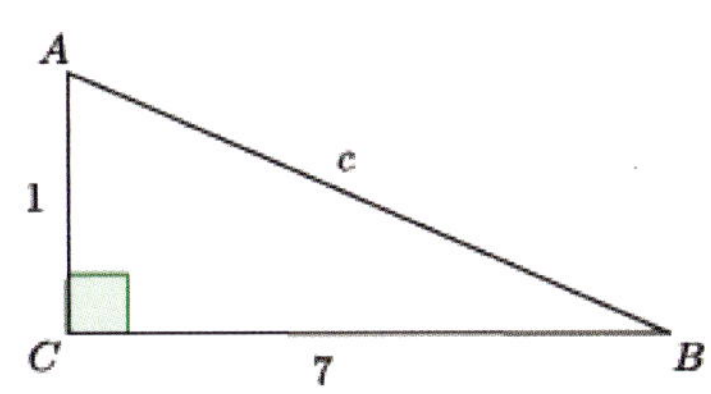

$$a^2 + b^2 = c^2$$
$$7^2 + 1^2 = c^2$$
$$49 + 1 = c^2$$
$$50 = c^2$$

5.

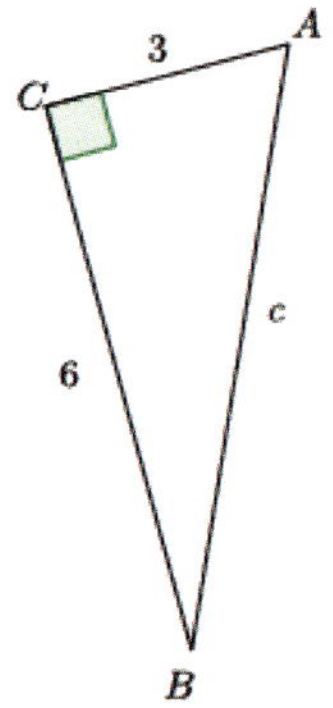

$$a^2 + b^2 = c^2$$
$$6^2 + 3^2 = c^2$$
$$36 + 9 = c^2$$
$$45 = c^2$$

6.

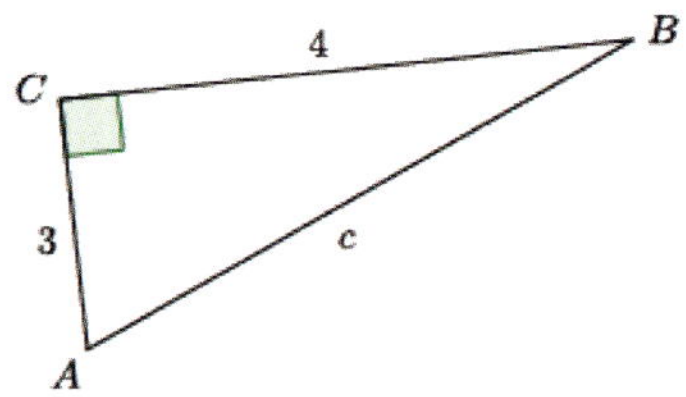

$$a^2 + b^2 = c^2$$
$$4^2 + 3^2 = c^2$$
$$16 + 9 = c^2$$
$$25 = c^2$$
$$5 = c$$

7.

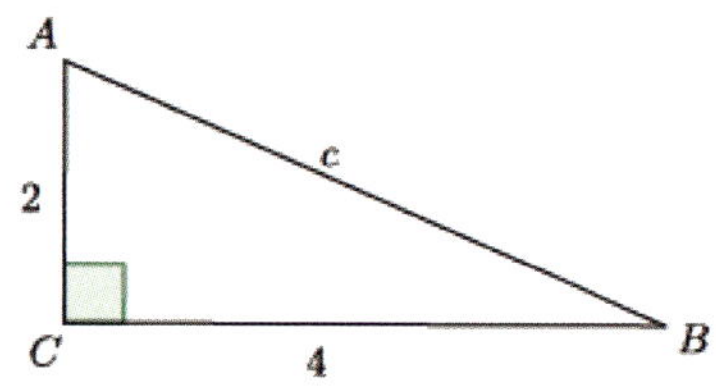

$$a^2 + b^2 = c^2$$
$$4^2 + 2^2 = c^2$$
$$16 + 4 = c^2$$
$$20 = c^2$$

8.

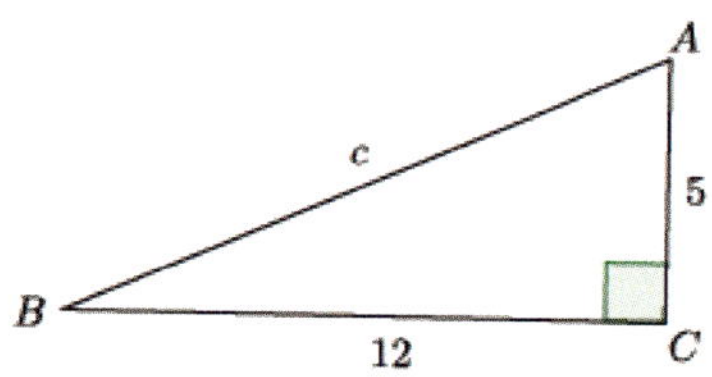

$$a^2 + b^2 = c^2$$
$$12^2 + 5^2 = c^2$$
$$144 + 25 = c^2$$
$$169 = c^2$$
$$13 = c$$

9.

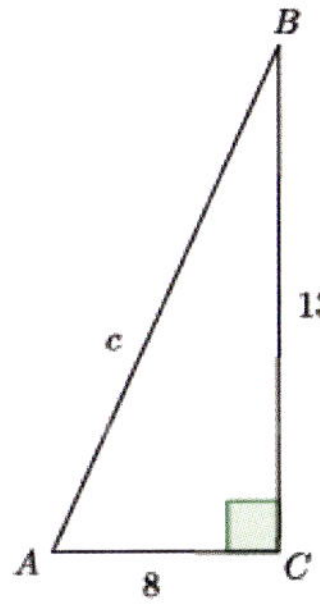

$$a^2 + b^2 = c^2$$
$$13^2 + 8^2 = c^2$$
$$169 + 64 = c^2$$
$$233 = c^2$$

10.

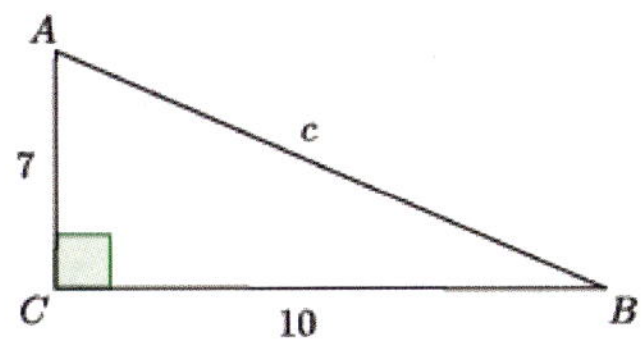

$$a^2 + b^2 = c^2$$
$$10^2 + 7^2 = c^2$$
$$100 + 49 = c^2$$
$$149 = c^2$$

11.

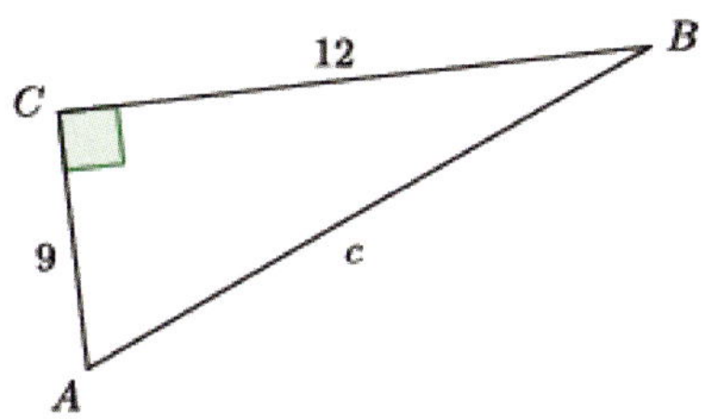

$$a^2 + b^2 = c^2$$
$$12^2 + 9^2 = c^2$$
$$144 + 81 = c^2$$
$$225 = c^2$$
$$15 = c$$

12.

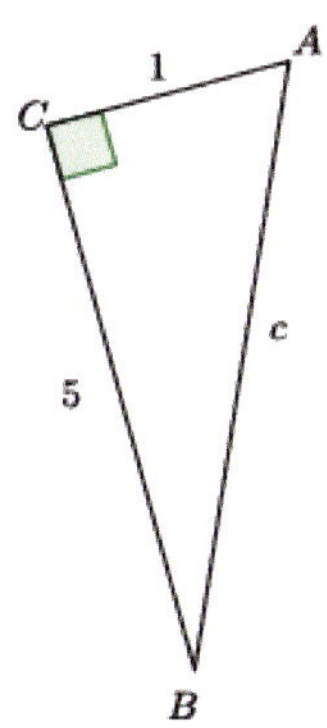

$$a^2 + b^2 = c^2$$
$$5^2 + 1^2 = c^2$$
$$25 + 1 = c^2$$
$$26 = c^2$$

Lesson 16: Applications of the Pythagorean Theorem

Student Outcomes

- Students use the Pythagorean theorem to determine missing side lengths of right triangles.

Lesson Notes

Since **8.G.B.6** and **8.G.B.7** are post-test standards, this lesson is designated as an extension lesson for this module. However, the content within this lesson is prerequisite knowledge for Module 7. If this lesson is not used with students as part of the work within Module 2, it must be used with students prior to beginning work on Module 7. Please realize that many mathematicians agree that the Pythagorean theorem is the most important theorem in geometry and has immense implications in much of high school mathematics in general (e.g., especially when studying quadratics and trigonometry). It is crucial that students see the teacher explain several proofs of the Pythagorean theorem and practice using it before being expected to produce a proof on their own.

Classwork

Example 1 (4 minutes)

The Pythagorean theorem as it applies to missing side lengths of triangles:

Example 1

Given a right triangle with a hypotenuse with length 13 units and a leg with length 5 units, as shown, determine the length of the other leg.

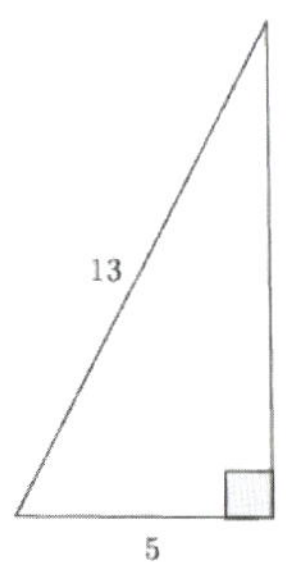

$$\begin{aligned} 5^2 + b^2 &= 13^2 \\ 5^2 - 5^2 + b^2 &= 13^2 - 5^2 \\ b^2 &= 13^2 - 5^2 \\ b^2 &= 169 - 25 \\ b^2 &= 144 \\ b &= 12 \end{aligned}$$

The length of the leg is 12 units.

- Let b represent the missing leg of the right triangle; then, by the Pythagorean theorem:
$$5^2 + b^2 = 13^2.$$
- If we let a represent the missing leg of the right triangle, then by the Pythagorean theorem:
$$a^2 + 5^2 = 13^2.$$
- Which of these two equations is correct: $5^2 + b^2 = 13^2$ or $a^2 + 5^2 = 13^2$?

- It does not matter which equation we use as long as we are showing the sum of the squares of the legs as equal to the square of the hypotenuse.
- Using the first of our two equations, $5^2 + b^2 = 13^2$, what can we do to solve for b in the equation?
  - *We need to subtract* 5^2 *from both sides of the equation.*

$$5^2 + b^2 = 13^2$$
$$5^2 - 5^2 + b^2 = 13^2 - 5^2$$
$$b^2 = 13^2 - 5^2$$

> *Scaffolding:*
>
> If students do not believe that they could use either equation, solve each of them, and show that the answer is the same.

- Point out to students that we are looking at the Pythagorean theorem in a form that allows us to find the length of one of the legs of the right triangle. That is, $b^2 = c^2 - a^2$.

$$b^2 = 13^2 - 5^2$$
$$b^2 = 169 - 25$$
$$b^2 = 144$$
$$b = 12$$

  - *The length of the leg of the right triangle is* 12 *units.*

Example 2 (4 minutes)

The Pythagorean theorem as it applies to missing side lengths of triangles in a real-world problem:

- Suppose you have a ladder of length 13 feet. Suppose that to make it sturdy enough to climb, you must place the ladder exactly 5 feet from the wall of a building. You need to post a banner on the building 10 feet above the ground. Is the ladder long enough for you to reach the location you need to post the banner?

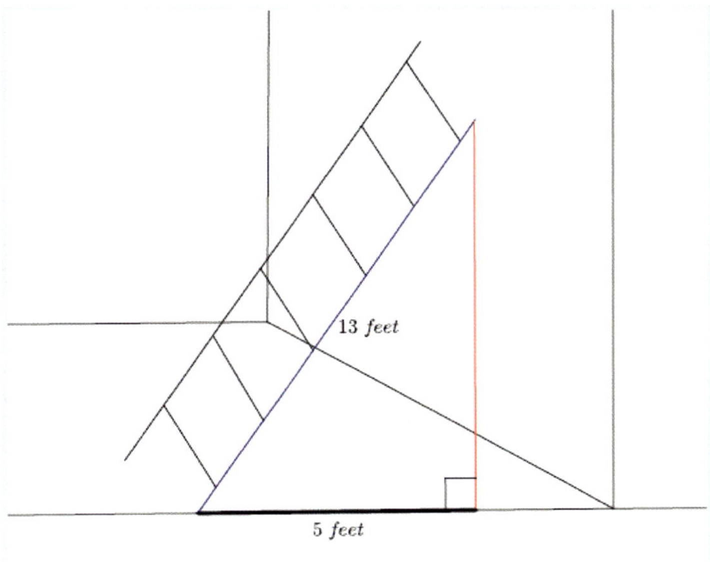

The ladder against the wall forms a right angle. For that reason, we can use the Pythagorean theorem to find out how far up the wall the ladder will reach. If we let h represent the height the ladder can reach, what equation will represent this problem?

  - $5^2 + h^2 = 13^2$ *or* $h^2 = 13^2 - 5^2$

- Using either equation, we see that this is just like Example 1. We know that the missing side of the triangle is 12 feet. Is the ladder long enough for you to reach the 10-foot banner location?
  - *Yes, the ladder allows us to reach 12 feet up the wall.*

Example 3 (3 minutes)

Pythagorean theorem as it applies to missing side lengths of a right triangle:

- Given a right triangle with a hypotenuse of length 15 units and a leg of length 9, what is the length of the other leg?

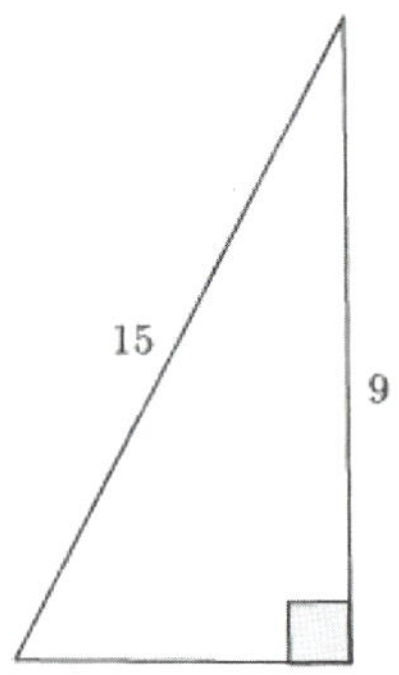

- If we let the length of the missing leg be represented by a, what equation will allow us to determine its value?
  - $a^2 + 9^2 = 15^2$ *or* $a^2 = 15^2 - 9^2$
- Finish the computation:

$$a^2 = 225 - 81$$
$$a^2 = 144$$
$$a = 12$$

  - *The length of the missing leg of this triangle is 12 units.*

Exercises 1–2 (5 minutes)

Students work on Exercises 1 and 2 independently.

Exercises 1–2

1. **Use the Pythagorean theorem to find the missing length of the leg in the right triangle.**

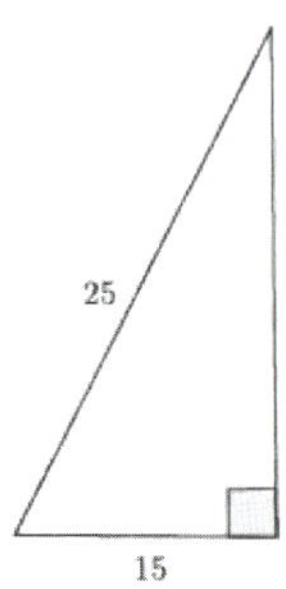

Let b represent the missing leg length; then,

$$15^2 + b^2 = 25^2$$
$$15^2 - 15^2 + b^2 = 25^2 - 15^2$$
$$b^2 = 625 - 225$$
$$b^2 = 400$$
$$b = 20.$$

The length of the leg is 20 units.

2. You have a 15-foot ladder and need to reach exactly 9 feet up the wall. How far away from the wall should you place the ladder so that you can reach your desired location?

Let a represent the distance the ladder must be placed from the wall; then,

$$a^2 + 9^2 = 15^2$$
$$a^2 + 9^2 - 9^2 = 15^2 - 9^2$$
$$a^2 = 225 - 81$$
$$a^2 = 144$$
$$a = 12.$$

The ladder must be placed exactly 12 feet from the wall.

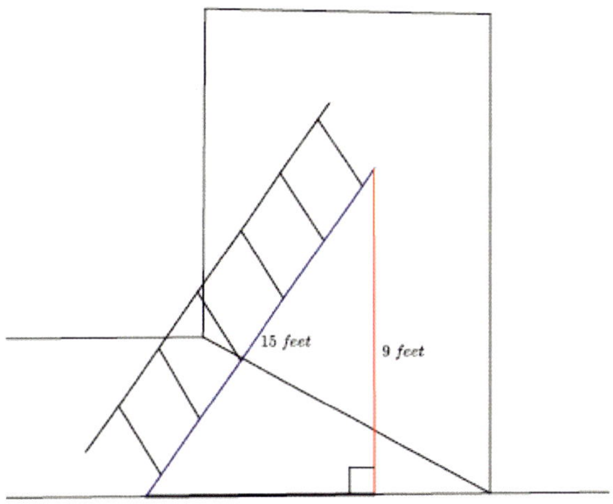

Example 4 (5 minutes)

The Pythagorean theorem as it applies to distances on a coordinate plane:

- We want to find the length of the segment AB on the coordinate plane, as shown.

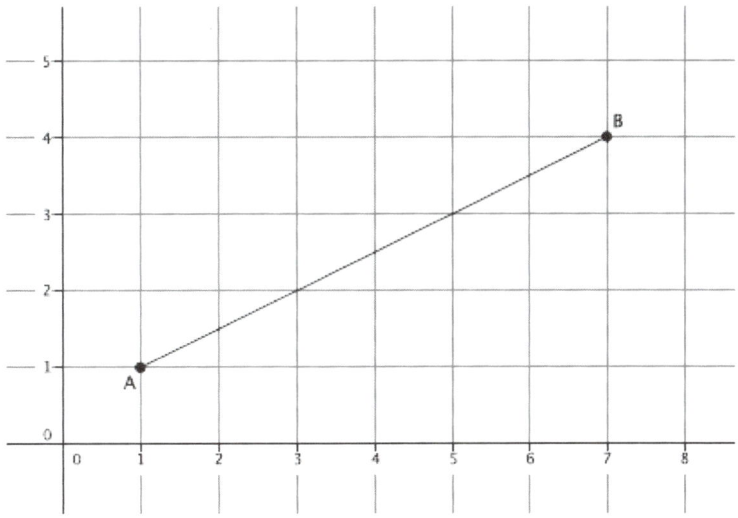

- If we had a right triangle, then we could use the Pythagorean theorem to determine the length of the segment. Let's draw a line parallel to the y-axis through point B. We will also draw a line parallel to the x-axis through point A.

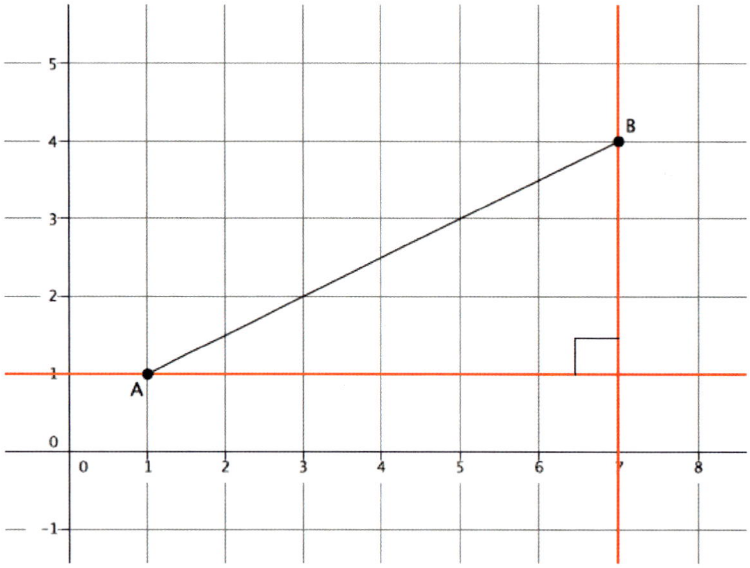

- How can we be sure we have a right triangle?
  - *The coordinate plane is set up so that the intersection of the x-axis and y-axis is perpendicular. The line parallel to the y-axis through B is just a translation of the y-axis. Similarly, the line parallel to the x-axis through A is a translation of the x-axis. Since translations preserve angle measure, the intersection of the two red lines is also perpendicular, meaning we have a $90°$ angle and a right triangle.*
- Now that we are sure we can use the Pythagorean theorem, we need to know the lengths of the legs. Count the units from point A to the right angle and point B to the right angle. What are those lengths?
  - *The base of the triangle is 6 units, and the height of the triangle is 3 units.*

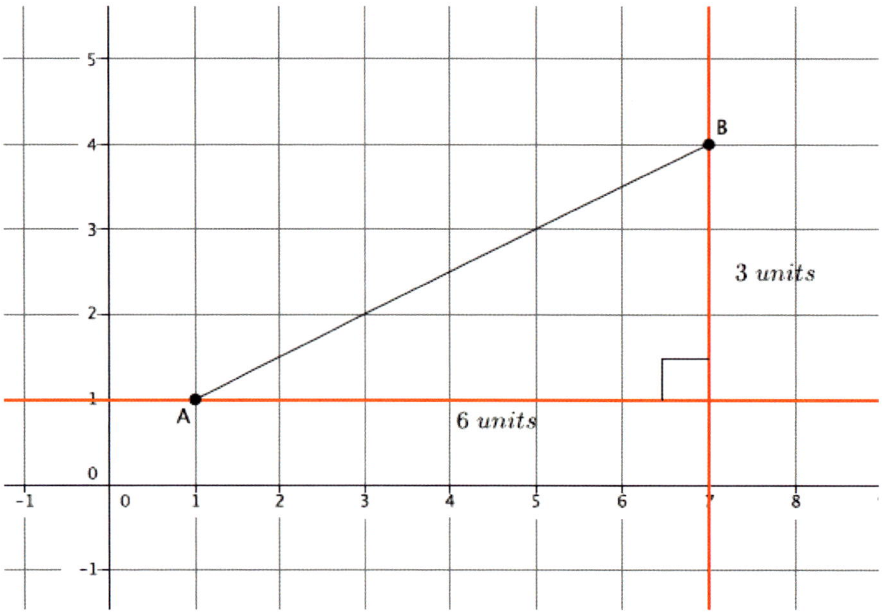

- What equation can we use to find the length of the segment AB? Let's represent that length by c.
  - $3^2 + 6^2 = c^2$
  - *The length of c is*

$$\begin{aligned} 3^2 + 6^2 &= c^2 \\ 9 + 36 &= c^2 \\ 45 &= c^2. \end{aligned}$$

- We cannot get a precise answer, so we will leave the length of c as $c^2 = 45$.

Example 5 (3 minutes)

The Pythagorean theorem as it applies to the length of a diagonal in a rectangle:

- Given a rectangle with side lengths of 8 cm and 2 cm, as shown, what is the length of the diagonal?

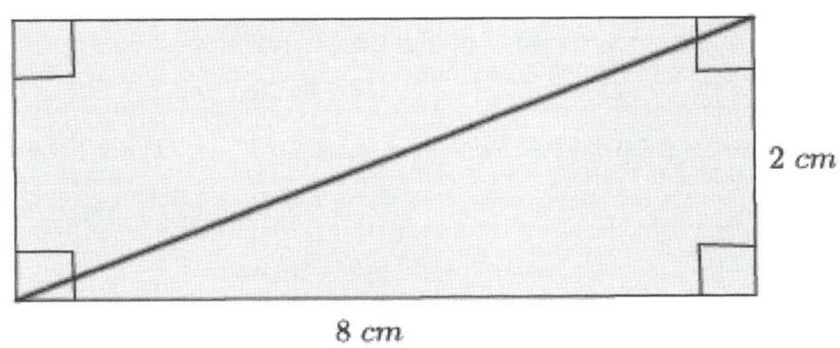

- If we let the length of the diagonal be represented by c, what equation can we use to find its length?
  - $2^2 + 8^2 = c^2$
  - *The length of the diagonal is given by the positive number c that satisfies any of the following equations:*

$$\begin{aligned} 2^2 + 8^2 &= c^2 \\ 4 + 64 &= c^2 \\ 68 &= c^2 \end{aligned}$$

- Since irrational square roots are not discussed until Module 7, until then, we state the answer as, "The measure of the length of the hypotenuse in centimeters is the positive number c that satisfies $c^2 = 68$."

Exercises 3–6 (11 minutes)

Students work independently on Exercises 3–6.

Exercises 3–6

3. **Find the length of the segment AB, if possible.**

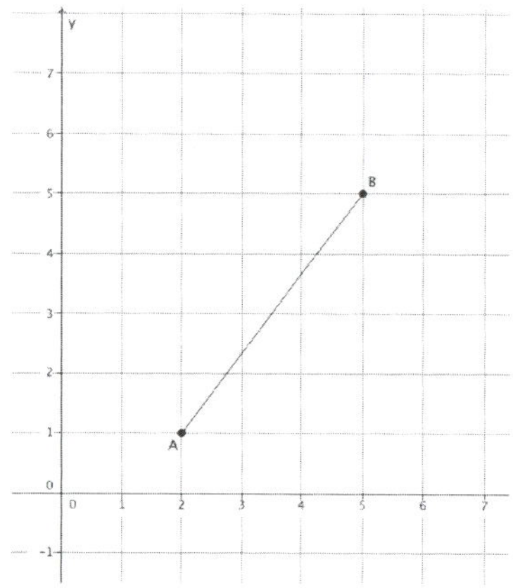

If we let the length of segment AB be represented by c, then

$$3^2 + 4^2 = c^2$$
$$9 + 16 = c^2$$
$$25 = c^2$$
$$5 = c.$$

The length of segment AB is 5 units.

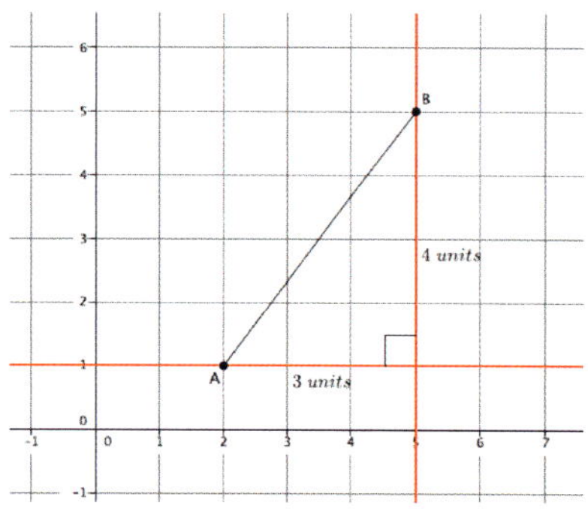

4. **Given a rectangle with dimensions 5 cm and 10 cm, as shown, find the length of the diagonal, if possible.**

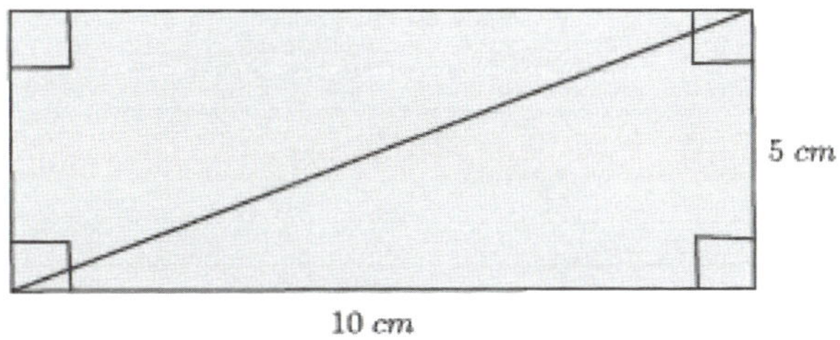

Let c represent the length of the diagonal; then,

$$c^2 = 5^2 + 10^2$$
$$c^2 = 25 + 100$$
$$c^2 = 125.$$

The measure of the length of the hypotenuse in centimeters is the positive number c that satisfies $c^2 = 125$.

5. **A right triangle has a hypotenuse of length 13 in. and a leg with length 4 in. What is the length of the other leg?**

If we let a represent the length of the other leg, then

$$a^2 + 4^2 = 13^2$$
$$a^2 + 4^2 - 4^2 = 13^2 - 4^2$$
$$a^2 = 13^2 - 4^2$$
$$a^2 = 169 - 16$$
$$a^2 = 153.$$

The measure of the length of the leg in inches is the positive number a that satisfies $a^2 = 153$.

6. **Find the length of b in the right triangle below, if possible.**

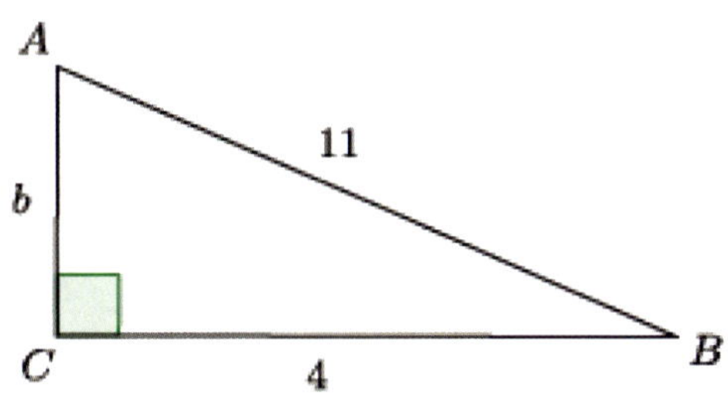

By the Pythagorean theorem,

$$4^2 + b^2 = 11^2$$
$$4^2 - 4^2 + b^2 = 11^2 - 4^2$$
$$b^2 = 11^2 - 4^2$$
$$b^2 = 121 - 16$$
$$a^2 = 105.$$

The length of side AC is the positive number b that satisfies $b^2 = 105$.

Closing (5 minutes)

Summarize, or have students summarize, the lesson.

- We know how to use the Pythagorean theorem to find the length of a missing side of a right triangle, whether it is one of the legs or the hypotenuse.
- We know how to apply the Pythagorean theorem to a real-life problem, like how high a ladder will reach along a wall.
- We know how to find the length of a diagonal of a rectangle.
- We know how to determine the length of a segment that is on the coordinate plane.

Lesson Summary

The Pythagorean theorem can be used to find the unknown length of a leg of a right triangle.

An application of the Pythagorean theorem allows you to calculate the length of a diagonal of a rectangle, the distance between two points on the coordinate plane, and the height that a ladder can reach as it leans against a wall.

Exit Ticket (5 minutes)

Name ______________________________ Date ______________

Lesson 16: Applications of the Pythagorean Theorem

Exit Ticket

1. Find the length of the missing side of the rectangle shown below, if possible.

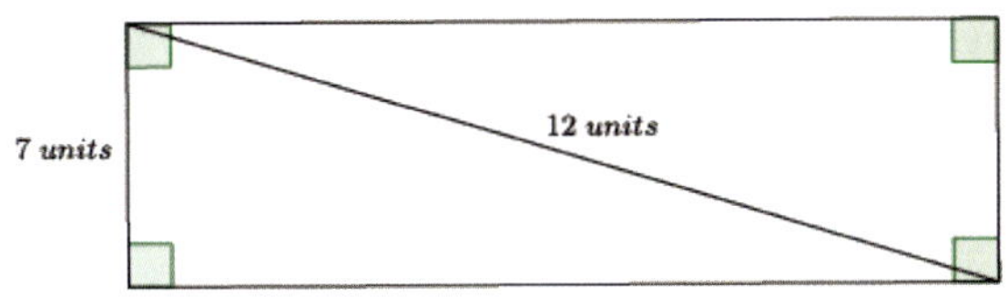

2. Find the length of all three sides of the right triangle shown below, if possible.

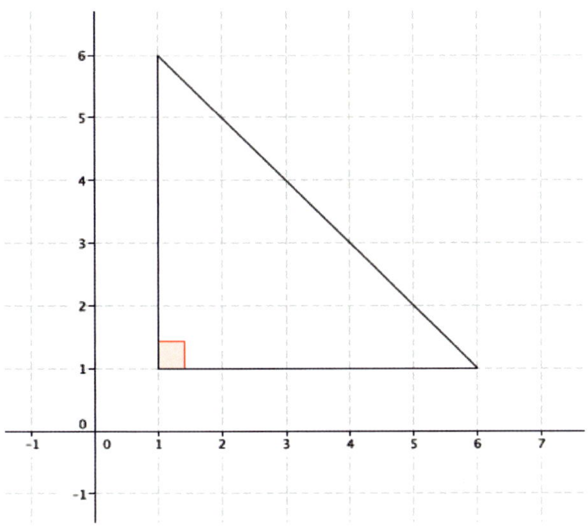

Exit Ticket Sample Solutions

1. Find the length of the missing side of the rectangle shown below, if possible.

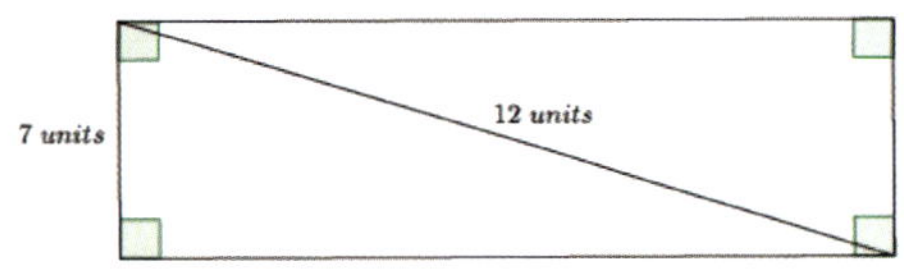

Let the length of the unknown leg be a units. Then,

$$a^2 + 7^2 = 12^2$$
$$a^2 + 7^2 - 7^2 = 12^2 - 7^2$$
$$a^2 = 12^2 - 7^2$$
$$a^2 = 144 - 49$$
$$a^2 = 95.$$

The number of units of the side is given by the positive number a that satisfies $a^2 = 95$.

2. Find the length of all three sides of the right triangle shown below, if possible.

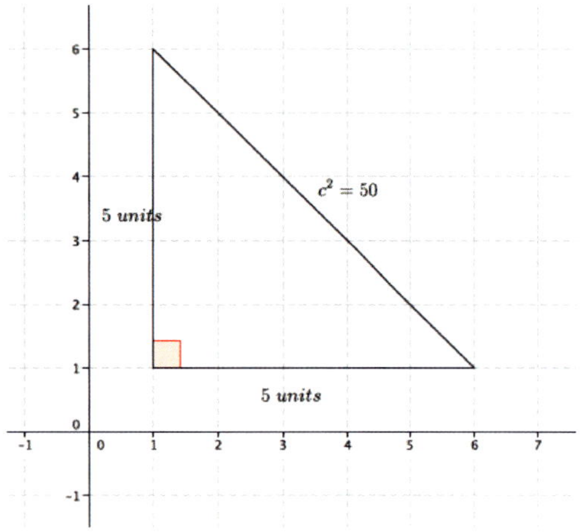

The two legs are each 5 units in length. If the hypotenuse is c units in length, then c is a positive number that satisfies

$$5^2 + 5^2 = c^2$$
$$25 + 25 = c^2$$
$$50 = c^2.$$

The length of the hypotenuse is given by the positive number c that satisfies $c^2 = 50$ units.

Problem Set Sample Solutions

Students practice using the Pythagorean theorem to find missing lengths in right triangles.

1. **Find the length of the segment AB shown below, if possible.**

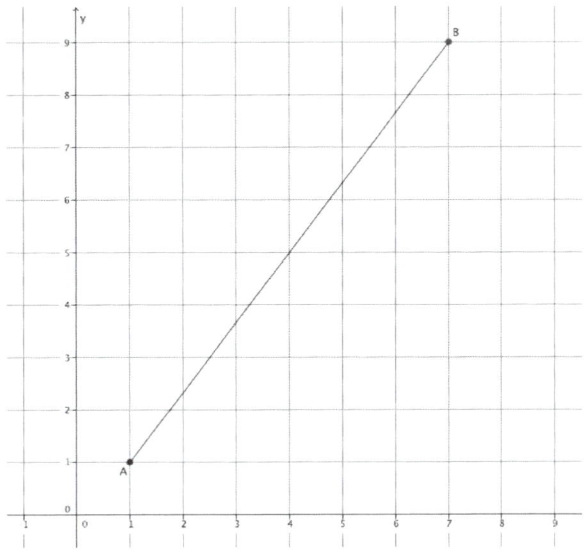

If we let the length of segment AB be represented by c units, then by the Pythagorean theorem

$$6^2 + 8^2 = c^2$$
$$36 + 48 = c^2$$
$$100 = c^2$$
$$10 = c$$

The length of the segment AB is 10 units.

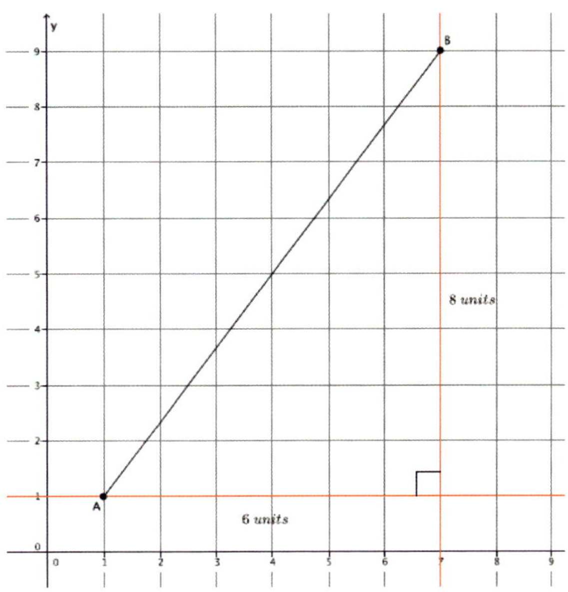

2. A 20-foot ladder is placed 12 feet from the wall, as shown. How high up the wall will the ladder reach?

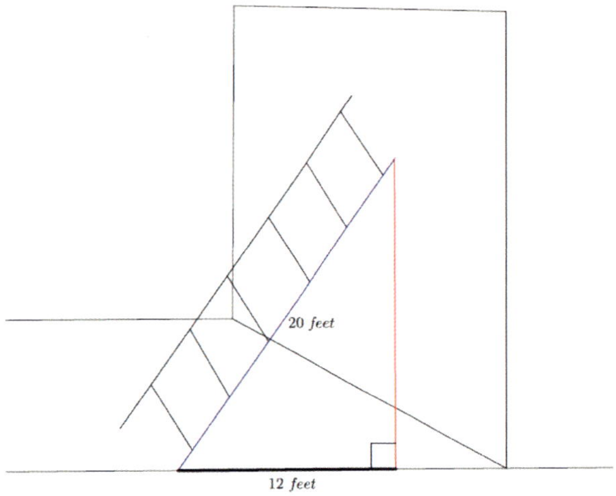

Let the height up the wall that the ladder will reach be a feet. Then,

$$a^2 + 12^2 = 20^2$$
$$a^2 + 12^2 - 12^2 = 20^2 - 12^2$$
$$a^2 = 20^2 - 12^2$$
$$a^2 = 400 - 144$$
$$a^2 = 256$$
$$a = 16.$$

The ladder will reach 16 feet up the wall.

3. A rectangle has dimensions 6 in. by 12 in. What is the length of the diagonal of the rectangle?

Let the length of the diagonal be c inches. Then c is a positive number that satisfies

$$6^2 + 12^2 = c^2$$
$$36 + 144 = c^2$$
$$80 = c^2.$$

The measure of the length of the diagonal in inches is the positive number c that satisfies $c^2 = 180$.

Use the Pythagorean theorem to find the missing side lengths for the triangles shown in Problems 4–8.

4. Determine the length of the missing side, if possible.

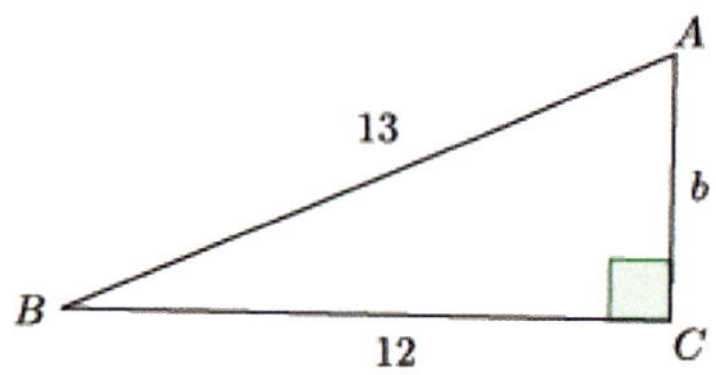

$$12^2 + b^2 = 13^2$$
$$12^2 - 12^2 + b^2 = 13^2 - 12^2$$
$$b^2 = 13^2 - 12^2$$
$$b^2 = 169 - 144$$
$$b^2 = 25$$
$$b = 5$$

The length of the missing side is 5 units.

5. **Determine the length of the missing side, if possible.**

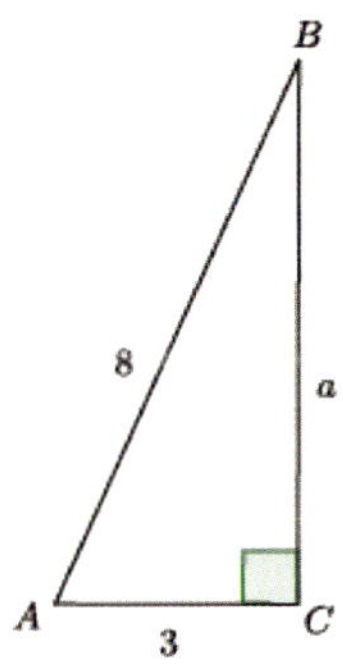

$$a^2 + 3^2 = 8^2$$
$$a^2 + 3^2 - 3^2 = 8^2 - 3^2$$
$$a^2 = 8^2 - 3^2$$
$$a^2 = 64 - 9$$
$$a^2 = 55$$

The number of units of the side is given by the positive number a that satisfies $a^2 = 55$.

6. **Determine the length of the missing side, if possible.**

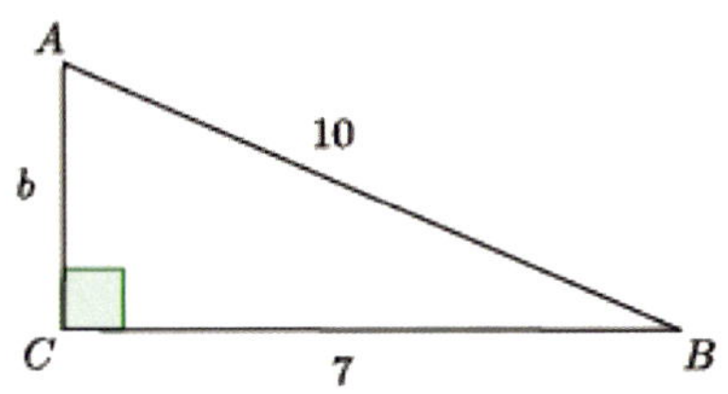

$$7^2 + b^2 = 10^2$$
$$7^2 - 7^2 + b^2 = 10^2 - 7^2$$
$$b^2 = 10^2 - 7^2$$
$$b^2 = 100 - 49$$
$$b^2 = 51$$

The number of units of the side is given by the positive number b that satisfies $b^2 = 51$.

7. **Determine the length of the missing side, if possible.**

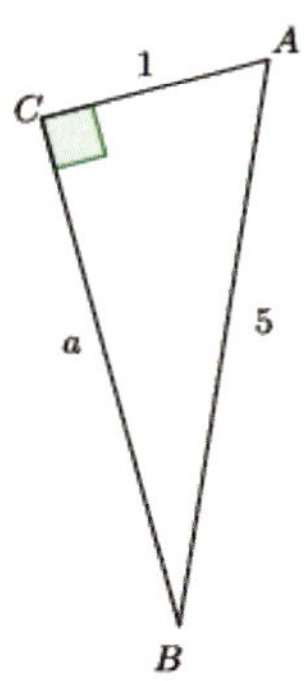

$$a^2 + 1^2 = 5^2$$
$$a^2 + 1^2 - 1^2 = 5^2 - 1^2$$
$$a^2 = 5^2 - 1^2$$
$$a^2 = 25 - 1$$
$$a^2 = 24$$

The number of units of the side is given by the positive number a that satisfies $a^2 = 24$.

8. **Determine the length of the missing side, if possible.**

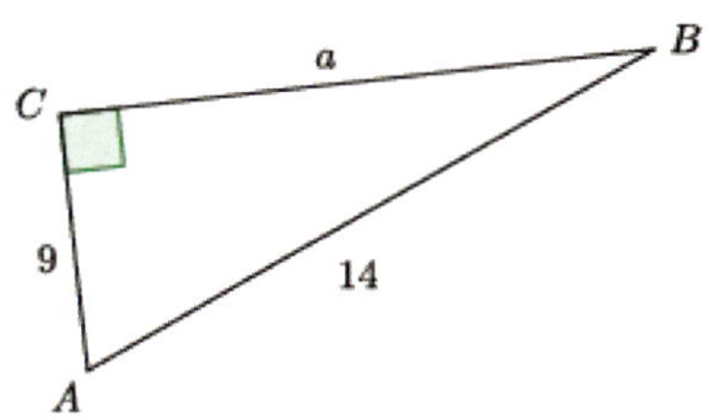

$$a^2 + 9^2 = 14^2$$
$$a^2 + 9^2 - 9^2 = 14^2 - 9^2$$
$$a^2 = 14^2 - 9^2$$
$$a^2 = 196 - 81$$
$$a^2 = 115$$

The number of units of the side is given by the positive number a that satisfies $a^2 = 115$.

This page intentionally left blank

This page intentionally left blank

This page intentionally left blank